THE CONTINENTAL SHELF
BEYOND 200 NAUTICAL MILES

The Continental Shelf Beyond 200 Nautical Miles

Rights and Responsibilities

JOANNA MOSSOP

Research supported by the New Zealand Law Foundation

UNIVERSITY PRESS

Great Clarendon Street, Oxford, OX2 6DP,
United Kingdom

Oxford University Press is a department of the University of Oxford.
It furthers the University's objective of excellence in research, scholarship,
and education by publishing worldwide. Oxford is a registered trade mark of
Oxford University Press in the UK and in certain other countries

© Joanna Mossop 2016

The moral rights of the author have been asserted

First Edition published in 2016

Impression: 1

All rights reserved. No part of this publication may be reproduced, stored in
a retrieval system, or transmitted, in any form or by any means, without the
prior permission in writing of Oxford University Press, or as expressly permitted
by law, by licence or under terms agreed with the appropriate reprographics
rights organization. Enquiries concerning reproduction outside the scope of the
above should be sent to the Rights Department, Oxford University Press, at the
address above

You must not circulate this work in any other form
and you must impose this same condition on any acquirer

Crown copyright material is reproduced under Class Licence
Number C01P0000148 with the permission of OPSI
and the Queen's Printer for Scotland

Published in the United States of America by Oxford University Press
198 Madison Avenue, New York, NY 10016, United States of America

British Library Cataloguing in Publication Data
Data available

Library of Congress Control Number: 2016954172

ISBN 978–0–19–876609–4

Printed and bound by
CPI Group (UK) Ltd, Croydon, CR0 4YY

Links to third party websites are provided by Oxford in good faith and
for information only. Oxford disclaims any responsibility for the materials
contained in any third party website referenced in this work.

To Esko, with love.

Preface

My interest in the law applying to the continental shelf beyond 200 nautical miles (nm) was sparked by a passing comment made at the 2006 meeting of the United Nations Ad Hoc Informal Working Group on Biodiversity beyond National Jurisdiction. A delegate referred to the need to preserve the coastal State's interests in the extended continental shelf. My curiosity was piqued and I became more intrigued as I began to look at the unique legal situation created by the continental shelf beyond 200 nm.

Considerable academic attention has been paid to the delineation process established in article 76 of the Law of the Sea Convention (LOSC). This is understandable because a large number of States have been engaged in gathering data about their continental shelves and preparing submissions for the Commission on the Limits of the Continental Shelf (CLCS). This work is important because it allows coastal States to delineate the outer limits of their continental shelf in a way that is final and binding. The question that motivated me to write on the topic, and to produce this book, was the question that I could imagine officials and governments asking: 'Now what? What do we do with this now we have it?' This book is an attempt to answer this question.

I acknowledge that it may be many years before individual coastal States are in a position to exploit the non-living resources of their extended continental shelves. But there are two reasons why it is important to explore this issue now. First, there are human activities currently taking place on some extended continental shelves. Exploration activity is being conducted on some of the more accessible extended continental shelves. Vessels targeting high seas fish species are conducting bottom trawling on some extended continental shelves. Secondly, coastal States are not required to wait until the CLCS issues recommendations in respect of the outer limits of their shelves. The inherent nature of sovereign rights to the shelf means that many States may already be in a position to exercise sovereign rights where there is clear evidence that their particular extended shelf exists. It is worthwhile for coastal States to put in place mechanisms to manage sustainable development for this area.

In writing this book I have tried to anticipate some of the questions that coastal States might ask in considering how to exercise the rights that they have over the extended continental shelf. Because there is almost no State practice in this area, it is inevitable that in the future new problems and questions will arise that I have not anticipated. However, I have attempted to consider the issues and answer some of these early questions. Although the issues facing coastal States and the exercise of sovereign rights over the continental shelf beyond 200 nm seem new, they are actually not—at least not completely. The 1958 Geneva Convention on the Continental Shelf achieved a legal situation similar to the extended continental shelf under the LOSC. Under the Convention on the Continental Shelf, coastal States had sovereign rights over the continental shelf beyond their territorial seas, but not over the superjacent high seas. Therefore, considerable guidance can be found in State practice under that Convention. This is particularly useful in evaluating how the high seas and continental shelf regimes should interact and the limits of coastal State jurisdiction under the LOSC.

Other issues are new in the LOSC, such as the article 82 obligation to share part of the production from resources on the continental shelf beyond 200 nm with the international community. Article 82 was the result of complex negotiations at the third United Nations Conference on the Law of the Sea. How this article, and others, will be implemented is among the important questions addressed in this book.

Acknowledgements

Writing a book is a solitary undertaking but I have received enormous support along the way. My first thanks must be to the New Zealand Law Foundation, which generously funded research assistance, as well as travel to allow me to ask questions of officials and academics in countries other than New Zealand. I am grateful for their support while the book was being written. Victoria University of Wellington has supported me with funding for my research. I appreciate the practical help provided by the Law Faculty administrative support team, especially Angela Funnell. I acknowledge the fine support by several research assistants over the years that contributed to chapters and articles that culminated in this book.

I am most obliged to the many government officials and academics who responded to my queries. At the time that I was researching this book, few governments had devoted much attention to their policies in relation to how they would exercise their sovereign rights, so I appreciate the interest and support I received. The people I spoke to are numerous, and please know I am deeply thankful for your help.

I must thank the colleagues, near and far, who read chapters and otherwise gave feedback or advice on parts of the book. Thanks go to Bevan Marten, Anastasia Telesetsky, Sir Kenneth Keith, Campbell McLachlan, Alberto Costi, Clive Schofield, Vasco Becker-Weinberg, Karen Scott and Kevin Baumert for their valuable wisdom. I am also grateful to the many other friends and colleagues who provided moral support.

I am obliged to the scientists who agreed to talk to me about their work and, in many cases, to correct my assumptions. Thanks to Dr Ray Wood, Dr Cornel de Ronde, Dr Malcolm Clark, and Dr Peter King. Of course, where I have got things wrong, it is my error, not theirs.

Finally, I am grateful to my family and friends for their patience and support during the writing of this book. They have heard more than they should have to about the continental shelf beyond 200 nm. In particular I thank my husband, Esko Wiltshire, for his enduring love and patience.

Contents

Table of Cases	xv
Table of Treaties	xvii
List of Abbreviations	xxiii

1. **Introduction**   1
   - A. Terminology   6
   - B. An Overview of Coastal States' Rights and Obligations on the Continental Shelf   7
   - C. Chapter Overview   14
   - D. Conclusion   17

2. **Resources and Human Activities on the Continental Shelf Beyond 200 Nautical Miles**   19
   - A. What is the Continental Shelf?   20
   - B. The Living Resources and the Seabed Environment of the Continental Shelf   21
   - C. Exploitation of Living Resources   27
   - D. The Non-living Resources of the Continental Shelf   33
   - E. Exploitation of Non-living Resources   36
   - F. Other Activities with a Direct Impact on the Seafloor   45
   - G. Conclusion   50

3. **The Development of Sovereign Rights to Continental Shelf Resources**   52
   - A. Overview of the Development of Continental Shelf Law   53
   - B. Sedentary Species   62
   - C. The Outer Limits of the Continental Shelf   67
   - D. The Status of Part VI of the Law of the Sea Convention in Customary International Law   81
   - E. Conclusion   91

4. **Living Resources and Protection of the Environment on the Continental Shelf Beyond 200 Nautical Miles**   93
   - A. Environmental Obligations in Relation to the Extended Continental Shelf   95
   - B. Bioprospecting   110
   - C. Fishing   119
   - D. Conclusion   120

5. **Non-living Resources on the Continental Shelf Beyond 200 Nautical Miles** — 123
   A. Mining and the Article 82 Obligation to Make Payments — 123
   B. Installations and Structures on the Extended Continental Shelf — 147
   C. Carbon Dioxide Sequestration — 149
   D. Conclusion — 150

6. **Marine Scientific Research** — 152
   A. Types of Marine Scientific Research — 154
   B. Coastal State Jurisdiction Over Marine Scientific Research Under the Law of the Sea Convention — 156
   C. The Scope of Coastal State Jurisdiction in Relation to Research Above the Extended Continental Shelf — 158
   D. Issues Arising from Article 246(6) — 164
   E. Conditions of Granting Consent for Marine Scientific Research Projects — 170
   F. Conclusion — 171

7. **The Intersection Between Coastal State Rights and High Seas Freedoms** — 173
   A. The Law of the Sea Convention and the Balancing of State Rights — 175
   B. A Framework for Balancing Rights on the Extended Continental Shelf — 187
   C. Conclusion — 201

8. **Enforcement Powers of Coastal States in Relation to the Continental Shelf Beyond 200 Nautical Miles** — 203
   A. Arguments Against an Enforcement Right — 203
   B. Arguments in Favour of an Enforcement Right — 205
   C. Conclusion — 215

9. **Cooperative Approaches to Regulating Activities on the Continental Shelf Beyond 200 Nautical Miles** — 216
   A. Example 1: Multilateral Regional Cooperation in the North-east Atlantic — 217
   B. Example 2: Bilateral Cooperation Between Mauritius and the Seychelles — 227
   C. Cooperation with International and Regional Organizations — 230
   D. Conclusion — 239

10. Looking Ahead	241
A. Disputed Areas	242
B. Developing States and the Information Gap	243
C. The Intersection Between the Extended Continental Shelf and an International Agreement for Biodiversity Beyond National Jurisdiction	244
D. Conclusion	248
Bibliography	251
Index	271

Table of Cases

INTERNATIONAL COURT OF JUSTICE/PERMANENT COURT OF INTERNATIONAL JUSTICE

Aegean Sea Continental Shelf (*Greece v Turkey*) (Interim Protection) [1976]
 ICJ Reports 3 .... 145, 161
Aegean Sea Continental Shelf (*Greece v Turkey*) (Judgment) [1978] ICJ Reports 3 161
Border and Transborder Armed Actions (*Nicaragua v Honduras*) (Jurisdiction) [1988]
 ICJ Reports 69 91
Certain Activities Carried out by Nicaragua in the Border Area and Construction of a Road in Costa
 Rica Along the San Juan River (*Costa Rica v Nicaragua*) (Merits) ICJ 16 December 2015 104
Continental Shelf (*Libyan Arab Jamahiriya v Malta*) (Judgment) [1985] ICJ Reports 13 ..... 59, 82, 84
Continental Shelf (*Tunisia v Libya Arab Jamahiriya*) (Judgment) [1982] ICJ Reports 18 ..... 54, 56, 59, 64, 82, 83, 84
Delimitation of the Continental Shelf between Nicaragua and Colombia Beyond 200
 Nautical Miles from the Nicaraguan Coast (*Nicaragua v Colombia*) (Preliminary
 Objections) ICJ 17 March 2016 .... 6, 78
Delimitation in the Maritime Boundary in the Gulf of Maine Area (*Canada v United States*)
 (Judgment) [1984] ICJ Reports 246 88
Fisheries Jurisdiction (*United Kingdom v Iceland*) (Jurisdiction) [1974] ICJ Reports 3 142
Gabčíkovo-Nagymaros Project (*Hungary v Slovakia*) (Judgment) [1997] ICJ Reports 7 102
Legality of the Threat or Use of Nuclear Weapons (Advisory Opinion) [1996]
 ICJ Reports 266 .... 102, 106
Maritime Delimitation and Territorial Questions Between Qatar and Bahrain (*Qatar v
 Bahrain*) (Merits) [2001] ICJ Reports 40 .... 82, 88
North Sea Continental Shelf (*Federal Republic of Germany v Denmark and the Netherlands*)
 (Judgment) [1969] ICJ Reports 3 .... 59, 139, 142
Nuclear Tests (*Australia v France*) (Merits) [1974] ICJ Reports 253; *(New Zealand v France)*
 (Merits) [1974] ICJ Reports 457 90
Pulp Mills on the River Uruguay (*Argentina v Uruguay*) (Judgment) [2010]
 ICJ Reports 14 .... 102, 103, 106
Reparation for Injuries Suffered in the Service of the United Nations (Advisory Opinion)
 [1949] ICJ Reports 174 147
Territorial Dispute (*Libyan Arab Jamahiriya v Chad*) (Judgment) [1994] ICJ Reports 6 135
Territorial and Maritime Dispute (*Nicaragua v Colombia*) (Judgment) [2012]
 ICJ Reports 624 .... 77–8, 84, 85, 88–9
Territorial and Maritime Dispute in the Caribbean Sea (*Nicaragua v Honduras*) (Judgment)
 [2007] ICJ Reports 659 .... 75–6, 77–8

INTERNATIONAL TRIBUNAL FOR THE LAW OF THE SEA

Arctic Sunrise (*Netherlands v Russian Federation*) (Provisional Measures) (2014) 53 ILM 607
 (*see also* International Arbitration) 205, 212
Delimitation of the Maritime Boundary between Bangladesh and Myanmar
 in the Bay of Bengal (*Bangladesh v Myanmar*) (Judgment) (2012)
 51 ILM 844 .... 2, 7, 70, 76–7
M/V 'Virginia G'(*Panama v Guinea-Bissau*) (Judgment) (2014) 53 ILM 1164 211

Request for an Advisory Opinion Submitted by the Sub-Regional Fisheries Commission 54
ILM 893 (2015) 98
Responsibilities and Obligations of States Sponsoring Persons and Entities
with Respect to Activities in the Area (Advisory Opinion) (2011)
50 ILM 458 .... 102, 103, 138

INTERNATIONAL ARBITRATION

Arctic Sunrise (*Netherlands v Russian Federation*) (Merits) PCA Case 2014-02, 14 August
2015 (*see also* ITLOS). 17, 185, 187, 189, 212–14
Barbados v Trinidad and Tobago (Award) (2006) 45 ILM 800 .... 2, 6–7
Chagos Marine Protected Area (*Mauritius v United Kingdom*) (Award) PCA Case No 2011-3,
18 March 2015. .... 183, 186, 189, 192
Delimitation of Maritime Areas between Canada and France (St Pierre and Miquelon)
(Award), 10 June 1992, 31 ILM (1992) 1145. 85
Guyana v Suriname (Arbitration) 139 ILR 566 (2007) 145, 163
Lac Lanoux Arbitration (*France v Spain*) (Award) 12 RIAA 306; 24 ILR 101 (1958). 142
South China Sea (*Philippines v China*) (Award) PCA Case No 2013-19,
12 July 2016. .... 97–8, 107
Trail Smelter (*United States v Canada*) (Award) 3 RIAA 1911 101

Table of Treaties

Agreement between the Government of Australia and the Government of the Democratic Republic of Timor-Leste relating to the Unitization of the Sunrise and Troubadour Fields (6 March 2003) 2483 UNTS 317.................. 140–1

Agreement between the Government of the United Kingdom of Great Britain and Northern Ireland and the Government of the Kingdom of Norway relating to the Delimitation of the Continental Shelf between the Two Countries (10 March 1965) 551 UNTS 214 141
  Art 4........................... 141

Agreement between United Kingdom of Great Britain and Northern Ireland and Norway Relating to the Exploitation of the Frigg Field Reservoir and the Transmission of Gas therefrom to the United Kingdom (10 May 1976) [1978] UNTS 4...................... 140

Agreement between the Government of the United States of America and the Government of the Union of Soviet Socialist Republics Relating to Fishing for King Crab (5 February 1965) 4 ILM 359 .... 66, 227, 240

Agreement between the United States of America and the United Mexican States Concerning Transboundary Hydrocarbon Reservoirs in the Gulf of Mexico (20 February 2012) TIAS 14-0718 ..... 141, 142, 143
  Art 6........................... 143
  Art 7........................... 143
  Art 8........................... 143
  Art 9........................... 143
  Art 13.......................... 144
  Art 14.......................... 143
  Art 18.......................... 143
  Art 19.......................... 143
  Art 24.......................... 143

Agreement for the Implementation of the Provisions of the United Nations Convention on the Law of the Sea of 10 December 1982 relating to the Conservation and Management of Straddling Fish Stocks and Highly Migratory Fish Stocks (opened for signature 4 December 1994, entered into force 11 December 2001) 2167 UNTS 88............. 226, 233, 244
  Art 7............................ 233
  Art 8(4) 226, 234

Agreement relating to the Implementation of Part XI of the United Nations Convention on the Law of the Sea of 10 December 1982 (opened for signature 28 July 1994, entered into force 28 July 1996) 33 ILM 1309 90

Convention for the Conservation of Antarctic Marine Living Resources (opened for signature 20 May 1980, entered into force 7 April 1982) 1329 UNTS 48 224

Convention for the Protection of the Marine Environment of the North-East Atlantic (opened for signature 22 September 1992, entered into force 25 March 1998) 32 ILM 1069.................. 80, 109, 218

Convention for the Protection of the World Cultural and Natural Heritage (opened for signature 16 November 1972, entered into force 17 December 1975) 1037 UNTS 151........................... 237

Convention for the Suppression of Unlawful Acts that Endanger the Safety of Navigation (opened for signature 4 March 1988, entered into force 1 March 1992) 1678 UNTS 222..................... 199

Convention on Biological Diversity
(opened for signature 5 June 1992,
entered into force 29 December
1993) 760 UNTS 79 .... 33, 93, 169
Art 1. 115
Art 3. . . . . 106, 115
Art 4. . . . . 99, 115
Art 5. 106
Art 6. . . . . 99, 116
Art 7. 116
Art 8. . . . . 99, 106, 116
Art 9. 116
Art 10. 116
Art 14. . . . . 99, 106, 116
Art 15. 115
Art 15(1) 116
Art 15(7) 116
Art 22. 117
Convention on the Conservation and
Management of Fishery Resources
in the South East Atlantic Ocean
(opened for signature 20 April
2001, entered into force 13 April
2003) 2221 UNTS 189 232
Convention on Fishing and the
Conservation of the Living
Resources of the High Seas (opened
for signature 29 April 1958, entered
into force 20 March 1966) 559
UNTS 286. 57
Convention on Future Multilateral
Cooperation in North-east Atlantic
Fisheries (opened for signature 18
November 1980, entered into force
17 March 1982, as amended in
2006) 1285 UNTS 129 217
Convention on Future Multilateral
Cooperation in the North-west
Atlantic Fisheries (opened for
signature 24 October 1978, entered
into force 1 January 1979) 1135
UNTS 369. . . . . 217, 232
Convention on the Continental Shelf
(opened for signature 29 April
1958, entered into force 10 June
1964) 499 UNTS 312 52,
94, 152, 175, 176, 205
Art 1. . . . . 57, 68, 83, 209
Art 2. . . . . 58, 97, 205
Art 2(4) 58, 63, 64

Art 3. . . . . 58, 177–9, 205
Art 5. . . . . 147, 177
Art 5(1) 16, 58, 177–9, 185, 205
Art 5(2) 58, 175
Art 5(3) 58, 175
Art 5(8) 152, 160–2
Art 6. 88
Convention on the High Seas (opened
for signature 29 April 1958, entered
into force 30 September 1962) 450
UNTS 82. 57, 152
Convention on the International
Regulations for Preventing
Collisions at Sea (opened for
signature 10 October 1972, entered
into force 15 July 1977) 1050
UNTS 18. 198
Convention on International Trade in
Endangered Species of Wild Flora
and Fauna (opened for signature 3
March 1973, entered into force 1
July 1975) 993 UNTS 243. 237
Convention on the Prevention of Marine
Pollution by Dumping of Wastes
and Other Matter 1972 (opened
for signature 29 December 1972,
entered into force 30 August
1975) 1046 UNTS 138 . . . . 46, 100
Convention on the Territorial Sea and
the Contiguous Zone (opened for
signature 29 April 1958, entered
into force 10 September 1964) 516
UNTS 206. 57
Exchange of Notes Constituting an
Agreement between the United
States of America and Japan Relating
to the King Crab Fishery in the
Eastern Bering Sea (25 November
1964) 1965 UNTS 32. . . . . 66, 240
International Convention for the
Prevention of Pollution from
Ships, as Modified by the Protocol
of 1978 (opened for signature 2
November 1971 and 17 February
1978, entered into force 2
October 1982) 1340 UNTS 62
(MARPOL 73/78) 100, 234
International Covenant on Civil and
Political Rights (opened for
signature on 19 December 1966,

entered into force 28 March 1979) 999 UNTS 171
  Art 19........................... 199
Nagoya Protocol on Access to Genetic Resources and the Fair and Equitable Sharing of Benefits Arising from Their Utilization to the Convention on Biological Diversity 116
  Art 6............................ 117
  Art 8............................ 117
  Art 10........................... 117
  Art 11........................... 117
Timor Sea Treaty between the Government of East Timor and the Government of Australia (20 May 2002) 2258 UNTS 3 146
Treaty between Australia the Government of Democratic Republic of Timor-Leste on Certain Maritime Arrangements in the Timor Sea (12 January 2006) 2483 UNTS 359..................... 146
Treaty between the Government of the United States of America and the Government of the United Mexican States on the Delimitation of the Continental Shelf in the Western Gulf of Mexico Beyond 200 Nautical Miles (9 June 2000) 2143 UNTS 417..................... 142
  Art 4............................ 143
  Art 5(1) 142
Treaty between Great Britain and Northern Ireland and Venezuela relating to the Submarine Areas of the Gulf of Paria (26 February 1942) 205 LNTS 121............... 54
Treaty Concerning the Joint Exercise of Sovereign Rights over the Continental Shelf in the Mascarene Plateau Region (13 March 2012) *Law of the Sea Bulletin* 79 (2012) 26..................... 228
Treaty Concerning the Joint Management of the Continental Shelf in the Mascarene Plateau Region (13 March 2012) *Law of the Sea Bulletin* 79 (2012) 41 228
  Art 4............................ 228

Art 5............................ 229
Art 12........................... 229
United Kingdom of Great Britain and Northern Ireland and Norway: Agreement Relating to the Exploitation of the Frigg Field Reservoir and the Transmission of Gas Therefrom to the United Kingdom (10 May 1976) 1978 UNTS 4....................... 140
United Nations Convention on the Law of the Sea (opened for signature 10 December 1982, entered into force 16 November 1994) 1834 UNTS 397............... 2, 52, 93, 152, 203
  Preamble 81
  Pt V 9, 95, 119
  Pt VI 9–11, 14, 15, 16, 52, 53, 81, 82, 92, 95, 96, 97, 108, 119, 121, 129, 175, 177, 180, 185, 193, 194, 204, 205, 215, 249
  Pt XI 90, 128, 130, 133, 139, 181, 246
  Pt XII............... 14, 97–8, 107, 121
  Pt XV........................... 138
  Art 1............................. 60
  Art 2......................... 8, 186
  Art 3.............................. 8
  Art 5.............................. 8
  Art 17............................. 9
  Art 18............................. 9
  Art 19............................. 9
  Art 21............................. 9
  Art 21(1) 153
  Art 22............................ 99
  Art 24............................. 9
  Art 25......................... 203–4
  Art 33........................... 203
  Art 42(2)........................ 181
  Art 44........................... 181
  Art 52........................... 181
  Art 53(2) 181
  Art 56................. 9, 182, 204, 211
  Art 56(1) 9, 95, 96, 174, 184, 204
  Art 56(2) 182, 183, 184, 186
  Art 56(3) 95, 180, 185, 193
  Art 57............................. 9
  Art 58............................. 9
  Art 58(2) 186
  Art 58(3) 180
  Art 59....................... 183, 191

Art 60. . . . . 13, 147, 148, 149, 150, 156, 174, 175, 193, 196, 200, 205, 213
Art 60(2) 212
Art 60(3) 182
Art 60(4) 212
Art 60(5) 149
Art 61. . . . . 95, 119
Art 62. 119
Art 62(2) 96
Art 68. 95
Art 73. . . . . 204–5, 212, 213, 214
Art 74(3) 80, 144–5, 163
Art 76. . . . . . 2, 10, 11, 15, 20, 50, 52–3, 62, 67, 69–71, 72, 74, 76–7, 79, 80, 82–85, 86, 87, 88, 89, 90–1, 123, 126, 129, 147, 151, 210, 249
Art 76(1) 69, 80, 82, 85, 87
Art 76(2) 82, 87
Art 76(3) 20, 69, 82, 85, 87
Art 76(4) 69, 79, 82, 85, 87
Art 76(5) 69, 79, 80, 82, 85, 87
Art 76(6) 69–70, 79, 80, 82, 85, 87
Art 76(7) 70, 80, 82, 85, 87
Art 76(8) 73, 78, 85, 86, 88
Art 76(9) 85, 86
Art 76(10) 70
Art 77. . . . . 11–12, 52, 76, 80, 82, 101, 105, 150, 157, 167, 205, 230, 242, 248
Art 77(1) 2
Art 77(2) 2, 73, 96, 129
Art 77(3) 73, 76
Art 77(4) 12, 27, 64, 67, 95, 121, 135, 165, 170, 210
Art 78. . . . . 12, 16, 82, 176, 177, 180, 181, 182, 185, 187, 188, 189, 191, 193, 202, 204, 214, 248
Art 78(1) 177, 205
Art 78(2) 110, 120, 149, 178, 185, 186–7, 189, 201, 202, 205, 227, 249
Art 79. . . . . 12, 182
Art 80. . . . . 13, 148, 150, 156, 174, 205
Art 81. . . . . 13, 150
Art 82. . . . . 2, 11, 13, 15, 16, 53, 61, 82, 84, 86–91, 123–5, 127–32, 133, 134, 136–9, 142, 143–4, 147, 150, 151, 157
Art 82(1) 130–2, 134
Art 82(2) 131, 132, 134, 150
Art 82(3) 131, 134–5, 144, 150
Art 82(4) 131, 134, 135–6, 138
Art 83. . . . . 13–14, 76, 79, 242
Art 83(3) 80, 144–5, 163
Art 84. 14
Art 85. 14
Art 87. 9
Art 87(1) 156
Art 87(2) 176, 182
Art 88. 129
Art 92. 204
Art 110. . . . . 204, 212
Art 111. . . . . 207, 212, 213, 215
Art 111(2) 205
Art 118. 194
Art 121. . . . . 82, 89
Art 121(1) 88
Art 121(2) 88, 89
Art 121(3) 88
Art 133. 246
Art 136. . . . . 9, 10, 60, 128
Art 137. 9
Art 139. 103
Art 140. . . . . 10, 60, 147
Art 141. 129
Art 142. . . . . 146, 182, 236
Art 142(2) 146, 236
Art 143. 62
Art 148. 182
Art 160(2) 136, 137, 182
Art 163(4) 182
Art 176. 147
Art 187. . . . . 137, 138
Art 188. 137
Art 189. . . . . 137, 138
Art 190. 137
Art 192. . . . . 97, 98, 99, 102, 113, 120
Art 193. . . . . 97, 101
Art 194. . . . . 98, 102, 106
Art 194(2) 101–2
Art 194(4) 181, 186
Art 204. 98
Art 206. . . . . 98, 99, 106, 113
Art 220. . . . . 204, 212
Art 221. 212
Art 226. 212
Art 239. 118
Art 240. . . . . 153, 156, 167
Art 244. 112
Art 246. . . . . 61, 113, 118, 153, 157, 158, 162, 171
Art 246(1) 162
Art 246(3) 117, 156, 168
Art 246(5) 13, 118, 156, 157, 166, 170, 174

Art 246(6) 3, 14, 16, 61, 110, 113, 128, 154, 156, 157, 162, 164–9, 171–2, 241, 249	Art 300. 127
Art 246(7) 157, 167	Annex II. 72, 73
Art 246(8) 157, 167	Art 1. 72
Art 249. 170	Art 2. 72
Art 249(1) 171	Art 4. 72
Art 249(2) 170	Annex III 130, 133
Art 256. 162	Art 13(12) 133
Art 257. .... 156, 162	Vienna Convention on the Law of Treaties (opened for signature 23 May 1969, entered into force 27 January 1980) 1155 UNTS 331. 127
Art 267. 182	
Art 279. 138	Art 31. .... 127, 164
Art 286. 138	Art 31(1) 135
Art 297(1) 138	Art 32. 127
Art 297(2) 118, 157, 158, 168	Art 34. .... 83, 226
Art 298(1) 212	Art 38. 83

List of Abbreviations

ABS	Access and benefit sharing
BBNJ	Biodiversity beyond national jurisdiction
CBD	Convention on Biological Diversity
CCAMLR	Convention for the Conservation of Antarctic Marine Living Resources
CLCS	Commission on the Limits of the Continental Shelf
Continental Shelf Convention	Geneva Convention on the Continental Shelf 1958
COP	Conference of the Parties
EBSA	Ecologically and Biologically Significant Area
EEZ	Exclusive Economic Zone
EIA	Environmental Impact Assessment
EMEPC	Task Group for the Extension of the Portuguese Continental Shelf
EU	European Union
IA	International Agreement on Biodiversity Beyond National Jurisdiction
ICCAT	International Commission for the Conservation of Atlantic Tunas
ICJ	International Court of Justice
ILA	International Law Association
ILC	International Law Commission
IMO	International Maritime Organization
ISA	International Seabed Authority
ITLOS	International Tribunal for the Law of the Sea
IWC	International Whaling Commission
JMA	Joint Management Area, Mauritius and the Seychelles
LOSC	Law of the Sea Convention
MOU	Memorandum of understanding
MPA	Marine Protected Area
MSR	Marine Scientific Research
NAFO	North-west Atlantic Fisheries Organization
NASCO	North Atlantic Salmon Conservation Organization
NEAFC	North-east Atlantic Fisheries Commission
OSPAR	Commission Established by the Convention for the Protection of the Marine Environment of the North-east Atlantic
PSSA	Particularly Sensitive Sea Area
SDC	Seabed Disputes Chamber of the International Tribunal for the Law of the Sea
SEAFO	South-east Atlantic Fisheries Organization
SPRFMO	South Pacific Regional Fisheries Management Organization
RFMO	Regional Fisheries Management Organization

List of Abbreviations

UNEP	United Nations Environment Programme
UNESCO	United Nations Educational, Scientific and Cultural Organization
UK	United Kingdom
US	United States of America
USSR	Soviet Union; Union of Soviet Socialist Republics
VCLT	Vienna Convention on the Law of Treaties
WHC	World Heritage Convention
WSSD	World Summit on Sustainable Development

1
Introduction

The continental shelf beyond 200 nautical miles (nm) is receiving increasing attention from States for several reasons. First, many States have now completed submissions to the Commission on the Limits of the Continental Shelf (CLCS) and some have received recommendations on the basis of which they can establish the outer limits of their continental shelf.[1] Secondly, human activities on the outer reaches of the continental shelf are increasing in scope and intensity and the rate of the increase will be exponential in future. Thirdly, an improving understanding of the diverse and sensitive ecosystems that can be found on the continental shelf beyond 200 nm means that coastal States will need to balance conservation and exploitation needs to ensure sustainable development.

This book focuses on the following questions. What does it mean to have sovereign rights over the resources of the continental shelf beyond 200 nm? What is the limit of permissible coastal State control over activities impacting on the seabed? Where coastal State interests intersect with the interests of flag States, how are these issues to be resolved? How can States protect their interests through regional or international organizations?

A number of valuable resources can be found on continental shelves, including oil and gas, minerals such as gold and copper, and living organisms that could yield valuable genetic resources for pharmaceuticals and other commercial products. Rather than being devoid of life, the continental shelf hosts strange and beautiful ecosystems. On some seamounts, cold-water corals that are hundreds, or even thousands, of years old may be found alongside sponges and sea lilies that are attached to the sea floor. These suspension feeders provide habitats for small, mobile invertebrates such as molluscs, crustaceans, and echinoderms. These ecosystems may be the basis of a food chain that includes valuable commercial fish species. Entire ecosystems may be based not on oxygen, but on the chemicals forced from the sea floor around hydrothermal vents and cold seeps. Ghost-like crabs crawl around tubeworms and enormous mats of bacteria form on the sea floor. Such resources and ecosystems may be found in a variety of locations on a State's continental shelf.

Continental shelves vary in length, depth and constitution. Some shelves will not extend far from the coastal area, descending rapidly to the deep sea. In other

---

[1] By 31 May 2016, the CLCS had received seventy-seven submissions and issued twenty-four recommendations. See www.un.org/depts/los.

places, the continental shelf may extend for hundreds of miles before beginning to slope down to meet the sea floor. Scientists refer to the shelf (a shallower area that extends from the coast to an edge or break), the slope (a relatively steep slope) and the rise of the slope (an area of gentle gradient) as the continental margin. Their research is centred around the physical extent of the margin, be it small or large. For international lawyers, the focus is on the 'juridical', or legal, shelf.[2] Under international law, the 'continental shelf' includes the shelf, slope and rise of the continental shelf.[3]

The United Nations Convention on the Law of the Sea (also called Law of the Sea Convention, or LOSC)[4] sets out the rights that coastal States may exercise over the juridical continental shelf, and their obligations. All coastal States have sovereign rights over the resources of the sea floor out to 200 nm (unless they are limited as a consequence of the claims of a neighbouring or opposite State). There are some cases where the physical continental shelf extends beyond 200 nm. The LOSC contains complicated rules to determine the precise extent of the continental shelf in these cases. Beyond 200 nm, a coastal State's rights and obligations in respect of the continental shelf must be exercised consistently with the rights and obligations of other States whose nationals might be navigating, fishing, researching or undertaking other activities near the continental shelf.

The first thing to note is that the rights of the coastal State to the resources of the continental shelf apply to the entire continental shelf, regardless of its distance from the coast. This has been confirmed by the International Tribunal for the Law of the Sea (ITLOS), which has stated that 'Article 76 embodies the concept of a single continental shelf'.[5] Article 77(1) and (2) provide that 'the coastal State exercises exclusive sovereign rights over the continental shelf in its entirety without any distinction being made between the shelf within 200 nm and the shelf beyond that limit'.[6] An arbitral tribunal in *Barbados v Trinidad and Tobago*, in the context of a delimitation case, stated that 'there is in law only a single "continental shelf" rather than an inner continental shelf and a separate extended or outer continental shelf'.[7]

The need for a separate examination of the legal issues arising from the rights of coastal States over the continental shelf beyond 200 nm arises for two reasons.

First, the LOSC itself contains specific articles that apply only in respect of the continental shelf beyond 200 nm. Article 82 requires coastal States to make

---

[2] See Figure 1.2.
[3] See ch 2 for more details.
[4] United Nations Convention on the Law of the Sea (opened for signature 10 December 1982, entered into force 16 November 1994) 1834 UNTS 397.
[5] Delimitation of the Maritime Boundary between Bangladesh and Myanmar in the Bay of Bengal *(Bangladesh v Myanmar)* (2012) 51 ILM 844, para. 361 (*Bay of Bengal* case). Article 76 of the LOSC establishes that a coastal State has jurisdiction over the continental shelf to the edge of the continental margin and imposes criteria by which the outer limits of the shelf should be determined.
[6] *Bay of Bengal* case, para. 361.
[7] *Barbados v Trinidad and Tobago* (2006) 45 ILM 800, para. 213 (*Barbados v Trinidad and Tobago* case).

payments or contributions in kind to the International Seabed Authority (ISA) representing a percentage of the production of non-living resources on the continental shelf beyond 200 nm.[8] Article 246(6) limits the ability of the coastal State to refuse consent for marine scientific research projects that may have significance for the exploration and exploitation of resources on the extended part of the continental shelf.

Secondly, the legal regime that applies to the water column above the shelf changes from the exclusive economic zone (EEZ) within 200 nm, to the high seas beyond 200 nm. Because activities on the extended continental shelf will almost always take place partly in the high seas, the intersection between the two regimes needs to be carefully examined. Within 200 nm, States typically have exclusive jurisdiction to regulate activities such as fishing, mining and the protection of the marine environment. This includes a right to enforce regulations against foreign-flagged vessels contravening those laws in the EEZ. Outside the 200 nm limit, although coastal States have the same substantive rights in relation to the continental shelf and its resources, activities in the water column above are governed by high seas freedoms. This will often change the way in which States can exercise their coastal State rights. For example, within its EEZ, a coastal State may restrict bottom fishing in a particular area to protect a vulnerable marine ecosystem on the sea floor. Beyond 200 nm, there is a question whether such a restriction may interfere with a foreign State's freedom of fishing.

The purpose of this book is to focus on the legal issues that require coastal States to regulate activities on and above the continental shelf beyond 200 nm differently to those within 200 nm. It is also important that States with vessels navigating and conducting activities in the high seas above a coastal State's continental shelf are aware of the extent to which coastal State regulations might appropriately limit such activities. The goal is to provide a framework for negotiating the potentially difficult problems that may arise when high seas freedoms and continental shelf rights are seen as being in conflict with one another.

The eminent scholar DP O'Connell opened his chapter on the continental shelf as follows:[9]

The doctrine of the continental shelf has been relegated at the Third Law of the Sea Conference to playing the minor and ancillary role of extending the exclusive rights of coastal States over the seabed, in a relatively few cases, to distances beyond the 200 miles of the EEZ. As an autonomous institution of international law, then, its importance is likely to be ephemeral, but as the device which mediated between the high seas and the territorial sea, or between the freedom of the high seas and coastal State sovereignty, its historic role has been of fundamental significance.

---

[8] The ISA is obliged to distribute the payments to State parties to the LOSC. At the time of writing, no State has undertaken exploitation on the continental shelf beyond 200 nm and so no payments or contributions have been made under article 82. See ch 5 for a discussion of article 82.

[9] DP O'Connell, *The International Law of the Sea: Volume I*, edited by Ivan Shearer, (Oxford: Clarendon Press, 1982): p. 467.

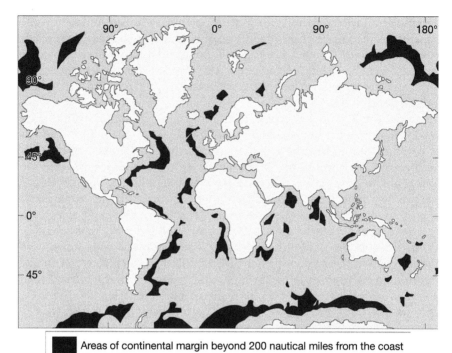

Figure 1.1 Potential for expanded continental shelf claims
Source: Based on Victor Prescott and Clive Schofield, *Maritime Political Boundaries of the World*, 2nd edn (Leiden: Martinus Nijhoff, 2004): p. 571.

O'Connell was correct that the EEZ has become the primary focus for States when considering the management of their rights to resources beyond the territorial sea. However, he might have been surprised to discover that, by January 2016, 66 coastal States had filed full or partial submissions with the CLCS and a further 14 States had submitted preliminary information but not a submission.[10] This amounts to more than 'a relatively few' States who believe that they may have rights to an area of the seabed beyond 200 nm from their coasts. Figure 1.1 gives a general indication of the extent of the potential expansion of coastal State jurisdiction. In light of this number and the increasing potential for human activities to take place on the outer parts of the continental shelf, it cannot be said today that the doctrine's importance is likely to be ephemeral.

It is also true that the traditional doctrine of the continental shelf was invaluable in mediating between the rights of the coastal State and the freedom of the high seas. For this reason, the analysis in this book is heavily influenced by the historical development of the continental shelf doctrine, including its enshrinement in the 1958 Geneva Convention on the Continental Shelf (Continental Shelf

---

[10] See www.un.org/depts/los. A total of forty-one States filed preliminary information with the Commission but subsequently twenty-seven of these have made full or partial submissions.

Convention). State practice under that Convention, as well as commentaries by the International Law Commission, will assist in the interpretation of how the continental shelf doctrine should apply to the continental shelf beyond 200 nm under the LOSC.

This book focuses on the nature of the coastal State's sovereign rights over the extended shelf and how these rights can be exercised consistently with the provisions of the LOSC. This book will not engage in any detail with the legal issues relating to the delineation of the continental shelf beyond 200 nm.[11] This is the subject of an excellent and comprehensive literature.[12] As will be established, the exercise of coastal State rights over the continental shelf beyond 200 nm is not dependent on the delineation of the outer limits. Where a continental shelf physically extends beyond 200 nm, the coastal State's sovereign rights are inherent and exist *ab initio*. Of course, the precise boundaries of the area in which the State can exercise these rights may be in doubt and, for this reason, coastal States should be extremely cautious in exercising rights in areas that may be disputed.

This book does not take a particular perspective on what policies a coastal State should take towards regulating activities on the continental shelf beyond 200 nm: this is a matter for States themselves. Rather, the focus is on using the scientific and technical information available to understand the scope of potential activities and consider the range of policies that States might adopt. The book aims to identify the appropriate limits of coastal State action. This will also, of course, be of interest to non-coastal States whose vessels may be operating in the vicinity of the coastal States' shelves.

In addition, the book only examines those issues relating to the extended shelf that are treated differently to the rest of the shelf because it is required due to either the provisions of the LOSC, or the fact that it lies under the high seas. For example, very few distinct issues are likely to arise about the laying of cables and pipelines on the extended shelf and so this book does not focus on that topic.

There are a number of places where one State's claim to a continental shelf beyond 200 nm is contested by another State. This book does not enter into detailed analysis of the delimitation of maritime boundaries.[13] There are duties under international law to seek to resolve disputes in good faith through peaceful means and it is

---

[11] A short overview of the process for delineation of the outer limits of the continental shelf is found in ch 3. Delineation refers to the process of establishing the outer limits of the continental shelf, which distinguishes between the shelf and the seabed beyond national jurisdiction. Delimitation refers to the process of determining the maritime boundaries between neighbouring or opposite States.

[12] See, e.g. Peter J Cook and Chris M Carleton (eds), *Continental Shelf Limits: The Scientific and Legal Interface* (Oxford: Oxford University Press, 2000); Suzette V Suarez, *The Outer Limits of the Continental Shelf: Legal Aspects of their Establishment* (Berlin: Springer, 2008); Øystein Jensen, *Commission on the Limits of the Continental Shelf: Law and Legitimacy* (Leiden: Martinus Nijhoff Publishers, 2014); Bjarni Már Magnússon, *The Continental Shelf Beyond 200 Nautical Miles* (Leiden: Brill Nijhoff, 2015).

[13] See ch 5 for a discussion of the issues arising regarding transboundary resources in situations of disputed boundaries.

possible for States with contested marine boundaries to work together for mutually beneficial outcomes pending a final resolution of the dispute. The rules of State responsibility will also be relevant where a State takes action that harms another State and is inconsistent with international law.

Finally, the book does not explore 'best practice' environmental management concepts that States could choose to apply to the governance of the continental shelf beyond 200 nm. Spatial management, integrated management, ecosystems-based management, adaptive management and the precautionary approach are all approaches that a coastal State can, and should, explore to order its domestic environmental management regime. This book largely focuses on the legal issues that apply particularly when the coastal State is exercising its rights and obligations on the continental shelf beyond 200 nm, and so those management issues are beyond the scope of the book.

A. Terminology

An initial problem arises when describing the continental shelf beyond 200 nm. At times, references have been made by judges and authors to the 'outer' or 'extended' part of the continental shelf. This approach is subject to criticism, primarily because of the point mentioned above: the continental shelf should properly be considered to be a single area. It may be tempting to treat the continental shelf as comprising an 'inner' shelf (within 200 nm) and an 'outer' shelf (outside 200 nm) because coastal States are subject to slightly different legal obligations beyond 200 nm. However, the LOSC is clear that the sovereign rights over the shelf apply to the entire shelf without distinction.[14] It has been suggested that referring to an 'outer' or 'extended' shelf may cause confusion about the coherency of the continental shelf.[15] In the ITLOS and some other arbitral tribunals, many judgments have not attempted to use a shortened form of referring to this area; instead, the 'continental shelf beyond 200 nm' has been used in full. The International Court of Justice (ICJ) also uses the full title.[16]

Despite this criticism, it must be acknowledged that the full title is a mouthful. Writers and others have sought a more succinct way to describe the part of the continental shelf beyond 200 nm without implying that there are two separate parts to the shelf. In a footnote, the Arbitral Tribunal in the *Barbados v Trinidad and Tobago*

---

[14] Article 77, LOSC.
[15] Frida Armas-Pfirter, 'Working Paper on Potential Options on Equitable Distribution of Payments and Contributions', Annex 6 in International Seabed Authority, *Implementation of Article 82 of the United Nations Convention on the Law of the Sea: Report of an International Workshop*, Technical Study No. 12 (2013): pp. 83–97, p. 86.
[16] In the recent *Nicaragua v Colombia* case on the continental shelf beyond 200 nm, the ICJ primarily uses 'continental shelf beyond 200 nautical miles', but occasionally refers to 'extended continental shelf' when referring to the parties' submissions. Delimitation of the Continental Shelf between Nicaragua and Colombia beyond 200 nautical miles from the Nicaraguan Coast *(Nicaragua v Colombia)* (Preliminary Objections) ICJ, 17 March 2016.

case rejected the use of 'extended continental shelf' on the basis that it might imply that the continental shelf was being extended, which is of course inconsistent with the single concept of the continental shelf.[17] Instead, in that case, the Tribunal preferred to use 'outer continental shelf'. However, this phrase also has the potential to cause confusion because the United States uses the 'outer continental shelf' to describe the part of the continental shelf beyond the control of states in the federal system (which usually control up to three nm from the coastline).[18] In addition, that wording could also imply that there are two different parts of the continental shelf.[19]

Indeed, neither of the more commonly used alternative descriptions of the continental shelf beyond 200 nm is ideal. However, this book does use 'extended continental shelf' to describe the continental shelf beyond 200 nm. The main reason is that this approach is less cumbersome than continually using the full description. This may not always be ideal but it is hoped that the context of the book will allay any concerns about a misunderstanding about the unitary concept of the continental shelf.

A second liberty taken in this book is that 'nm' is used to refer to nautical miles. In scientific literature, 'nm' refers to nanometres and 'N' refers to nautical miles. However, in the legal literature it is common for authors to shorten the reference in this way. It is also used in some international legal cases[20] and so this book has adopted this abbreviated format.

B. An Overview of Coastal States' Rights and Obligations on the Continental Shelf

(1) The development of the law of the sea

It has been over 400 years since Hugo Grotius published his work *Mare Liberum*.[21] Although he was not the first to make the argument, his thesis famously was that the oceans could not be appropriated because they were limitless and not capable of occupation. Everyone could enjoy navigation and fishing on the oceans. The concept of the freedom of the seas was attractive because it allowed all States to operate on the oceans and share in their resources. This became the organizing principle for the law of the sea.

---

[17] *Barbados v Trinidad and Tobago* case, p. 165.
[18] Outer Continental Shelf Lands Act 1953 (United States) 43 United States Code, Subchapter III, from §1331.
[19] Clive Schofield, 'New Marine Resource Opportunities, Fresh Challenges' *University of Hawaii Law Review* 35 (2013): pp. 715–33, p. 718.
[20] See, e.g. Delimitation of the Maritime Boundary between Bangladesh and Myanmar in the Bay of Bengal *(Bangladesh v Myanmar)* (2012) 51 ILM 844, para. 42; *Barbados v Trinidad and Tobago* case, para. 42.
[21] Hugo Grotius, *The Freedom of the Seas or the Right Which Belongs to the Dutch to Take Part in the East Indian Trade*, translated by Ralph Van Deman Mogoffin (Oxford: Oxford University Press, 1916).

The freedom of the seas was challenged in a limited way by the expansion of control of the waters around the coastline by coastal States. This expansion of coastal State jurisdiction was initially motivated by the desire to ensure the security of the State from foreign naval attack, but also became popular for the protection of fisheries resources.[22] Thus the notion of the territorial sea developed, over which the coastal State exercised a level of control similar to that over its territory.[23]

The law of the sea has been characterized by a tension between the dominant theory of freedom of the high seas and a gradually expanding notion of coastal State jurisdiction over activities in the oceans adjacent to the coast.[24] Over time, coastal State jurisdiction has gradually made inroads into the initial idea of freedom of the seas. This has been described as the phenomenon of 'creeping jurisdiction'.[25] This caused great instability in the law of the sea, with States making inconsistent claims to jurisdiction throughout the twentieth century. The first United Nations conferences produced four Conventions on the law of the sea in 1958. However, these Conventions were undermined by ongoing dissatisfaction among coastal States about the extent of the territorial sea and fisheries zones. In 1982, the LOSC settled many of these disputes in a 'package deal', which meant that States either had to accept the whole of the Convention, or nothing. The negotiations were conducted on the basis of achieving consensus, in the hope that all States would come to accept the new Convention as reflecting a balance of all interests.[26]

One of the major achievements of the LOSC was the agreement on the rights of States in a range of maritime zones. The relevant zones for this book are the territorial sea, the EEZ, the continental shelf, the high seas and the Area.[27] As the maritime zones becomes more distant from the coast, coastal State jurisdiction reduces and the strength of high seas freedoms increases.

In the territorial sea, the coastal State has sovereignty over the ocean up to twelve nm from its baselines.[28] Foreign vessels have the right of innocent passage, which

---

[22] Donald R Rothwell and Tim Stephens, *The International Law of the Sea* (Oxford: Hart, 2010): p. 4.

[23] For an excellent discussion of the history of the law of the sea, see DP O'Connell, *The International Law of the Sea: Volume I*, edited by Ivan Shearer (Oxford: Clarendon Press, 1982).

[24] H Lauterpacht, 'Sovereignty over Submarine Areas' *British Yearbook of International Law* 27 (1950): pp. 376–433, p. 378.

[25] See, e.g. Barbara Kwiatkowska, 'Creeping Jurisdiction Beyond 200 Miles in the Light of the 1982 Law of the Sea Convention and State Practice' *Ocean Development and International Law* 22 (1991): pp. 153–87.

[26] Allott described this as involving 'a partial loss of autonomy on the part of the members of the group in exchange for a share in the enhanced power of the group', resulting in 'cumulative causation ... as each member seeks to promote its own interest and thereby contributes to the communal benefit of the group'. Philip Allott, 'Power Sharing in the Law of the Sea' *American Journal of International Law* 77 (1983): pp. 1–30, p. 6.

[27] This is clearly a simplification of the complex maritime areas created by the LOSC. Other important zones in the LOSC, but which are not important to this book, include internal waters, straits and archipelagic waters. The complexity of the legal situation under the LOSC is reflected in the fact that Allott identified fifty-eight different 'legal sea areas' in the LOSC. Allott, 'Power Sharing in the Law of the Sea', p. 28.

[28] Articles 2 and 3, LOSC. The normal baseline is the low-water line along the coast as marked on large-scale charts officially recognized by the coastal State. Article 5, LOSC.

must not be hampered by the coastal State.²⁹ The coastal State can regulate aspects of innocent passage to prevent infringement of its interests.³⁰

The EEZ was first established in the LOSC and can extend to a distance of no more than 200 nm from a State's coast.³¹ In the EEZ, the coastal State has certain sovereign rights, as distinct from the sovereignty exercised in the territorial sea. The sovereign rights allow the coastal State to explore, exploit, conserve and manage the natural resources of the water and the seabed. It also has jurisdiction in relation to other matters including the establishment of artificial islands, installations and structures, marine scientific research and the protection and preservation of the marine environment.³² In other respects, the freedoms of the high seas prevail so long as they are not incompatible with the exercise of coastal State jurisdiction.³³ Therefore, freedom of navigation is preserved in the EEZ.

The continental shelf regime owes its separate existence in Part VI of the LOSC to the historical development of the concept from the 1940s.³⁴ It was the object of a separate regime in the 1958 Continental Shelf Convention and the LOSC continued its character as a maritime zone with rights distinct from those in the water column above it.³⁵ Some confusion can result from the fact that the EEZ regime formally includes rights over the seabed and subsoil, as well as the water column.³⁶ It is common for States and commentators to refer to the EEZ regime as the predominant legal regime for the water column and the seabed between twelve and 200 nm. However, the negotiators of the LOSC kept a clear separation between the water column and the seabed by declaring that the rights with respect to the seabed and subsoil must be exercised in accordance with Part VI (the continental shelf) rather than Part V (the EEZ). Therefore, the two regimes are related, but are independent of one another.³⁷ As a consequence, Part VI is the primary source of legal rights and obligations for the continental shelf under the LOSC. The provisions of Part VI are discussed below.

On the high seas, all States enjoy the freedoms of navigation and overflight, fishing, laying of cables and pipelines, construction of installations and scientific research, so long as they are exercised consistently with the LOSC.³⁸

Another new development in the LOSC was the regime applying to the deep seabed beyond national jurisdiction, referred to in the LOSC as 'the Area'. The Area and its resources are declared to be the common heritage of mankind.³⁹ This means that the rights to the resources of the Area are vested in mankind as a whole.⁴⁰

---

²⁹ Ibid., articles 17–19 and 24. ³⁰ Ibid., article 21.
³¹ Ibid., article 57. See generally Part V of the LOSC. ³² Article 56, LOSC.
³³ Ibid., article 58.
³⁴ See ch 3 for a more detailed description of the history of the continental shelf doctrine.
³⁵ It is likely that the ambition of many coastal States to extend their jurisdiction beyond 200 nm played a part in keeping the continental shelf and EEZ regimes separate. RR Churchill and AV Lowe, *The Law of the Sea*, 3rd edn (Manchester: Manchester University Press, 1999): p. 166.
³⁶ Article 56(1), LOSC.
³⁷ Attard described the two regimes as autonomous but intimately linked. David Attard, *The Exclusive Economic Zone in International Law* (Oxford: Clarendon Press, 1987): pp. 139–40.
³⁸ Article 87, LOSC. ³⁹ Ibid., article 136. ⁴⁰ Ibid., article 137.

The ISA was created to act on behalf of mankind. The key feature of the Area is that activities in the Area must be carried out for the benefit of all States and the ISA is instructed to share the economic benefit of those activities equitably among States.[41] As described below, the creation of the Area governed by the common heritage of mankind played an important role in the shaping of Part VI of the LOSC in relation to the continental shelf.

(2) Part VI of the Law of the Sea Convention

The history and development of the sovereign rights and obligations of coastal States in relation to the continental shelf beyond 200 nm are explored in Chapter 3. However, it is useful to summarize the important provisions of Part VI of the LOSC, which sets out rights and obligations of States in relation to the continental shelf.

It must be remembered that Part VI of the LOSC is part of a larger convention that was negotiated as a package deal. One element of the package deal related to the outer limits of the continental shelf. The LOSC introduced the common heritage of mankind as the organizing principle for exploitation of resources in the Area.[42] The concept of the common heritage of mankind was supported by developing countries who were opposed to applying the freedom of the high seas to the deep seabed. They were concerned that they did not have the technical ability to exploit resources and would miss out on the benefits of the resources that lay in areas controlled by no State. These States generally wished to maximize the area of seabed covered by the common heritage regime, and argued that coastal State jurisdiction over the continental shelf should be limited. Among these States, the prevailing argument was that coastal State rights over the continental shelf should not extend beyond 200 nm. On the other hand, coastal States with continental shelves that physically extended beyond 200 nm (broad margin States) argued that the continental shelf doctrine meant that they should have rights over the resources of the entire shelf, no matter how far from the coast it stretched. The effect of this approach was to minimize the area to which the common heritage principle would apply and reduce the possible benefits that would return to the international community.[43]

Within Part VI of the LOSC, there is evidence of the compromise made to resolve these differences. On the one hand, broad margin States were allowed to exercise rights over the resources of the continental shelf in cases where it physically extends beyond 200 nm. On the other hand, article 76 established criteria that would be used to determine the outer limits of the continental shelf. This was important to provide clarity as between the seabed included in the Area (to which the principle of common heritage applied) and the continental shelf (over which coastal States had sovereign rights). A process was created to ensure the criteria

---

[41] Ibid., article 140.   [42] Ibid., article 136.
[43] SC Vasciannie, *Land-locked and Geographically Disadvantaged States in the International Law of the Sea* (Oxford: Clarendon Press, 1990).

Coastal States' Rights and Obligations on the Continental Shelf 11

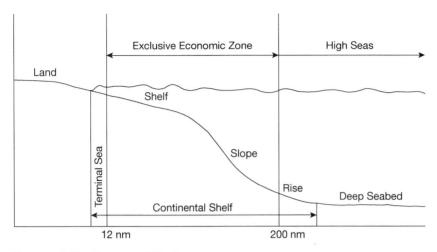

Figure 1.2 The Continental Shelf

were followed by encouraging States to submit their proposed outer limits to the Commission on the Limits of the Continental Shelf. A second major concession made by the broad margin States was that some revenue would be shared with the international community when resources were exploited on the extended part of the continental shelf.[44] This was to recognize the fact that the international community had given up on the possibility of applying the common heritage of mankind to those resources.

Beyond those important concessions, most of the rest of Part VI is based on rights established in the Continental Shelf Convention. These rights had, by and large, been acknowledged as customary international law prior to the negotiations for the LOSC, so there was little appetite to make significant changes.

The outer limits of the continental shelf are established in article 76 of the LOSC. The details of this are discussed in Chapter 3. Essentially, a coastal State may claim sovereign rights over the resources of the seabed and subsoil out to 200 nm (except where a boundary between two or more States must be established due to overlapping claims). In cases where the physical continental shelf extends beyond 200 nm, article 76 sets out the criteria by which such claims should be evaluated (see Figure 1.2). The article also establishes the Commission on the Limits of the Continental Shelf (CLCS), which evaluates State submissions and issues recommendations in relation to the delineation of the extended shelf. When a coastal State sets its outer limits on the basis of the CLCS's recommendations, the limits 'shall be final and binding'.

Article 77 sets out the basis for the sovereign rights a coastal State exercises over the continental shelf and so is fundamental to the inquiry of this book.

[44] Article 82, LOSC.

Article 77 is discussed in several chapters of this book. The coastal State has 'sovereign rights' for the purposes of exploring and exploiting the resources of the continental shelf, meaning that the rights are exclusive and no other State can exploit them without permission from the coastal State. The rights of the coastal State do not depend on occupation or proclamation. As discussed in Chapter 3, the rights of the coastal State on the extended continental shelf do not depend on establishing the outer limits of the shelf according to recommendations by the CLCS. States have these rights if they have a continental shelf that physically extends beyond 200 nm.

The resources over which a coastal State has jurisdiction are mineral resources and other non-living resources, as well as living resources belonging to sedentary species. Article 77(4) defines sedentary species as 'organisms which, at the harvestable stage, either are immobile on or under the seabed or are unable to move except in constant physical contact with the seabed or subsoil'. The definition and development of the sedentary species category is discussed in Chapter 3. Living resources are discussed in Chapter 4 and non-living resources in Chapter 5.

On the extended continental shelf, the relationship between the high seas and the continental shelf rights is particularly complex. Article 78 establishes that the rights of the coastal State over the resources of the continental shelf do not affect the legal status of the superjacent waters or airspace above. The exercise of the rights of the coastal State over the continental shelf must not infringe or result in any unjustifiable interference with navigation and other rights and freedoms of other States. However, questions remain about how these principles should be implemented in practice. To what extent must a coastal State tolerate high seas activities that may have a detrimental impact on sedentary species on the sea floor? Can the coastal State require that vessels operating in the high seas curtail their activities if they may infringe on the coastal State's interests in the seabed resources? Within 200 nm, the coastal State has more control over activities in the water column such as fishing, which it does not in the case of the extended continental shelf. Chapter 7 explores article 78 in some detail and discusses it in light of potential issues on the extended continental shelf.

Article 79 sets out the rights and interests of the coastal State in relation to submarine cables and pipelines on the continental shelf. All States are entitled to lay submarine cables and pipelines on the continental shelf. Although the coastal State may take reasonable measures for the exploration and exploitation of the continental shelf and the control of pollution from pipelines, it may not otherwise impede the laying or maintenance of cables and pipelines. In relation to pipelines, the course of the pipeline is subject to the consent of the coastal State. This does not apply to cables. Despite these provisions, the coastal State retains the right to establish conditions for cables or pipelines entering its territorial sea, as well as for cables and pipelines used in connection with the exploration and exploitation of its natural resources. In addition, all States must have due regard to existing cables and pipelines when laying their own equipment. They must not prejudice the ability to repair the existing cables or pipelines.

Coastal States' Rights and Obligations on the Continental Shelf 13

Article 80 applies the provisions of article 60 to artificial islands, installations and structures ('structures') on the continental shelf *mutatis mutandis*.[45] Article 60 sets out detailed instructions for the status and operation of structures over which the coastal State has exclusive rights. Coastal State rights include the construction of the structures and exclusive jurisdiction over them, 'including jurisdiction with regard to customs, fiscal, health, safety, and immigration laws and regulations'. Coastal States must not establish structures where interference may be caused to the use of recognized sea lanes essential to navigation. Coastal States must give due notice of the construction of structures, and remove abandoned or disused structures to ensure safety of navigation. One of the most important rights for the coastal State is the ability to establish reasonable safety zones of no more than 500 metres[46] around structures, in which it can take measures to ensure the safety of navigation and of the structures themselves. All ships must respect the safety zones and comply with international standards regarding navigation in the vicinity.

The coastal State has the exclusive right to authorize and regulate drilling on the continental shelf for all purposes under article 81. This is not confined to the exploration or exploitation of resources—it also applies to other types of drilling. Article 246(5)(b) gives coastal States the authority to refuse consent for marine scientific research on the continental shelf if it involves drilling on the continental shelf. Drilling for other purposes is similarly restricted.

Article 82 is particularly important to this book, and is discussed in detail in Chapter 5. As mentioned above, this article is the result of the compromise during the LOSC negotiations between those States wishing to exercise control over the full extent of their continental shelves, and those States that insisted that all of the seabed beyond 200 nm should be included in the Area and subject to the common heritage of mankind. It applies to exploitation of the continental shelf beyond 200 nm and requires coastal States to make payments or contributions in kind in respect of a proportion of the production situated there. Payments begin in the sixth year of operation and rise from one per cent of the value or volume of production in year six, to seven per cent by the twelfth year. Payments are to be made to the International Seabed Authority, which will distribute them to State Parties to the LOSC on the basis of equitable sharing criteria. Because there has been no State practice on this point, there is much speculation on how this article will be applied in practice.

Article 83 deals with the delimitation of the continental shelf between States with opposite or adjacent coasts. Delimitation is a matter for agreement between

---

[45] *Mutatis mutandis*: with any necessary changes. It does not appear that this requires any substantive changes between article 60 and article 80. The authors of *UNCLOS: A Commentary* suggest that article 80 applies in two circumstances: when the coastal State has not established an EEZ, and where the continental shelf extends beyond 200 nm. See Myron H Nordquist (ed), *United Nations Convention on the Law of the Sea 1982: A Commentary, Volume II* (Leiden: Martinus Nijhoff Publishers, 1993): p. 919.

[46] Although there is provision for larger safety zones to be established if permitted by generally accepted international standards or the recommended competent international organization, this has not yet been agreed to by the International Maritime Organization (IMO) or other organizations.

the States involved but may involve dispute settlement if no agreement can be reached in a reasonable period of time.

One of the more interesting aspects of article 83 is the requirement that, pending agreement, the relevant States shall make every effort to enter into provisional arrangements of a practical nature and not jeopardize or hamper the reaching of the final agreement. This book will not be dealing with issues relating to the delimitation of boundaries with adjacent or opposite States, except for consideration of issues relating to transboundary resources, discussed in Chapter 5.

Article 84 requires coastal States to deposit charts showing the outer limits of the continental shelf, as well as a list of coordinates, with the Secretary-General of the United Nations and with the Secretary-General of the International Seabed Authority.

Finally, article 85 permits the coastal State to exploit the subsoil through tunnelling under the seabed, irrespective of the depth of the water above the subsoil. This transfers a right found in the Continental Shelf Convention but which is unlikely to be relevant to the shelf beyond 200 nm, as such tunnels usually began on land.

Beyond Part VI, there are a number of general provisions that also apply to the continental shelf, such as the obligations to protect and preserve the marine environment in Part XII, which are discussed in Chapter 4. Additionally, in article 246(6) there is an explicit restriction on the rights of the coastal State to refuse permission for marine scientific research into the resources of the extended continental shelf. This restriction will be discussed in detail in Chapter 6.

C. Chapter Overview

This book is structured in three stages. The first section provides context: an overview of the resources of the continental shelf and the development of the continental shelf regime in international law.[47] The second section examines particular legal issues arising in relation to different types of activities.[48] The final section explores the options for States to take unilateral and multilateral actions to protect their interests in the continental shelf beyond 200 nm.[49]

Chapter 2 examines the physical characteristics of the continental shelf—describing both living and non-living resources that may be located in and on the shelf. The purpose of the chapter is to provide context for the subsequent discussion about coastal State rights. The chapter describes the types of resources likely to be found on the continental shelf beyond 200 nm, activities that might occur in relation to the resources and the environmental impact of those activities.

Although in most cases the area of the continental shelf beyond 200 nm is likely to be very deep as well as distant from the shore, this will not prevent human

---

[47] Chapters 2 and 3.   [48] Chapters 4, 5 and 6.   [49] Chapters 7, 8 and 9.

activities from occurring there. Evidence shows that oil and gas exploration and exploitation is already starting to happen on some extended continental shelves, fishing vessels can and do target seamounts and marine scientific research is occurring in those areas. Other potential activities that could affect the sea floor beyond 200 nm include dumping of waste and storage of carbon dioxide in the deep ocean or in sub-floor deposits. Although the amount of activity is relatively low compared to parts of the continental shelf closer to shore, it is timely for coastal States to consider how to effectively manage and protect their sovereign rights to the resources of the continental shelf beyond 200 nm.

Chapter 3 outlines the development of the continental shelf regime and addresses some fundamental questions about the nature of coastal State rights. It examines the historical development of the continental shelf regime, from the Truman Declaration, to the Continental Shelf Convention, to the LOSC. It also sets out the historical basis for the inclusion of sedentary species in Part VI of the LOSC. This historical perspective assists in understanding how the provisions in the LOSC should be interpreted. The chapter then gives a brief overview of the process for the delineation of the outer limits of the continental shelf and the operation of the Commission on the Limits of the Continental Shelf (CLCS). The chapter argues that coastal States have rights to the resources of the continental shelf beyond 200 nm independently of the issuance of any recommendations from the CLCS. Finally, the chapter evaluates the status of articles 76 and 82 of the LOSC under customary international law.

Chapter 4 explores the issues that arise from coastal State rights over living resources on the continental shelf. The words 'living resources' used in the LOSC reflect the fact that interest was primarily in exploiting living creatures for economic purposes. However, in subsequent years the international community has placed increasing importance on the conservation and sustainable use of marine biodiversity. The chapter describes the range of environmental obligations on coastal States to preserve and protect the marine environment and to take procedural steps to ensure that activities on the extended continental shelf do not cause harm to another State or to the global commons. These obligations provide the context for discussing the issues involved with exploiting both living and non-living resources on the continental shelf beyond 200 nm.

Chapter 4 also discusses issues arising from bioprospecting and fishing. Recent attention has been focused in the international community on living resources as the source of genetic resources that may have value for biotechnology. Scientists have discovered abundant ecosystems on seamounts and hydrothermal vents. Some species, including some species that will be considered to be sedentary species, may lead to the development of pharmaceutical, cosmetic and other industrial products. Some of the issues that may arise for coastal States include what legal regime to apply to activities involving bioprospecting on the extended continental shelf and how to identify sedentary species collected for biotechnological purposes. With regard to fishing, one of the key concerns will not be the exploitation of sedentary species, but rather the protection of sedentary species such as corals and sponges from harm from vessels fishing for high seas species.

Chapter 5 focuses on non-living resources. Although coastal States have rights to the resources of the entire shelf, article 82 establishes an unusual system whereby coastal States are required to make payments or contributions in kind when exploitation occurs on the extended continental shelf. This chapter discusses some of the issues that have been raised about how article 82 is to be interpreted. In some cases, there may be difficulties when a deposit of oil and gas or minerals is located across a boundary or partly on the continental shelf and partly in the Area governed by the International Seabed Authority. The chapter suggests approaches for dealing with these difficulties. It will not be long until hydrocarbon exploitation occurs on an extended continental shelf and these issues will become very important at that point.

Chapter 6 addresses marine scientific research. Coastal States have different rights in relation to marine scientific research on the extended continental shelf than they do in relation to the EEZ and continental shelf within 200 nm. First, given that the vessels conducting research above the extended continental shelf will be in the high seas, how is it possible to distinguish between research in the high seas and research 'on' the continental shelf? Secondly, article 246(6) creates a particular limitation on coastal State rights to refuse consent for marine scientific research on the extended continental shelf, and the chapter discusses how States might interpret this provision.

Chapter 7 explores the intersection between the rights of the coastal State to the continental shelf beyond 200 nm and the rights and freedoms of other States in the high seas that lie above the shelf. This is an issue that is likely to create significant disagreement if coastal States attempt to limit the exercise of high seas freedoms. As in the EEZ, it is clear that there must be a balance struck between coastal States' rights and flag State rights to high seas freedoms. However, Part VI takes a different approach to balancing rights on the continental shelf as opposed to the EEZ. Under the EEZ regime, States are required to have 'due regard' to each other's interests. In Part VI, article 78 requires coastal States not to 'infringe or unjustifiably interfere' with high seas freedoms. The consequences of this difference are examined with reference to the negotiating history of article 78 and its predecessor, article 5(1) of the Continental Shelf Convention. The chapter then outlines a proposed framework for decision-making to determine whether a coastal State's regulations or actions violate the terms of article 78. The framework is illustrated with two examples: limitations on bottom fishing on the extended continental shelf and limitations on navigating close to vessels undertaking seismic surveys. Essentially, this discussion explores the extent to which coastal States can unilaterally regulate the exercise of high seas freedoms.

Chapter 8 explains why coastal States have a right to enforce their regulations in relation to the continental shelf beyond 200 nm against foreign vessels on the high seas. Although there is no explicit provision in the LOSC allowing for enforcement jurisdiction in relation to the continental shelf, the historical development of the continental shelf doctrine demonstrates that 'sovereign rights' was understood to include both legislative and enforcement rights. This approach is supported by limited State practice, and was confirmed in the *Arctic Sunrise* arbitration.[50]

Chapter 9 suggests that, rather than dealing with activities on the continental shelf beyond 200 nm through unilateral action by the coastal State, there is value in adopting bilateral and multilateral approaches to pursue coastal State interests. International organizations, regional fisheries management authorities and regional environmental regimes may offer solutions that minimize the risk of potential disputes. The chapter examines two examples of State practice in this regard. First, Portugal has worked with the regional environmental and fishing regimes in the North-east Atlantic to gain protection for some areas on its extended continental shelf. The chapter examines the lessons from that experience and also the applicability of the example as a model for other parts of the world. Secondly, the Seychelles and Mauritius provide a model for cooperative development of the resources of the continental shelf beyond 200 nm. Rather than focusing on maritime delimitation, the two States prepared a joint submission to the CLCS and have subsequently created a process underpinned by treaties which allows for joint decision-making and equal sharing of profits generated in the area. The treaty that establishes the process is an excellent example of a sustainable development approach that balances exploitation and environmental protection.

Chapter 10 draws some conclusions and lessons from the previous chapters, and makes recommendations for future approaches by coastal States and high seas user States. It also discusses the role that extended continental shelf issues might play in the international negotiations for the protection of marine biodiversity in areas beyond national jurisdiction.

D. Conclusion

Some might ask: why is it necessary for coastal States to turn their minds to the regulation of a part of the continental shelf that is so far from the coast, and where activities such as oil and gas exploration may not occur for many years? In many cases, coastal States do not yet have recommendations from the CLCS, and may not for many years.

The answer is twofold. First, although it may take a long time before most States undertake oil and gas exploration and exploitation on their extended continental shelf, in some places around the world exploration activities are taking place beyond 200 nm. In addition, it is likely that fishing activity is occurring on extended shelves around the world, on slopes and on seamounts.[51] Secondly, even though many

---

[50] Arctic Sunrise *(Netherlands v Russian Federation)* (Merits) PCA Case No. 2014-02, 14 August 2015.
[51] See ch 2.

States may not have recommendations from the CLCS in relation to the outer limits of their continental shelf, this does not prevent them from exercising their sovereign rights over the extended continental shelf. They will need to exercise caution where there is some doubt over the outer limits but, legally, coastal States already have the right to exercise, and protect, their sovereign rights.[52] The issue is not simply an academic exercise, but a reality for many States.

---

[52] See ch 3.

2
Resources and Human Activities on the Continental Shelf Beyond 200 Nautical Miles

A variety of geomorphological conditions and seabed habitats give rise to potential living and non-living resources on the extended continental shelf. There is no special character to the area of the shelf beyond 200 nm, in that similar conditions and resources are also likely to be present where shelves do not extend beyond 200 nm. However, it is valuable to consider the variety of resources that may be located beyond 200 nm and the potential human activities that may need to be regulated by the coastal State.

The range of likely activities on the continental shelf and the extent of those activities have expanded since the negotiation of the United Nations Convention on the Law of the Sea (LOSC). Although oil and gas exploitation remains the primary interest relating to non-living resources, increasing attention is being paid to mineral deposits on the sea floor. Sedentary species are still harvested, and developments in fishing technology allow access to ever deeper areas of the oceans. In addition, new scientific understanding of the resources of the continental shelf indicates that more complex ecological systems than previously thought are present in the deeper waters of the continental shelf.[1] Therefore, as technology improves, it is expected that there will be greater interest in exploring and exploiting the resources of the extended continental shelf. At the same time, States have obligations in relation to the marine environment and they will be seeking to find ways to balance economic activities with environmental protection.

This chapter summarizes information about the types of sea floor ecosystems and resources that might be found on the continental shelf and some of the environmental impacts that exploitation of those resources might have. This information is intended to provide context for the discussions in later chapters, which address both the regulation of such activities under the LOSC and the obligations on States to protect the marine environment.

Three points should be borne in mind when considering the impact of human activities on the sea floor environment. First, all human activities (whether based on land or at sea) have an impact on the environment, and in some cases the impact of

---

[1] See Laurence P Madin et al., 'The Unknown Ocean' in LK Glover and SA Earle (eds), *Defying Ocean's End: An Agenda for Action* (Washington, DC: Island Press, 2004): pp. 213–36, p. 213.

ocean-based activities may be less than their land-based equivalents.[2] The effectiveness of environmental regulation and mitigation measures will influence the level of impact the activity might have. Secondly, some activities have an impact on a large scale, whilst others will have more limited impact. For example, the footprint of bottom trawling will usually be much larger than that of most mining operations that might be conducted at hydrothermal vents. Finally, to put the impact of human activities in context, it should be noted that marine ecosystems are subject to natural disturbances, including volcanic eruptions and variations in benthic currents. These natural events can have effects similar to those of some human activities. In addition, human-induced climate change is likely to have a much larger impact on marine environments when compared with activities with a localised impact. Ultimately, coastal States will have to balance these factors when making determinations about uses of the extended continental shelf.

A. What is the Continental Shelf?

The rights to a 'continental shelf' as set out in the LOSC are over the continental margin that includes the shelf, the continental slope and the continental rise.[3] In the LOSC, the words 'continental shelf' describe all three parts of the continental margin. Therefore the legal, or juridical, concept of the continental shelf is different to that of scientists, who treat the slope, shelf and rise as separate ecological regions.[4] Even between disciplines such as geomorphology and geology there can be disagreement about how to interpret concepts such as the edge of the continental rise and the natural prolongation of the land territory.[5] Additionally, geologists often refer to the 'continental margin' as the zone separating the thin oceanic crust of the deep ocean basins from the thicker continental crust.[6] Despite attempts to connect the legal definition of the continental shelf with the scientific notion of the continental margin, in practice the two may not completely overlap. For example, article 76 imposes outer limits to the coastal State's claims when the physical continental margin may exceed these limits.[7]

The geomorphology and composition of the sea floor of the continental shelf will depend on the geological processes of plate tectonics, sedimentation patterns and ocean circulatory systems. Continental margins can be passive (where the continent

---

[2] James R Hein et al., 'Deep-ocean Mineral Deposits as a Source of Critical Metals for High- and Green-technology Applications: Comparison with Land-based Resources' *Ore Geology Reviews* 51 (2013): pp. 1–14, p. 10.
[3] Article 76(3). See figure 1.2, p. 11.
[4] Betsy B Baker, 'Law, Science, and the Continental Shelf: The Russian Federation and the Promise of Arctic Cooperation' *American University International Law Review* 25 (2010): pp. 251–81, p. 265. See also CLCS Scientific and Technical Guidelines, para. 1.3. Unless the context indicates otherwise, references in this chapter to the continental shelf should be read as including the entire continental margin.
[5] Philip A Symonds et al., 'Characteristics of Continental Margins' in Peter Cook and Chris M Carleton (eds), *Continental Shelf Limits: The Scientific and Legal Interface* (New York: Oxford University Press, 2000): pp. 25–63, pp. 27–9.
[6] Ibid., p. 25.   [7] See ch 3 for a discussion of article 76.

has been split in two and is moving apart) or active (where two tectonic plates collide or move past each other). These geological processes lead to a range of features on the sea floor such as volcanoes, trenches and seamounts.[8] It is not always easy to separate a continental shelf into a shelf, slope and rise. These terms are largely based on the types of continental shelf found in the Atlantic Ocean, which are passive and have little volcanic activity.[9] In active regions such as the Pacific Ocean, continental margins can be highly complex. It has been suggested that the areas of extended continental shelf with the greatest resource potential are more likely to be found in passive zones, although some may be found in active margins.[10]

According to Symonds et al., continental margins underpin approximately 28 per cent of the total area of the oceans:[11]

The continental shelf generally slopes gently (gradient of less than 1:1000) away from the shoreline to a relatively well-defined shelf edge or shelf break ... at an average depth of about 130 m.... [The shelf] can range in width from a few kilometres to more than 400 km.... Seaward of the shelf break, water depth increases rapidly from 100–200 m to 1500–3500 m over the continental slope ... The slope zone is usually less than 200 km in width and is relatively steep (gradients greater than 1:40, averaging about 4°, but can be as high as 35–90° ...), with abrupt boundaries with the adjacent, flatter continental shelf and rise provinces. The continental rise lies between the slope and the deep ocean basin and varies in width from 100 km to 1000 km, with gentle gradients (1:100 to 1:700) and low local relief ... Many margins do not have continental rises for a variety of tectonic and depositional reasons.

This description makes it clear that coastal States with extended shelves may have rights over the seabed to thousands of metres below sea level. Over the breadth of the continental shelf a range of geomorphological and oceanographic characteristics will be found at different depths, leading to a diversity of both non-living and living resources. Both mineral and biological resources can be of interest to coastal States.

B. The Living Resources and the Seabed Environment of the Continental Shelf

The period following the negotiation of the LOSC saw a plethora of new discoveries about the living and non-living resources of the sea floor, including those located on continental shelves. It is now understood that diverse ecosystems can be found on continental shelves in areas previously thought to be barren. Levin and Dayton poetically describe characteristics of the continental margin as follows:[12]

Water masses with distinct hydrographic characteristics overly [sic] the bottom, creating strong gradients in pressure, temperature, oxygen, food supply and substrate stability that

---

[8] E. Ramirez-Llodra et al., 'Deep, Diverse and Definitely Different: Unique Attributes of the World's Largest Ecosystem' *Biogeosciences* 7 (2010): pp. 2851–99, 2857–8. Active margins will be characterized by greater levels of volcanic activity.
[9] Symonds et al., 'Characteristics of Continental Margins', p. 29.   [10] Ibid., p. 32.
[11] Ibid., pp. 29–31.
[12] Lisa A Levin and Paul K Dayton, 'Ecological Theory and Continental Margins: Where Shallow Meets Deep' *Trends in Ecology and Conservation* 24 (2009): pp. 606–17, p. 606.

rival in intensity any on the planet. Canyons, gulleys, mounds and banks interact with currents to create flow conditions suitable for reefs of corals, sponges, cnidarians, and giant, agglutinated protozoans, while plate subduction squeezes methane-rich fluids from the crust, supporting large beds of sibroglinid worms, clams and mussels. These structure-forming species in turn create crucial habitat for numerous protozoan and invertebrate species.

The continental margin has been described as the most geologically diverse component of the deep-ocean floor, containing a range of habitats for biological resources, many of which will be located on the continental shelf beyond 200 nm.[13] Two factors influencing the presence of these habitats are the geomorphology of the sea floor and the depth of the water.[14] Continental margin ecosystems play a role in the overall health of the ocean, providing essential ecosystem functions.[15] For example, the continental margins act as a sink for 90 per cent of the ocean's carbon dioxide.[16]

One of the characteristics of many continental shelf ecosystems deeper than 200 metres is that they are generally not extensively investigated by scientists and relatively little is known about them.[17] The distances from shore and the depths involved make scientific investigation difficult and expensive. This can pose a challenge for coastal States in identifying them.

This chapter looks at two particular examples of ecosystems that can be located on extended continental shelves: seamount ecosystems and ecosystems found at hydrothermal vents and cold seeps.

(1) Seamount ecosystems

Seamounts are undersea mountains, and there may be as many as 30,000 seamounts with an elevation of 1000 metres or higher from the sea floor.[18] These seamounts can be found on continental shelves, as well as in the deep sea. There are also numerous features (knolls and hills) on the sea floor that are between 100 and 1000 metres high and potentially host similar ecosystems to those of the larger seamounts.[19] More information about these smaller features is being

---

[13] Ramirez-Llodra et al., 'Deep, Diverse and Definitely Different', p. 2857.
[14] Lenaick Menot et al., 'New Perceptions of Continental Margin Biodiversity' in Alasdair D McIntyre (ed), *Life in the World's Oceans: Diversity, Distribution, and Abundance* (Chichester: Wiley-Blackwell, 2010): pp. 79–102.
[15] Roberto Danovaro, 'Exponential Decline of Deep-sea Ecosystem Functioning Linked to Benthic Biodiversity Loss' *Current Biology* 18(1) (2008): pp. 1–8.
[16] Levin and Dayton, 'Ecological Theory and Continental Margins'.
[17] Ramirez-Llodra et al., 'Deep, Diverse and Definitely Different'. See the table on p. 2856 of this article for the estimated area, and proportion investigated, of a range of deep-sea habitats.
[18] Report of the Secretary General of the United Nations, A/60/63/Add.1 (2005) para. 24; Chris Yesson et al., 'The Global Distribution of Seamounts Based on 30 Arc Seconds Bathymetry Data' *Deep-Sea Research* 1 (2011): pp. 442–53, p. 446.
[19] References to seamounts in this book include all features rising more than 100 metres above the floor. This approach is consistent with that in J Anthony Koslow et al., 'Biological Communities on Seamounts and Other Submarine Features Potentially Threatened by Disturbance', Chapter 51, *First*

collected with improvements in satellite and other technology, but estimates are that seamounts and knolls could cover approximately 21 per cent of the sea floor.[20] Smaller features are often remnants of volcanic plugs and submarine volcanoes. The harder substrate of the features makes it possible for corals and sponges to attach to the sea floor.[21] Hills, knolls and seamounts provide barriers to water circulation above the sea floor, causing nutrients to rise from deeper waters. These nutrients can support high levels of biomass[22] on or above the seamount.[23] Seamounts that are closer to the ocean's surface are more likely to support the presence of organisms because the circulation of plankton from the surface can provide additional food sources on such seamounts.[24] In some cases, seamounts may act as 'biological hotspots', attracting an abundance of top-level predators as a result of the ocean currents concentrating nutrients, plankton, shrimp and fish above the feature.[25]

Scientists have discovered cold-water corals, sponges, crinoids (sea lilies) and molluscs living on seamounts deep below the sea surface.[26] These organisms are usually the dominant species found on seamounts.[27] Deep-water corals have been discovered that create reef-like structures that act as habitats for other species, both sessile (immobile) and mobile.[28] These corals can cover large areas of seamounts but are also found on other geomorphological features such as in canyons and on continental slopes.[29]

Global Integrated Marine Assessment (United Nations, 2015), www.un.org/depts/los/global_reporting/WOA_RegProcess.htm.

[20] Koslow et al., 'Biological Communities on Seamounts and Other Submarine Features Potentially Threatened by Disturbance', p. 1. For information about seamount distribution, see Paul Wessel, 'Seamount Characteristics' in Tony J Pitcher et al., (eds), *Seamounts: Ecology, Fisheries and Conservation* (Oxford: Blackwell Publishing, 2007): pp. 3–25.

[21] M Clark et al., 'The Ecology of Seamounts: Structure, Function and Human Impacts' *Annual Review of Marine Science* 2 (2010): pp. 253–78.

[22] The total quantity or weight of organisms in a given area.

[23] Ashley A Rowden et al., 'A Test of the Seamount Oasis Hypothesis: Seamounts Support Higher Epibenthic Megafaunal Biomass than Adjacent Slopes' *Marine Ecology* 31 (Supp. 1) (2010): pp. 95–106, p. 96.

[24] Amatzia Genin and John F Dower, 'Seamount Plankton Dynamics' in Tony J Pitcher et al., (eds), *Seamounts: Ecology, Fisheries and Conservation* (Oxford: Blackwell Publishing, 2007): pp. 85–100.

[25] Ramirez-Llodra et al., 'Deep, Diverse and Definitely Different', p. 2863; T Koslow *The Silent Deep: The Discovery, Ecology and Conservation of the Deep Sea* (Sydney: UNSW Press, 2007): pp. 121, p. 125; Filipe M Porteiro and Tracey Sutton, 'Midwater Fish Assemblages and Seamounts' in Tony J Pitcher et al., (eds), *Seamounts: Ecology, Fisheries and Conservation* (Oxford: Blackwell Publishing, 2007): pp. 101–16.

[26] Ramirez-Llodra et al., 'Deep, Diverse and Definitely Different', p. 2863; Koslow *The Silent Deep*, p. 125; Clark et al., 'The Ecology of Seamounts'; Tony J Pitcher et al., (eds), *Seamounts: Ecology, Fisheries and Conservation* (Oxford: Blackwell Publishing, 2007).

[27] Clark et al., 'The Ecology of Seamounts'.

[28] F Althaus et al., 'Impacts of Bottom Trawling on Deep-coral Ecosystems of Seamounts are Long-lasting' *Marine Ecology Progress Series* 397 (2009): pp. 279–94, pp. 279–80. These include fish, shrimp, lobsters and other crustaceans, including amiphods and starfish. See also Alex D Rogers et al., 'Corals on Seamounts' in Tony J Pitcher et al., (eds), *Seamounts: Ecology, Fisheries and Conservation* (Oxford: Blackwell Publishing, 2007): pp. 141–69, p. 147.

[29] Ramirez-Llodra et al., 'Deep, Diverse, and Definitely Different', p. 2864.

Seamounts can attract large numbers of commercially valuable fish for spawning or feeding.[30] This puts pressure on States to allow bottom fishing, which can be damaging for the benthic ecosystems.[31] Deep-water species are typically characterized by slow growth, low fecundity and late maturity.[32] Because of the length of time these benthic invertebrates and fish take to mature, this makes them particularly vulnerable to overfishing and recovery from disturbance can take a long time.[33]

It is apparent that continental shelf ecosystems vary between (and even within) ocean basins and so species found in one area may not be found in another.[34] There is some debate about the level of endemism of species found on seamounts. For example, it has been estimated that nine to thirty-five per cent of species discovered on seamounts may not be found elsewhere in the ocean.[35] However, these sorts of estimates have been challenged as reflecting a lack of data, or limited sampling technology, rather than a biological reality.[36] On average, levels of ten to twenty per cent endemism appear realistic.[37] What is clear is that benthic communities differ across many spatial scales, both in terms of the composition of the ecosystem and the types of species found.[38]

(2) Ecosystems at hydrothermal vents and cold seeps

Other types of ecosystems can be found on continental margins, including chemosynthetic ecosystems. Species in these ecosystems do not rely on photosynthesis from the sun for energy; rather, they are dependent for their energy on chemicals.[39] Species with chemosynthetic qualities are found at hydrothermal vents and cold seeps. Hydrothermal vents are found where hot water laden with chemicals is expelled from the sea floor as part of large, sub-sea floor hydrothermal systems, driven by heat derived from underlying magma activity. Commonly, sulfide

---

[30] Malcolm R Clark, 'Are Deepwater Fisheries Sustainable? The Example of Orange Roughy (Hoplostethus atlanticus) in New Zealand' *Fisheries Research* 51 (2000): pp. 123–35; Malcolm Clark 'Deep-sea Seamount Fisheries: A Review of Global Status and Future Prospects' *Latin American Journal of Aquatic Research* 37(3) (2009): pp. 501–12.
[31] See below Section C(1). [32] Clark, 'Deep-sea Seamount Fisheries', p. 507.
[33] Sarah Samadi, Thomas Schlacher and Bertrand Richer de Forges, 'Seamount Benthos' in Tony J Pitcher et al., (eds), *Seamounts: Ecology, Fisheries and Conservation* (Oxford: Blackwell Publishing, 2007): pp. 119–40.
[34] Koslow, *The Silent Deep*, p. 69.
[35] Gregory Stone et al., 'Seamount Biodiversity, Exploitation and Conservation' in LK Glover and SA Earle (eds), *Defying Ocean's End: An Agenda for Action* (Washington DC: Island Press, 2004): pp. 43–70, p. 50.
[36] Clark et al., 'The Ecology of Seamounts', p. 255; Samadi et al., 'Seamount Benthos', p. 123; Koslow et al., 'Biological Communities on Seamounts and Other Submarine Features Potentially Threatened by Disturbance', p. 2.
[37] Karen I Socks and Paul JB Hart, 'Biogeography and Biodiversity of Seamounts' in Tony J Pitcher et al., (eds), *Seamounts: Ecology, Fisheries and Conservation* (Oxford: Blackwell Publishing, 2007): pp. 255–81, p. 263.
[38] Clark et al., 'The Ecology of Seamounts', p. 258.
[39] Madin et al., 'The Unknown Ocean', p. 223.

deposits form around the vent, providing a mineral-rich substrata that is of potential economic interest.[40]

Hydrothermal vents are associated with volcanic activity, which primarily occurs along tectonic plate boundaries. In the Western Pacific there are numerous subduction zones where the tectonic plates are converging. These are often located within the national jurisdiction of coastal States, including New Zealand, Papua New Guinea, Japan and other nations along the 'Pacific Ring of Fire'. In New Zealand, numerous active hydrothermal vents are located along the Kermadec Arc, including on the extended continental shelf.[41] By contrast, mid-ocean ridges are formed by volcanic activity where plates are diverging, and hydrothermal vents may be found there too. However, mid-ocean ridges are typically far from land, although a few States have extended continental shelves that are closer to the ridges and may find hydrothermal vents under their jurisdiction. For example, in Portugal the Rainbow hydrothermal vents are located beyond 200 nm,[42] whilst in Iceland the Mid-Atlantic Ridge bisects the country.

Species found at hydrothermal vents and cold seeps often display similar characteristics. They are frequently sessile, filter-feeding, long-lived and slow-growing.[43] They include sponges, corals, anemones, squat lobsters, hydroids, brittle stars and sea cucumbers.[44] Juveniles are dispersed as planktonic larvae that circulate in the water, which means that some species can be found at several vent locations.[45] These benthic communities are typically high in biomass and low in diversity compared to species found in the vicinity, although new species are constantly being discovered.[46] The nature of the communities changes depending on their proximity to the hydrothermal vent. In the central vent zone (ranging from ten to 100 metres from the vent), large communities of invertebrates can be found, some of which live in symbiosis with chemosynthetic bacteria. For example, tubeworms have no mouth or gut but live in symbiosis with sulfur-oxidizing bacteria stored inside their

---

[40] Nadine Le Bris et al., 'Chapter 45: Hydrothermal Vents and Cold Seeps' *First Integrated Global Marine Assessment* (New York: United Nations, 2015) www.un.org/depts/los/global_reporting/WOA_RegProcess.htm. See p. 2 of this source for a useful image showing the locations of vents around the world.
[41] Cornel EJ de Ronde et al., 'Intra-oceanic Subduction-related Hydrothermal Venting, Kermadec Volcanic Arc, New Zealand' *Earth and Planetary Science Letters* 193 (2001): pp. 359–69; Cornel EJ de Ronde et al., 'Evolution of a Submarine Magmatic-Hydrothermal System: Brothers Volcano, Southern Kermadec Arc, New Zealand' *Economic Geology* 100 (2005): pp. 1097–133.
[42] Jeff A Ardron, 'The Challenge of Assessing Whether the OSPAR Network of Marine Protected Areas is Ecologically Coherent' *Hydrobiologia* 606 (2008): pp. 45–53, p. 47.
[43] RE Boschen, AA Rowden, MR Clark, JPA Gardner, 'Mining of Deep-sea Seafloor Massive Sulfides: A Review of the Deposits, Their Benthic Communities, Impacts from Mining, Regulatory Frameworks and Management Strategies' *Ocean and Coastal Management* 84 (2013): pp. 54–67, p. 57.
[44] CL Van Dover, *The Ecology of Deep-sea Hydrothermal Vents* (Princeton: Princeton University Press, 2000); Boschen et al., 'Mining of Deep-sea Seafloor Massive Sulfides', p. 57.
[45] International Seabed Authority (ISA), 'Environmental Management of Deep-sea Chemosynthetic Ecosystems: Justification of and Considerations for a Spatially-based Approach', Technical Study No. 9 (2011): p. 3.
[46] Boschen et al., 'Mining of Deep-sea Seafloor Massive Sulfides', p. 57; ISA, Technical Study No. 9, p. 2.

bodies.[47] Other creatures such as shrimp may eat the bacteria. In many cases, the invertebrates are so dependent on the chemosynthetic bacteria that they cannot survive away from the vents.[48] Species that are not dependent on the hydrothermal vent are more prevalent as the distance from the vent increases. The phenomenon where a hydrothermal vent has a higher density of fauna when compared to the surrounding area has been described as a 'halo' effect.[49]

Cold seeps are another form of chemosynthetic ecosystem, found where gases (especially methane) and chemicals related to the compaction of large accumulations of sediment are slowly released from the sea floor. These seeps are often supported by sub-surface hydrocarbon reservoirs.[50] Cold seeps host concentrations of species including bivalves, tubeworms and crustaceans.[51] Symbiont-bearing invertebrates are an important feature of cold seeps. A variety of creatures live on and within the cold-seep organisms.[52]

Microbial mats can cover large areas of the sea floor, including at hydrothermal vents and cold seeps and are of great interest to scientists.[53] These microbes act as a food source for the high numbers of invertebrates found at such sites.[54]

Because both hydrothermal vents and cold seeps are chemosynthetic communities based on compounds such as sulfide and gases such as methane, some species found at each can be similar.[55] However, critical differences do exist, owing to different sources of compounds, the nature of the geological structure of the seabed and the duration of habitats and life-spans of dominant species.[56]

Ecosystems found at these locations can be unstable because the cessation of the release of chemicals at a cold seep, or the ending of volcanic activity at a hydrothermal vent, will usually mean the death of the ecosystem.[57] It is clear that hydrothermal vents are temporary in nature and the species found at the vent will die off or move if the vent becomes inactive. However, inactive vents may also support species known as background, or halo, fauna and may host unique communities associated with inactive systems.[58]

Extensive surveys have been carried out on some hydrothermal vent systems, but limited biological research has been done on most known vents. Seeps are even less

---

[47] TA Lösekann et al., 'Endosymbioses Between Bacteria and Deep-sea Siboglinid Tubeworms from an Arctic Cold Seep (Haakon Mosby Mud Volcano, Barents Sea)' *Environmental Microbiology* 10 (2008): pp. 3237–54, p. 3237–8.
[48] Patrick Colman Collins et al., 'A Primer for the Environmental Impact Assessment of Mining at Seafloor Massive Sulphides' *Marine Policy* 42 (2013): pp. 198–209, p. 201.
[49] Boschen et al., 'Mining of Deep-sea Seafloor Massive Sulfides', p. 57.
[50] Le Bris et al., 'Chapter 45: Hydrothermal Vents and Cold Seeps', p. 1
[51] Ramirez-Llodra et al., 'Deep, Diverse and Definitely Different', p. 2862.
[52] Ann Vanreusel et al., 'Biodiversity of Cold Seep Ecosystems Along the European Margins' *Oceanography* 22(1) (2009): pp. 110–27, p. 117. Vanreusel et al., 'Biodiversity of Cold Seep Ecosystems', p. 117.
[53] Ibid., p. 122.
[54] Le Bris et al., 'Chapter 45: Hydrothermal Vents and Cold Seeps', p. 3.
[55] Vanreusel et al., 'Biodiversity of Cold Seep Ecosystems', p. 115.
[56] ISA, Technical Study No. 9, p. 3.
[57] Vanreusel et al., 'Biodiversity of Cold Seep Ecosystems', p. 117.
[58] Boschen et al., 'Mining of Deep-sea Seafloor Massive Sulfides', p. 58.

studied. It is therefore very difficult to determine the extent to which species are distributed among hydrothermal vents and cold seeps. It appears that some species are broadly distributed among hydrothermal vents (approximately fifteen per cent of species are shared between vents of a region or between basins), whilst at some cold seeps there may be little distribution between sites.[59]

C. Exploitation of Living Resources

Both human and non-human activities and events can have an impact on benthic ecosystems. Non-human-related events, including storms, turbidity flows, mass wasting, volcanic eruptions, and upwelling-induced hypoxia, disturb the marine environment.[60] At hydrothermal vents, ecosystems are naturally affected if the vent ceases expelling hot water, which occurs frequently as a result of changes in the underlying volcanic system. Even seasonal changes can affect the composition of the ecosystem.[61]

Human activities also have an impact on the marine environment. An understanding of the kinds of effects that may occur on the sea floor environment is useful when considering the types of regulation that coastal States may decide are necessary to protect their interests. The following discussion relates only to human activity directed at exploiting living resources. The impacts of hydrocarbon exploitation and mining on the sea floor environment are discussed later in the chapter.

(1) Fishing

Fishing activities can have a negative impact on the benthic marine environment. The coastal State has two interests related to fishing on the continental shelf beyond 200 nm. First, the coastal State has exclusive jurisdiction over sedentary species and may regulate the fishing of those resources. Sedentary species are 'organisms which, at the harvestable stage, either are immobile on or under the seabed or are unable to move except in constant physical contact with the seabed or subsoil'.[62] Secondly, the seabed environment (including sedentary species) can be negatively impacted by fishing that targets non-sedentary species located in the water column. The following discussion focuses on the potential impact that these activities may have on sea floor ecosystems.

As species traditionally fished in shallow zones near the coasts were put under pressure, interest turned to species located in deeper waters. Since the late 1960s, fishing vessels have targeted deep-water fish such as armourhead, cardinalfish,

---

[59] Craig R McClain and Sarah Mincks Hardy, 'The Dynamics of Biogeographic Ranges in the Deep Sea' *Proceedings of the Royal Society* 277 (2010): pp. 3533–46, p. 3534.
[60] Levin and Dayton, 'Ecological Theory and Continental Margins', p. 607.
[61] Ibid., p. 608. The authors describe huge mats of bacteria that form on the Chilean continental margin in summer but are replaced by annelids in winter.
[62] Article 77(4), LOSC.

orange roughy, roundnose grenadier and oreos through bottom trawling on seamounts.[63] Fishing of deep-water species found below 500 metres (those most likely to be found on, or above, the continental shelf beyond 200 nm) began in earnest in the 1980s.[64] The attractiveness of an area for bottom fishing will often be based in part on the bathymetry of the sea floor. Seamounts, banks and continental slopes are known to be productive fishing sites because upwellings of nutrients from deeper waters can result in extensive benthic ecosystems that attract commercially valuable fish species for feeding or spawning.[65] Cold seeps may also attract valuable fish species owing to the habitat and food sources located there.[66]

A variety of fishing methods can be used to target species located close to the sea floor. Bottom trawling, in which large and heavy nets are dragged along the sea floor, is the most destructive for sea-floor environments.[67] As technology develops, increasingly deeper parts of the seabed are becoming accessible to bottom trawling, which can currently be undertaken at depths of up to 1500 metres.

Koslow describes the equipment used in bottom trawling:[68]

> Modern deepwater trawls are particularly large and heavy, enabling them to sink rapidly to depth ... The footrope of a net like this is between 50 and more than 100 meters long and weighs up to 4800 kilograms ... Whatever is not dragged up by such a behemoth is generally left crushed in its wake ... Given the 100-meter footrope and a trawl speed of 3 knots, a single trawler will disturb approximately 10 square kilometers of seafloor each day.

The doors of the net alone can weigh over 2000 kilograms and can penetrate the seabed thirty centimetres or more.[69] The impact of the trawl doors and sweeps can create sediment clouds, which can reach two to four metres high and may be 120–150 metres in width, depending on the trawl equipment.[70] These sediment clouds reduce light levels, smother small animals, corals and sponges, and affect feeding and metabolic rates of organisms. Churning of the sea floor can mix surface organic material with sub-surface material, changing the chemical composition of the sea floor and the sediment layers.[71] Organisms living in the top parts of the sediment can be crushed.

The consequences of bottom trawling on the sea-floor environment can be severe. Bottom trawling can remove approximately 90 per cent of the native corals and sponges, usually within the early stages of an area being targeted by fishers.[72]

---

[63] Malcolm R Clark et al., 'Large-scale Distant-Water Trawl Fisheries on Seamounts' in Tony J Pitcher et al., (eds), *Seamounts: Ecology, Fisheries and Conservation* (Oxford: Blackwell Publishing, 2007): pp. 361–99, p. 362.

[64] JA Koslow et al., 'Continental Slope and Deep-sea Fisheries: Implications for a Fragile Ecosystem' *ICES Journal of Marine Science* 57 (2000): pp. 548–57, p. 549.

[65] Ibid.   [66] ISA, Technical Study No. 9, p. 7.

[67] Koslow, *The Silent Deep*, p. 220.   [68] Ibid.

[69] Malcolm R Clark et al., 'The Impacts of Deep-sea Fisheries on Benthic Communities: a Review' *ICES Journal of Marine Science* 73 (Supp. 1) (2016): doi: 10.1093/icesjms/fsv123, p. i57.

[70] Ibid.

[71] M Clark and T Koslow, 'Impacts of Fisheries on Seamounts' in Tony J Pitcher et al., (eds), *Seamounts: Ecology, Fisheries and Conservation* (Oxford: Blackwell Publishing, 2007): p. 417; Clark et al., 'The Impacts of Deep-sea Fisheries on Benthic Communities: a Review', p. i57.

[72] Elliott A Norse et al., 'Sustainability of Deep-sea Fisheries' *Marine Policy* 36 (2012): pp. 307–20, p. 315.

Studies have shown that bottom trawling tends to result in high levels of sedentary species such as coral being caught as by-catch. For example, in the first year of trawling for orange roughy on the South Tasman Rise (1997–1998), vessels landed 1.6 tonnes of coral every hour and in total took over 1100 tonnes.[73] Between 1990 and 2002, Alaskan fishers landed approximately 4186 tonnes of corals and sponges, mostly as a result of bottom trawling.[74] Comparisons between fished and unfished seamounts have demonstrated significant differences in the diversity of invertebrates such as coral and sponges and the overall biomass.[75] This has led some to suggest that bottom trawling is the marine equivalent of forest clear-cutting.[76] Because deep-water corals are slow-growing, the impact of trawling can be long term, with little recovery of corals in the short to medium term.[77]

Other types of fishing gear can be used to target species on the sea floor. These include bottom longlines, anchor gill nets, pots and traps and tangle nets.[78] Longlines may also impact on sea-floor ecosystems as the weights on the line can be dragged through coral communities.[79] However, the impact is exponentially lower than the effects of bottom trawling.[80]

The direct impacts on seabed organisms fall into three basic types.[81] Blunt impacts occur when a broad object, such as a ground rope, trawl doors or mesh, travels through the benthic community. This can result in the dislodgement or crushing of species such as coral, especially larger types. Line shear occurs when a narrow object, such as a longline, travels across the sea floor. This can shear off such organisms. Hooking occurs when hooks, such as those attached to longlines, catch non-target animals or organisms. This direct impact can remove creatures that create habitat for other species, cause a decline in abundance or diversity and change the structure of the benthic community.

A further concern is the impact on benthic ecosystems of fishing gear that is lost or abandoned. This gear can continue to catch species on the sea floor, sometimes

---

[73] Owen F Anderson and Malcolm R Clark, 'Analysis of Bycatch in the Fishery for Orange Roughy, *Hoplostethus atlanticus*, on the South Tasman Rise' *Marine and Freshwater Research* 54 (2003): pp. 643–52.
[74] Norse et al., 'Sustainability of Deep-sea Fisheries', p. 315.
[75] Malcolm R Clark and J Anthony Koslow, 'Impacts of Fisheries on Seamounts' in Tony J Pitcher et al., (eds), *Seamounts: Ecology, Fisheries and Conservation* (Oxford: Blackwell Publishing, 2007): pp. 413–41, p. 423.
[76] Les Watling and Elliott A Norse, 'Disturbance of the Seabed by Mobile Fishing Gear: A Comparison to Forest Clearcutting' *Conservation Biology* 12 (1998): pp. 1180–97.
[77] Althaus et al., 'Impacts of Bottom Trawling on Deep-coral Ecosystems of Seamounts', pp. 279–94; Alan Williams et al., 'Seamount Megabenthic Assemblages Fail to Recover from Trawling Impacts' *Marine Ecology* 31 (Supp. 1) (2010): pp. 183–99, p. 194.
[78] Clark and Koslow, 'Impacts of Fisheries on Seamounts' in Pitcher et al., (eds), p. 414.
[79] Ramirez-Llodra et al., 'Deep, Diverse and Definitely Different', p. 2883; E Ramirez-Llodra et al., 'Man and the Last Great Wilderness: Human Impact on the Deep Sea' *PLoS One* 6(7) (2011): e22588. Doi:10.1371/journal.pone.0022588. p. 10.
[80] Christopher K Pham et al., 'Deep-water Longline Fishing Has Reduced Impact on Vulnerable Marine Ecosystems' *Scientific Reports* 4 (2014): DOI: 10.1038/srep04837; Clark et al., 'The Impacts of Deep-sea Fisheries on Benthic Communities: a Review'.
[81] These impacts are described by Clark et al., 'The Impacts of Deep-sea Fisheries on Benthic Communities: a Review', p. i53.

called 'ghost fishing'.[82] As well as seamounts, vent and seep ecosystems can also be damaged by fishing gear. The impact of lost nets can be much higher on deep-sea ecosystems than in shallower waters because nets lost in water shallower than 100 metres lose their catching efficiency much faster than nets lost in deep water with weaker currents.[83]

It is becoming possible for fishing vessels to target demersal fish (those living near the seabed) using mid-water trawl gear that is navigated very close to the sea floor—in some cases between one and five metres from the bottom. Although this technique does not have as severe an impact on the benthic environment, it can still have detrimental effects.[84] For example, a net towed a few metres above the sea floor may catch the tops of taller corals and sponges. The passage of the net can create pressure waves, which disturb organisms and create sediment clouds. There is also the problem of accidental impact with the seabed when towing nets close to the seabed. Even bottom trawl nets can be pushed off course by the strength of currents around a seamount.[85]

Large-scale fishing for deep-sea species risks following a 'boom and bust' pattern, whereby the majority of the biomass is harvested in the first few years of exploitation, following which the population falls to low levels.[86] In deep waters the fish species grow slowly and reach maturity later in life, making them more vulnerable to over-fishing.[87] In some cases the population does not recover, even once fishing has stopped.[88] Bottom fishing can result in the environment of the sea floor and the surrounding waters being fundamentally altered.[89] In 2003 the International Council for the Exploration of the Seas suggested that most deep-water stocks were probably outside safe biological limits.[90]

The concern about the impact of destructive fishing methods on the seabed environment is reflected in United Nations General Assembly calls for States

---

[82] Ramirez-Llodra et al., 'Deep, Diverse and Definitely Different', p. 2883.

[83] Andrew J Davies, J Murray Roberts and Jason Hall-Spencer, 'Preserving Deep-sea Natural Heritage: Emerging Issues in Offshore Conservation and Management' *Biological Conservation* 138 (2007): pp. 299–312, p. 302.

[84] Conversation with Malcolm Clark. See SJ Baird, BA Wood and NW Bagley, 'Nature and Extent of Fishing Effort on or Near the Seafloor within the New Zealand 200 Nautical Mile Exclusive Economic Zone 1989–90 to 2004–05' NZ Aquatic Environment and Biodiversity Report No. 73 (Wellington: Ministry of Fisheries, 2011).

[85] Koslow, *The Silent Deep*, p. 121.   [86] Clark et al., 'The Ecology of Seamounts', p. 263.

[87] Elliott A Norse et al., 'Sustainability of Deep-sea Fisheries' *Marine Policy* 36 (2012): pp. 307–20, p. 309.

[88] Clark et al., 'The Ecology of Seamounts', p. 266.

[89] It has been observed on some seamounts that little regeneration is observed five to ten years after the discontinuation of trawling, and 'recovery may require centuries to millenia'. Koslow et al., 'Biological Communities on Seamounts and Other Submarine Features Potentially Threatened by Disturbance', p. 9. The detrimental impact on species from bottom trawling may extend as far as 70 kilometres from the fishing area. DM Bailey et al., 'Long-term Changes in Deep-water Fish Populations in the Northeast Atlantic: a Deeper Reaching Effect of Fisheries?' *Proceedings of the Royal Society of Britain* 276 (2009): doi: 10.1098/rspb.2009.0098. See also Clark et al., 'The Impacts of Deep-sea Fisheries on Benthic Communities: a Review', p. i58.

[90] Koslow et al., 'Biological Communities on Seamounts and Other Submarine Features Potentially Threatened by Disturbance', p. 8.

and regional fisheries management organizations (RFMOs) to protect vulnerable marine ecosystems.[91] These resolutions primarily relate to areas beyond national jurisdiction and are without prejudice to State rights over the extended continental shelf.[92] However, the request that States identify and protect benthic habitats from the impacts of fishing indicates a high level of concern about the potential impacts of fishing on sensitive benthic communities.

(2) Bioprospecting

Bioprospecting is part of a process in which commercially useful products are derived from living resources. Generally, samples of living organisms are gathered, returned to the laboratory and then analysed. Marine genetic resources of use include nucleic acid sequences, chemical compounds produced by marine organisms and unrefined materials extracted from marine biomass.[93] Potentially useful qualities may then be identified and ultimately developed into an output, such as a pharmaceutical or other substance. This has been described as occurring in four stages.[94] In stage one, samples are collected. In stage two, scientists will attempt to isolate, characterize and culture microbes. In stage three, the sample will be screened for useful qualities that may result in a commercial product. Stage four involves the development of the product, including securing intellectual property rights, trials and sales and marketing. Only an extremely small number of organisms sampled (perhaps less than one per cent) will lead to a commercial product.[95] In some cases, the success of a product derived from organisms will depend on whether the substance can be chemically synthesized in the laboratory because harvesting of sufficient organisms can be difficult.[96]

Marine living resources have become increasingly important to the biotechnology industry as comparatively little of the total marine biodiversity has been explored, compared to terrestrial biodiversity.[97] It has been estimated that the success rate in finding previously undescribed active chemicals in marine organisms is 500 times higher than in terrestrial organisms.[98] Deep-sea organisms exist in harsh,

---

[91] For example, see General Assembly Resolution 59/25 (17 November 2004), paras. 66–68; General Assembly Resolution 61/105 (6 March 2007), para. 83; and General Assembly Resolution 64/72 (19 March 2010), paras. 119–124.

[92] General Assembly Resolution 64/72 (19 March 2010), para. 115.

[93] Michael Banks et al., 'Chapter 29: Use of Marine Genetic Resources' *First Global Integrated Marine Assessment* (United Nations, 2015) http://www.un.org/depts/los/global_reporting/WOA_RegProcess.htm.

[94] See Julia Jabour-Green and Dianne Nichol, 'Bioprospecting in Areas Outside National Jurisdiction: Antarctica and the Southern Ocean' *Melbourne Journal of International Law* 4 (2003): pp. 76–111, p. 85; Leary, *International Law and the Genetic Resources of the Deep Sea*, p. 164.

[95] Jabour-Green and Nichol, 'Bioprospecting in Areas Outside National Jurisdiction', p. 87.

[96] JF Imhoff, A Labes and J Wiese, 'Bio-mining the Microbial Treasures of the Ocean: New Natural Products' *Biotechnology Advances* 29 (2011): pp. 468–82, p. 470.

[97] D Leary et al., 'Marine Genetic Resources: A Review of Scientific and Commercial Interest' *Marine Policy* 33 (2009): pp. 183–94, p. 185.

[98] JM Arrieta, S Arnaud-Haond and CM Duarte, 'What Lies Underneath: Conserving the Ocean's Genetic Resources' *Proceedings of the National Academy of Sciences* 107 (2010): pp. 18318–24, p. 18320.

extreme environments with conditions such as low (or no) oxygen, high pressure, no light and strong acidity. The ability of these organisms to survive such conditions is of interest to the marine biotechnology industry.[99]

Although the marine biotechnology industry is in its relative infancy, a growing range of products has been derived from marine genetic resources. A study of applications for patents in relation to genes isolated in marine organisms found that pharmacology and human health uses represented the largest number, accounting for around 55 per cent of applications.[100] Marine genetic resources have been the source of products that have anti-inflammatory, anti-cancer or other medical properties.[101] Other uses include cosmetics, agriculture, enzymes that are used in industrial processes, and marine nutraceuticals (compounds that are found to be beneficial to humans and embedded in food).[102] However, commentators have cautioned that marine bioprospecting is not guaranteed to produce commercially valuable results and the time frame needed to develop such results can amount to decades.[103]

Individual pharmaceuticals can generate revenue in the billions of dollars if successful for the treatment of widespread diseases such as cancer.[104] Millions of dollars are earned annually from treatments for herpes and AIDS derived from marine organisms.[105] Cosmetics are also highly profitable. One enzyme from a hydrothermal vent used in a facial cream generated annual sales of approximately US$150 million.[106] Inevitably, such figures can only be estimates. However, it is clear that the potential rewards are high, providing incentives for companies to undertake the costly and uncertain process of looking for useful compounds in marine genetic resources.

In the future, scientists may have less need to conduct research on the sea floor. Leary and Juniper have pointed out that, increasingly, biotechnology companies are relying less on physical samples and more on information about genetic resources that is contained on databases resulting from general scientific research.[107] In addition, the development of synthetic biology, where novel organisms are created by researchers, may decrease the need for sampling of marine biodiversity.[108]

---

[99] Report of the Secretary General of the UN, A/60/63/Add.1, para. 79.
[100] Arrieta et al., 'What Lies Underneath', p. 18320.
[101] Leary et al., 'Marine Genetic Resources', pp. 185–6; Imhoff et al., 'Biomining the Microbial Treasures of the Ocean', p. 471; David Leary 'International Law and the Genetic Resources of the Deep Sea' in Davor Vidas (ed), *Law, Technology and Science for Oceans in Globalisation: IUU Fishing, Oil Pollution, Bioprospecting, Outer Continental Shelf* (Leiden: Martinus Nijhoff, 2010): pp. 353–69, p.359.
[102] Leary et al., 'Marine Genetic Resources', p. 191. See also Imhoff et al., 'Bio-mining the Microbial Treasures of the Ocean'; Banks et al., 'Chapter 29: Use of Marine Genetic Resources', p. 3.
[103] Banks et al., 'Chapter 29: Use of Marine Genetic Resources', pp. 2–3.
[104] Lyle Glowka, 'The Deepest of Ironies: Genetic Resources, Marine Scientific Research, and the Area' *Ocean Yearbook* 12 (1996): pp. 154–78, p. 160.
[105] Leary et al., 'Marine Genetic Resources', p. 192. [106] Ibid.
[107] David Leary and S Kim Juniper, 'Addressing the Marine Genetic Resources Issue: Is the Debate Headed in the Wrong Direction?' in Clive Schofield, Seokwoo Lee and Moon-Sang Kwon (eds), *The Limits of Maritime Jurisdiction* (Leiden: Martinus Nijhoff Publishers, 2014): pp. 769–85, p. 777.
[108] Ibid, pp. 778–80.

In some cases, marine species might need to be harvested to produce commercial quantities of a product. For example, the marine nutraceutical market can use marine biomass, including harvested algae, to produce health food products or cosmetics.[109] However, in general, scientists do not need a large sample to evaluate its genetic qualities—even 500 grams can be sufficient.[110] The main environmental problem would be from damage to vulnerable marine ecosystems during sample collection, especially where the site is well studied and frequently visited.[111]

One of the risks for coastal States has been the concern that researchers may remove samples of species from the continental shelf and use the sample to develop a commercial product but fail to share the benefits of that product with the coastal State. Indeed, this concern was behind the development of the Convention on Biological Diversity in 1992, although the motivation for that convention was primarily terrestrial resources and the abuse by companies of indigenous knowledge.[112] The Convention on Biological Diversity established an approach based on 'access and benefit sharing'.[113]

D. The Non-living Resources of the Continental Shelf

Many States have committed significant resources to exploring their extended continental shelves for the purposes of making a submission to the CLCS. A driver for this activity is undoubtedly the expectation that an extended continental shelf is likely to provide coastal States with rich rewards from the exploitation of the non-living resources of the shelf including minerals and hydrocarbons. Although resource exploitation on extended continental shelves has been limited to date, it is probable that this will change in the next decade. Factors driving this development include the clarification of legal rights to the extended continental shelf, rising demand for energy and minerals, the plateauing of terrestrial and near-shore reserves of hydrocarbons and minerals, and improving technology allowing safer and more economic extraction in deep waters.[114]

---

[109] Banks et al., 'Chapter 29: Use of Marine Genetic Resources', p. 3.
[110] Edgar J Asebey and Jill D Kempenaar, 'Biodiversity Prospecting: Fulfilling the Mandate of the Biodiversity Convention' *Vanderbilt Journal of Transnational Law* 28 (1995): pp. 703–54, p. 706.
[111] Banks et al., 'Chapter 29: Use of Marine Genetic Resources', p. 5.
[112] James O Odek, 'Bio-Piracy: Creating Proprietary Rights in Plant Genetic Resources' *Journal of Intellectual Property Law* 2 (1994–1995): pp. 141–81; Asebey and Kempenaar, 'Biodiversity Prospecting'.
[113] See ch 4.
[114] Clive Schofield, 'Securing the Resources of the Deep: Dividing and Governing the Extended Continental Shelf' Paper from Securing the Ocean for the Next Generation, Conference held at the Korea Institute of Ocean Science and Technology, 21–24 May, Seoul, Korea. See also *IAEA World Energy Outlook 2014* (12 November 2014).

(1) Hydrocarbons

Oil and gas reservoirs are formed where organic matter is trapped in layers of sediment under sufficient pressure and heat. Oil and gas are less dense than water, and so, once formed, they migrate upwards through relatively porous rocks. Most of the oil and gas escapes to the surface, but some is trapped by impermeable rocks through which they cannot flow, forming reservoirs. The location of oil and gas reservoirs is determined by the history of sedimentation and tectonics, and the existence of trapping rock (reservoir) formations.[115] When the hydrocarbon reservoir is drilled, the pressure of the hydrocarbons forces them to the surface.[116] In some situations water is pumped into the reservoir to aid the hydrocarbon's extraction at the surface.

The width of the shelf 'is not, in itself, an index of favourability for petroleum or other resources'.[117] Factors that determine incidence of petroleum occurrence are purely geological and can vary from region to region.[118] In order for large quantities of hydrocarbons to form, sediments must be at least one or two kilometres deep, and usually three or four kilometres deep.[119] Passive continental margins have the greatest potential for hydrocarbons because of the areas of large sediment accumulation that can be located there.[120]

It has been estimated that approximately one-third of global crude oil is located offshore.[121] A number of extended continental shelves have sediment thickness in excess of one kilometre and are likely to contain hydrocarbons.[122] The International Seabed Authority (ISA) has estimated that the top ten extended continental shelf areas could have combined oil and gas resources of 185.6 billion barrels of oil equivalence.[123] If present, the recoverability of these resources will depend on factors such as depth and environmental conditions (e.g. ice cover).

(2) Minerals

Several types of mineral deposits may occur on the continental shelves of States in amounts suitable for exploitation. It is useful to observe at the outset that the

---

[115] See Knut Bjørlykke, 'Introduction to Petroleum Geology' in Knut Bjørlykke (ed), *Petroleum Geoscience: From Sedimentary Environments to Rock Physics* (Berlin: Springer-Verlag, 2010): pp. 1–26.
[116] Vasco Becker-Weinberg, *Joint Development of Hydrocarbon Deposits in the Law of the Sea* (Heidelberg: Springer, 2014): p. 9.
[117] Lewis G Weeks, 'Petroleum Resources Potential of Continental Margins' in Creighton A Burk and Charles L Drake (eds), *The Geology of Continental Margins* (New York: Springer-Verlag, 1974): pp. 953–64, p. 953.
[118] Weeks, 'Petroleum Resources Potential of Continental Margins', p. 954.
[119] International Seabed Authority, '*Global Non-living Resources on the Extended Continental Shelf: Prospects at the Year 2000*', Technical Study No. 1 (2001): p. 38.
[120] Bruce C Heezen, 'Atlantic-Type Continental Margins' in Creighton A Burk and Charles L Drake (eds), *The Geology of Continental Margins* (New York: Springer-Verlag, 1974): pp. 13–24, p. 18.
[121] Clive Schofield and Robert van de Poll, 'Exploring the Outer Continental Shelf', Annex 5 to International Seabed Authority, *Implementation of Article 82 of the United Nations Convention on the Law of the Sea: Report of an International Workshop*, Technical Study No. 12 (2013).
[122] ISA, Technical Study No. 1, p. 44.
[123] ISA, Technical Study No. 1, p. 46. World annual production in 2014 was approximately 93 million barrels of oil. US Energy Information Administration www.eia.gov.

mix of mineral deposits on the continental shelf is largely similar to those found in equivalent territorial areas but some studies have indicated that ocean deposits may be richer than existing territorial deposits.[124] However, some minerals are more likely to be present on the extended continental shelf than others.

Manganese nodules and crusts are sources of manganese, nickel, copper, cobalt and other minerals. They can occur in a variety of environments, including continental shelves, seamounts and abyssal plains and are distributed widely throughout the world.[125] Ferromanganese crusts rich in cobalt are formed on hard rock substrate, where there is very little sediment at depths of 400–7000 metres, especially on seamounts. Manganese nodules and ferromanganese crusts form incredibly slowly over millions of years from the precipitation of minerals from seawater. The ISA has tentatively estimated that as much as 12 billion tonnes of manganese nodules and crusts could be located on the extended continental shelves of coastal States.[126]

Gas hydrates are crystalline compounds composed of gas molecules such as methane in a solid ice form. Gas formed from organic matter in sediments rises until it 'freezes' under certain pressure and temperature conditions.[127] Methane hydrates are of the most commercial interest and may provide an alternative to traditional sources of gas.[128] Gas hydrates have been identified on many continental shelves around the world in areas where the sediment thickness is greater than 500 metres. However, significant risks are involved in exploiting gas hydrates, including the potential for destabilization of the seabed.[129]

Polymetallic sulfides, also known as sea floor massive sulfides, contain gold, copper and other precious and base metals such as silver, lead and zinc.[130] Sea-floor massive sulfides can be formed both on, and beneath, the sea floor in the oceanic crust, when metal-rich hydrothermal fluids of up to 400°C mix with seawater.[131] The most attractive deposits are those that form metal-rich chimneys and mounds around hydrothermal vent sites.[132] Those most accessible are the shallow sites that are found in back-arcs as in Papua New Guinea, or on volcanoes, as in those along the Kermadec Arc.[133] Some deposits can be large, ranging from several thousand to

---

[124] Michael J Cruickshank, 'Mineral Resources Potential of Continental Margins' in Creighton A Burk and Charles L Drake (eds), *The Geology of Continental Margins* (New York: Springer-Verlag, 1974): pp. 965–1000, p. 974; Hein et al., 'Deep-ocean Mineral Deposits', p. 9.
[125] ISA, Technical Study No. 1, p. 32. See also Koslow, *The Silent Deep*, p. 170.
[126] ISA, Technical Study No. 1, p. 37. This figure is not an estimate of economically viable reserves.
[127] ISA, Technical Study No. 1, p. 47.   [128] Koslow, *The Silent Deep*, p. 174.
[129] ISA, Technical Study No. 1, p. 52.
[130] See RE Boshen et al., 'Mining of Deep-sea Seafloor Massive Sulphides: A Review of the Deposits, their Benthic Communities, Impacts from Mining, Regulatory Frameworks, and Management Strategies' *Ocean and Coastal Management* 84 (2013): pp. 54–67, p. 56; M Hannington et al., 'The Abundance of Seafloor Massive Sulfide Deposits' *Geology* 39 (2011): pp. 1155–8; Koslow, *The Silent Deep*, p. 171.
[131] Boshen et al., 'Mining of Deep-sea Seafloor Massive Sulphides', p. 55.
[132] Jochen Halfar and Rodney M Fujita, 'Precautionary Management of Deep-sea Mining' *Marine Policy* 26 (2002): pp. 103–6, p. 104.
[133] Porter Hoagland et al., 'Deep-sea Mining of Seafloor Massive Sulfides' *Marine Policy* 34 (2010): pp. 728–32, p. 728.

about 100 million tonnes, and it has been estimated that the potential for seabed sulfide deposits rivals that of land-based sources.[134] Although up to 40 per cent of known sites occur within 200 nm of the coast, it is unclear how many deposits are in areas of the extended continental shelf.[135] It should be remembered that hydrothermal vents are frequently host to unusual ecosystems, leading to potential tensions between conservation and economic objectives.[136]

In addition, large quantities of rare earth elements and yttrium have been found in mud from various deep-sea locations in the Pacific around 2000 kilometres from mid-ocean ridges.[137] Currently, all sources of commercial exploitation are on land and it is highly unlikely that extensive extraction will occur at sea in the medium term. Rare earth elements are used in petroleum fluid cracking catalysts, metallurgical additives and alloys, glass polishing, ceramics and magnets. It is estimated that 18 kilograms of rare earth metals are used in each hybrid car produced.[138]

Demand for minerals is increasing over time. Rapid adoption of emerging technologies (e.g. smart phones, laptops, solar panels, wind turbines and electric cars) requires mass consumption of mineral resources at unprecedented rates. For example, a cell phone is on average twenty-five per cent metal. In 2010, approximately 1.5 billion cell phones were sold worldwide, requiring about fifty-one tonnes of gold, 525 tonnes of silver and 24,000 tonnes of copper in their production (plus many other metals).[139] This demand, together with the exhaustion of land-based sources over time, means that ocean-based sources of these metals may be very valuable in the future.[140]

E. Exploitation of Non-living Resources

(1) Hydrocarbon exploration and exploitation

Oil and gas exploration and exploitation on continental shelves is being conducted in water as deep as 3000 metres.[141] Currently such exploitation is primarily conducted within States' territorial seas and EEZs but it is feasible that

---

[134] Halfar and Fujita, 'Precautionary Management of Deep-sea Mining', p. 103; M Hannington et al., 'The Abundance of Seafloor Massive Sulfide Deposits', p. 1155.

[135] ISA, Technical Study No. 1, p. 27; Hoagland, 'Deep-sea Mining of Seafloor Massive Sulfides', p. 728.

[136] Cindy Lee Van Dover, 'Tighten Regulations on Deep-sea Mining' *Nature* 470 (3 February 2011): pp. 31–33, p. 32.

[137] Yasuhiro Kato, 'Deep-sea Mud in the Pacific Ocean as a Potential Source for Rare-Earth Elements' *Nature Geoscience* 3 July 2011: doi: 10.1038/NGEO1185.

[138] National Mining Association, *40 Common Minerals and Their Uses* http://www.nma.org/index.php/minerals-publications/40-common-minerals-and-their-uses.

[139] Hein et al., 'Deep-ocean Mineral Deposits', p. 3.

[140] Ibid.; Richard A Kerr, 'The Coming Copper Peak' *Science* 343 (2014): pp. 722–4.

[141] Ramirez-Llodra et al., 'Man and the Last Great Wilderness', p. 12; National Commission on the BP Deepwater Horizon Oil Spill and Offshore Drilling, 'Deep Water: The Gulf Oil Disaster and the Future of Offshore Drilling' Report to the President, 2011, p. 41 http://www.gpo.gov/fdsys/pkg/GPO-OILCOMMISSION/pdf/GPO-OILCOMMISSION.pdf.

in the future, as pressures on oil and gas supplies increase, companies will look to exploit reserves found on the extended parts of their continental shelves. There are significant concerns about the impact of hydrocarbon exploration and exploitation on the environment. Regular oil exploration and exploitation can result in accidental discharges that negatively impact on ecosystems. Incidents such as the explosion on the *Deepwater Horizon* rig in the Gulf of Mexico can have negative consequences for the environment.[142] However, any consideration of the environmental impact of oil and gas exploration should take into account that hydrocarbons leak into the oceans naturally where their upward migration is not prevented by a cap of denser rock. Estimates of the worldwide seepage of crude oil in the environment are around 600,000 tonnes a year (between 200,000 and two million tonnes).[143] Oil seepage in the Gulf of Mexico alone is considered to be between 40 and 110 million tonnes a year.[144] This can be compared with the almost 16,000 tonnes of oil equivalent expelled into the Gulf of Mexico by the *Deepwater Horizon* spill. Of course, a rapid release of oil into the marine environment carries different risks to slow release of naturally occurring seeps.

There are several stages of hydrocarbon exploitation. First, general information about an area's subsurface geology is obtained using a variety of methods, including seismic surveys, gravity surveys and magnetic surveys. Where surveys indicate the possibility of hydrocarbon deposits, exploration wells are drilled to analyse the geology of the sediments and to determine whether commercially viable accumulations of hydrocarbons are present. If successful, appraisal wells are then drilled to obtain further information about the reservoir, such as the quantity and quality of the resource and the percentage of oil that can be recovered.[145]

Drilling usually involves using rotary drill bits and attaching sections of drill pipe as the drill hole deepens. Drilling mud—a mixture of water, clay and chemicals—is pumped into the well to prevent uncontrolled entry of water or hydrocarbons into the well.[146] Finally, the well is lined with steel, which is cemented in place to prevent collapse. In a production well, the pipe at the reservoir level is perforated to allow the hydrocarbon to enter the well. The pressure in the reservoir enables the hydrocarbons to be moved to the surface of the well.[147] Modern wells can be drilled horizontally as well as vertically, which means that more of the resource can be extracted. Another modern trend is to create subsea developments in which connected systems of pipelines on the sea floor direct hydrocarbons from individual wells to a single platform. This sort of development makes deep-water oil exploitation more economic by reducing the size and number of platforms

---

[142] Ramirez-Llodra et al., 'Man and the Last Great Wilderness', p. 13.
[143] National Research Council, *Oil in the Sea III: Inputs, Fates and Effects* (Washington, DC: National Academies Press, 2003): p. 70.
[144] IR MacDonald et al., 'Transfer of Hydrocarbons from Natural Seeps to the Water Column and Atmosphere' *Geofluids* 2 (2002): pp. 95–107.
[145] Hussein K Abdel-Aal and Mohammed A Alsahlawi, *Petroleum Economics and Engineering*, 3rd edn (Boca Raton: CRC Press, 2014): p. 229; Bjørlykke, 'Introduction to Petroleum Geology', p. 21.
[146] Bjørlykke, 'Introduction to Petroleum Geology', p. 22.   [147] Ibid.

needed.[148] In these cases the footprint of the seabed infrastructure extends well beyond the base of the platform.

The ability to explore and exploit hydrocarbon resources at greater depths is growing. This increases the possibility that hydrocarbon deposits on the extended continental shelf could become exploitable. However, the technical difficulty of extracting the resource increases with depth; and areas of the continental shelf beyond 200 nm are likely to be very deep.[149] The distance from shore also increases the economic cost of getting the product to market and so companies usually prefer to focus on inshore reserves first. It is likely that the near future will see only a small number of extended shelves being explored, and only where the reserves are large and easily accessed. This may change in the longer term, depending on demand and the price of oil and gas.

Some coastal States are already exploring for hydrocarbons on their extended shelf.[150] At least four countries have already commenced exploration activities and up to sixteen countries have 'issued and/or are offering' offshore oil and gas exploration concession licences on their extended continental shelf.[151] The Canada–Newfoundland and Labrador Offshore Petroleum Board has issued seven current exploration licences in respect of its extended continental shelf in the Grand Banks region. These licences have resulted in two validation wells in the area to date.[152] Two significant discovery licences have also been issued for that area.[153] In the United States, exploration leases have been issued in respect of approximately 92,000 acres in the northern part of the Western Gap, an area of the extended shelf that was the subject of a boundary agreement with Mexico.[154] Norway has issued exploration licences in respect of three relatively small areas that either straddle or are beyond the 200 nm EEZ limit.[155] New Zealand has issued a petroleum exploration permit for the New Caledonia Basin, much of which is located on its continental shelf beyond 200 nm.[156] Initial seismic surveying began in 2016.

---

[148] Two examples in the Gulf of Mexico are: the Perdido development http://www.shell.com/about-us/major-projects/perdido.html; and the Independence Hub platform http://www.offshore-technology.com/projects/independence/.

[149] 'Deep Water: The Gulf Oil Disaster and the Future of Offshore Drilling', Report to the President, p. 91.

[150] Ramirez-Llodra et al., 'Man and the Last Great Wilderness', p. 12.

[151] Clive Schofield, 'New Marine Resource Opportunities, Fresh Challenges' *University of Hawaii Law Review* 35 (2013): pp. 715–33, pp. 723–4.

[152] Canada–Newfoundland and Labrador Offshore Petroleum Board, 'Current Exploration Licences' (13 August 2014) www.cnlopb.ca/pdfs/elgbr.pdf.

[153] Canada–Newfoundland and Labrador Offshore Petroleum Board, 'Current Significant Discovery Licences' (30 September 2013) www.cnlopb.ca/pdfs/sdlgbr.pdf.

[154] Personal communication, Bureau of Ocean Energy Management. See Treaty with Mexico on Delimitation of Continental Shelf, 9 June 2000.

[155] Harald Brekke, 'Licensing System in Norway', presentation to ISA Workshop, (26–30 November 2012) www.isa.org.jm/sites/default/files/workshop/presenters-2012.pdf.

[156] Neil Ritchie, 'Anadarko Determined to Stay in New Zealand' *Oil and Gas Australia* http://energy-pubs.com.au/oil-gas-australia/anadarko-determined-to-stay-in-nz/.

There are many environmental impacts of hydrocarbon exploitation. The day-to-day production of offshore oil and gas results in the discharge of water contaminated with chemicals and deposition on the seabed of drilling muds and cuttings from the well.[157] These deposits can be substantial, but are usually localized. For example, in one study the authors noted that the largest pile of contaminated cuttings in the North Sea had a volume of 66,816 cubic metres and the tallest pile reached twenty-six metres above the seabed.[158] Impacts on the immediate environment include the smothering of ecosystems by sediment, as well as increased levels of toxic substances in surrounding organisms, leading to increased mortality.[159] A strong inverse relationship between the concentrations of oil in the sediments and the biodiversity at the site has been identified.[160] However, the measured impact varies according to the types of chemicals released into the environment, the hydrographic conditions and the particle size of the discharged materials.[161] In most cases the effect on the benthic community is limited to a relatively small area around each platform.[162] Where oil-based drilling mud has been used and discharged, the toxic effects are greater than where water-based drilling mud is used.[163] Environmental effects may be greater in deeper waters compared with shallower waters,[164] but the effects will vary depending on the ocean currents and other environmental conditions.[165]

Other concerns include leaks from pipelines. These can be caused by a range of factors, including corrosion, damage from storms, damage from ship anchors and

---

[157] Peter Harris et al., 'Chapter 21: Offshore Hydrocarbon Industries', in *United Nations World Ocean Assessment*, pp. 10–12, www.un.org/depts/los/global_reporting/WOA_RPROC/Chapter_21.pdf; Charles H Peterson et al., 'Ecological Consequences of Environmental Perturbations Associated with Offshore Hydrocarbon Production: A Perspective on Long Term Exposures in the Gulf of Mexico' *Canadian Journal of Fisheries and Aquatic Sciences* 53 (1996): pp. 2637–54, p. 2638; Paul F Kingston, 'Impact of Offshore Oil Production Installations on the Benthos of the North Sea' *ICES Journal of Marine Science* 49 (1992): pp. 45–53.

[158] Alastair Grant and Andrew D Briggs, 'Toxicity of Sediments From Around a North Sea Oil Platform: Are Metals or Hydrocarbons Responsible for Ecological Impacts?' *Marine Environmental Research* 53(1) (2002): pp. 95–116, p. 96.

[159] Ibid. However, in some studies the impact on the local ecosystem was transient: see DR Currie and Leanne R Isaacs, 'Impact of Exploratory Offshore Drilling on Benthic Communities in the Minerva Gas Field, Port Campbell, Australia' *Marine Environmental Research* 59(3) (2005): pp. 217–33.

[160] Kingston, 'Impact of Offshore Oil Production Installations on the Benthos of the North Sea', p. 48.

[161] Frode Olsgard and John S Gray, 'A Comprehensive Analysis of the Effects of Offshore Oil and Gas Exploration and Production on the Benthic Communities of the Norwegian Continental Shelf' *Marine Ecology Progress Series* 122 (1995): pp. 277–306, p. 300.

[162] Kingston, 'Impact of Offshore Oil Production Installations on the Benthos of the North Sea', p. 47.

[163] Olsgard and Gray, 'A Comprehensive Analysis of the Effects of Offshore Oil and Gas Exploration and Production', p. 278.

[164] Adrian Glover and Craig R Smith, 'The Deep-sea Floor Ecosystem: Current Status and Prospects of Anthropogenic Change by the Year 2025' *Environmental Conservation* 30(3) (2003): pp. 219–41, p. 228.

[165] Andrew J Davies, J Murray Roberts and Jason Hall-Spencer, 'Preserving Deep-sea Natural Heritage: Emerging Issues in Offshore Conservation and Management' *Biological Conservation* 138 (2007): pp. 299–312, p. 303.

from fishing gear. Damage from anchors and fishing gear generally causes the greatest amount of oil leakage.[166]

The impacts of oil and gas platforms are not always detrimental. Platforms, which are typically located on soft sea floors, can provide a hard surface for species such as mussels, barnacles and anemones.[167] These can then provide a food source for other local species including fish. When the mussels fall from the platform during storms, they provide food for benthic creatures such as crabs, but also create a shell mound that can be 8.5 metres high and up to 70 metres across, creating a new form of habitat.[168] Decommissioned rigs are often left in place or toppled to create artificial reefs.[169]

There is naturally concern about a major accident at deep-sea oil wells. The *Deepwater Horizon* incident in 2010 brought international attention to the potentially catastrophic consequences of a blowout at oil wells in deep waters.[170] A blowout occurs when the pressure in the well is not balanced and gas, oil or water flows to the surface, often at very high pressures. These substances can then ignite, causing intense fires. Blowout preventers are valves installed in the well that are designed to prevent fluids and other material from escaping and can be used to seal the wellhead in an emergency. In the *Deepwater Horizon* incident, the rig was drilling an exploratory well in about 1500 metres of water to depths of 5600 metres below the sea floor. On 20 April 2010, explosions occurred on the rig, killing eleven workers and causing it to sink. Pressure had built up in the drill pipe while members of the crew were preparing to seal the well for a temporary abandonment. The blowout preventer failed to prevent the release of large amounts of gas and, subsequently, mechanisms designed to seal the well did not operate. The official investigation found that a series of poor decisions had contributed to the disaster.[171] As a result of the destruction of the pipes, oil continued to leak into the ocean. Approximately five million gallons (about nineteen million litres) of oil from the well entered the Gulf of Mexico before the well could be capped eighty-seven days later.[172] Around

---

[166] Harris et al., 'Chapter 21: Offshore Hydrocarbon Industries', p. 15.

[167] RE Bomkamp, HM Page and JE Dugan, 'Role of Food Subsidies and Habitat Structure in Influencing Benthic Communities of Shell Mounds at Sites of Existing and Former Offshore Oil Platforms' *Marine Biology* 146 (2004): pp. 201–11.

[168] Bomkamp et al., 'Role of Food Subsidies and Habitat Structure in Influencing Benthic Communities of Shell Mounds', p. 202.

[169] For an excellent discussion about the ecological and social impacts of decommissioning platforms, see Donna M Schroeder and Milton S Love, 'Ecological and Political Issues Surrounding Decommissioning of Offshore Oil Facilities in the South California Bight' *Ocean and Coastal Management* 47 (2000): pp. 21–48.

[170] See 'Black Storm Rising: The Gulf of Mexico Oil Spill' *The Economist* 395.8681, 8 May 2010, pp. 69–71. It has been estimated that an accident (usually much smaller than the *Deepwater Horizon* incident) occurs every 17 years, although there is considerable uncertainty surrounding that rate. See Harris et al., 'Chapter 21: Offshore Hydrocarbon Industries', p. 20.

[171] Deep Water: The Gulf Oil Disaster and the Future of Offshore Drilling: Report to the President, ch 4.

[172] Bruce C Glavovic, 'Disasters and the Continental Shelf: Exploring New Frontiers of Risk' in Myron H Nordquist et al., (eds), *The Regulation of Continental Shelf Development: Rethinking International Standards* (Leiden: Martinus Nijhoff, 2013): pp. 225–58, p. 232.

seven million litres of oil dispersants were introduced into the Gulf to speed up the rate of oil dispersion, although these too have their own potential environmental effects.[173]

The blowout at the *Deepwater Horizon* rig had significant environmental consequences, reflecting the experience with other significant oil spills. Onshore, approximately 430 miles of marsh shorelines were exposed to oil, leading to high mortality for many species.[174] Declines in abundance of certain pelagic fisheries were observed in the year following the disaster.[175] Some coral communities less than twenty kilometres from the well site were affected, although they improved over two years of study.[176]

Oil and gas platforms located at great distances from shore and in great depths are vulnerable to severe weather such as hurricanes, and it is much more difficult to respond to disasters offshore.[177] Areas such as the Gulf of Mexico can be prone to severe storms in which waves can be between seven and fifteen metres high. Freak waves of up to thirty metres have been recorded.[178]

Over time, oil from a spill will be dispersed naturally. The rate at which the environment recovers depends on a variety of factors but recovery usually takes between two and ten years.[179] After major incidents, it can take even longer for the effects to disperse. Following the *Exxon Valdez* disaster, in which forty-two million litres of crude oil entered the Alaskan environment, studies showed that effects were still found eight to ten years after the spill. For example, species that were associated with sediments and certain shorebirds showed long-term elevated contamination and mortality. These losses caused indirect decline in other species and ecosystems.[180]

Owing to the potential economic benefits from hydrocarbon exploitation, many States are keen to promote such developments. There are a number of international environmental law obligations that coastal States will need to observe. The public is demonstrating a greater interest in, and some cases opposition to, decisions to

---

[173] Danielle M DeLeo et al., 'Response of Deep-water Corals to Oil and Chemical Dispersant Exposure' *Deep Sea Research Part II: Topical Studies in Oceanography* 129 (2016): pp. 137–47.

[174] Irving A Mendelssohn et al., 'Oil Impacts on Coastal Wetlands: Implications for the Mississippi River Delta Ecosystem After the Deepwater Horizon Oil Spill' *Bioscience* 62(6) (2012): pp. 562–74, p. 568.

[175] Jay R Rooker, 'Spatial, Temporal and Habitat-related Variation in Abundance of Pelagic Fishes in the Gulf of Mexico: Potential Implications of the Deepwater Horizon Oil Spill' *PLOS One* 10 October 2013, doi: 10.1371/journal.pone.0076080.

[176] Pen-Yuan Hsing et al., 'Evidence of Lasting Impact of the Deepwater Horizon Oil Spill on a Deep Gulf of Mexico Coral Community' *Elementa: Science of the Anthropocene* (2013): doi: 10.12952/journal.elementa.000012. See also DeLeo, 'Response of Deep-water Corals to Oil and Chemical Dispersant Exposure'.

[177] 'Deep Water: The Gulf Oil Disaster and the Future of Offshore Drilling' Report to the President, p. 50.

[178] Morgan Downey, *Oil 101* (New York: Wooden Table Press, 2009): pp. 117–18.

[179] Paul F Kingston, 'Long-term Environmental Impact of Oil Spills' *Spill Science and Technology Bulletin* 7 (2002): pp. 53–61.

[180] Charles H Peterson et al., 'Long-term Ecosystem Response to the Exxon Valdez Oil Spill' *Science* 302(5653) (19 December 2003): pp. 2082–6.

explore and exploit offshore resources. Despite this, it is likely that there will be exploitation on some extended shelves in the short to medium term where conditions are favourable to such activities.

(2) Mineral exploration and exploitation

Seabed mining for minerals such as diamonds, tin and aggregate minerals is currently taking place in shallow water, for example off Namibia's coast. Several deep-sea mining initiatives are in development but, at the time of writing, none was in full production.[181] Minerals found on and below the seabed have a wide range of important uses in modern economies. For example, manganese is used in iron and steel construction.[182] Nickel is an alloy in stainless steel and has a wide range of uses in electrical equipment, transportation and industrial machinery. Copper is used in a range of products, including building construction, electronic products, roofing and chemical and pharmaceutical machinery. Cobalt is used primarily in super alloys, chemicals and magnets. Rare earth elements and yttrium are used in new electronic technologies.

Production of gas hydrates has been demonstrated onshore in Canada and offshore in Japan. Significant risks are associated with exploiting gas hydrates, including concerns about the possibility that mining the hydrates may undermine the sea floor. In addition, gas hydrates can become destabilized, without direct human interference, as a result of climate change, which can cause large landslides. Changes in water temperature could destabilize the hydrates, turning the crystalline compound into gas and releasing large quantities of methane into the water.[183]

Extensive effort has been expended on exploring for manganese nodules in the deep seabed beyond national jurisdiction (the Area). The ISA has established environmental protection measures and issued prospecting licences for companies to develop the potential for mining of these nodules. However, commercial quantities of manganese nodules have not yet been recovered from the sea floor and there is currently no industrial-scale process to extract the minerals. The bulk of the effort has been focused on the Clarion–Clipperton Fracture Zone in the Area in the Pacific.

Mining of cobalt-rich crusts on the seabed is much more difficult than other types of mining because they are formed on the steep sides of seamounts, making

---

[181] Nautilus Minerals is advanced in developing the technology to exploit a mining site offshore of Papua New Guinea and exploitation should begin in the next few years. See the discussion below regarding the Solwara 1 site.

[182] The description of mineral uses in this paragraph are from National Mining Association, *40 Common Minerals and Their Uses* http://www.nma.org/index.php/minerals-publications/40-common-minerals-and-their-uses.

[183] Matthew T Reagan and George J Moridis, 'Modeling of Oceanic Gas Hydrate Instability and Methane Release in Response to Climate Change' *Proceedings of the 6th International Conference on Gas Hydrates* (2008) https://www.netl.doe.gov/File%20Library/Research/Oil-Gas/methane%20hydrates/ICGH_5803_ESD07-014.pdf.

them harder to access.[184] Removal of the crusts involves digging or scraping the top section of the sea-floor substrate, which will affect the associated benthic communities.[185] Mining of cobalt-rich crusts is in its very early stages, even in areas near to the coast.

The most advanced offshore mining operation is for sea floor massive sulfides and is managed by Nautilus Minerals in Papua New Guinea's territorial sea. Papua New Guinea has issued a twenty-year mining lease to Nautilus Minerals to extract minerals from the Solwara 1 site at depths of 1600 metres.[186] Estimates are that 20–30 cubic metres of ore will be removed from an area of approximately 0.112 square kilometres of sea floor.[187] The proposed approach to mining at Solwara 1 involves crushing the ore on the seabed, lifting it hydraulically to a surface vessel within a riser pipe, dewatering the ore on the ship and pumping the fluid back into the water column a few metres above the sea floor to minimize the extent of any plume created.[188] Nautilus has identified potential sites for sea-floor massive sulfides in Tonga, and has applied for prospecting licences in the Solomon Islands, Fiji, Vanuatu and New Zealand. A second company, Neptune Minerals, also has interests in New Zealand, Vanuatu and Solomon Islands waters.

Mineral exploitation on a commercial scale will have an impact on ecosystems in the surrounding area, although the overall effect will depend on the scale of the operation and the geographic area covered. Several potential environmental effects of mining sea-floor massive sulfides have been identified.[189] The primary direct effect will be the removal of species from the area of mining. The environmental consequences of this removal will depend on the ecosystem and the process. The ability for a vent ecosystem to recover following mining may depend on the proximity of other vent communities that can be a source of organisms. So long as reserve sites are preserved from mining activity, Van Dover has argued that the impact on the ecosystems around vents could be equivalent to the impact of natural sea-floor eruptions.[190] Where the vent is still active following the mining, chimneys can regrow in a matter of months in some cases.[191] If organisms are able to be

---

[184] Samantha Smith and Robert Heydon, 'Processes Related to the Technical Development of Marine Mining' in Elaine Baker and Yannick Beaudoin, *Cobalt-rich Ferromanganese Crusts: A Physical, Biological, Environmental, and Technical Review* (Arendal: SPC-Grid Arendal, 2013): pp. 41–6.

[185] Malcolm R Clark and Samantha Smith, 'Environmental Management Considerations' in Elaine Baker and Yannick Beaudoin (eds), *Cobalt-rich Ferromanganese Crusts: A Physical, Biological, Environmental, and Technical Review* (Arendal: SPC-Grid Arendal, 2013): pp. 23–40, p. 32.

[186] See www.nautilusminerals.com/irm/content/png.aspx?RID=258. Solwara 1 is the first site to be developed and others may also be exploited in future.

[187] PC Collins, R Kennedy and CL Van Dover, 'A Biological Survey Method Applied to Seafloor Massive Sulphides (SMS) with Contagiously Distributed Hydrothermal-vent Fauna' *Marine Ecology Progress Series* 452 (2012): pp. 89–107. doi: 10.3354/meps09646; Van Dover, 'Tighten Regulations on Deep-sea Mining', p. 32.

[188] Hoagland, 'Deep-sea Mining of Seafloor Massive Sulfides', p. 731; Collins et al., 'A Biological Survey Method Applied to Seafloor Massive Sulphides', p. 91.

[189] See Clark and Smith, 'Environmental Management Considerations'.

[190] Cindy Lee Van Dover, 'Mining Seafloor Massive Sulphides and Biodiversity: What is at Risk?' *ICES Journal of Marine Science* (2010): p. 5, doi:10.1093/icesjms/fsq086.

[191] Boschen et al., 'Mining of Deep-sea Seafloor Massive Sulfides', p. 59.

recruited from nearby reserve sites, some species will be able to recover after mining in a short time frame, between several months and five years.[192]

A further possible effect of mining is caused by the disposal of sediment and waste rock. At Solwara 1, it has been estimated that approximately 130,000 tonnes of sediment and 115,000 tonnes of waste rock will be disposed of, disturbing the sea floor and creating plumes of sediment that will affect water quality and creatures in the vicinity.[193] The impact will depend on the concentration and nature of the suspended material, how close to the sea floor it is released and how quickly it disperses.

Koslow has suggested that mining inactive hydrothermal vent sites could have fewer environmental effects compared with active sites.[194] However, locating the inactive sites is a significant challenge for scientists.[195] Active vent sites are more easily identified owing to the 'black smoke' created by the chemical plumes expelled into the oceans as the hydrothermal fluids discharged from the vents mix with the much colder, ambient seawater.[196]

As a result of the risks posed by seabed mining, the ISA has developed regulations governing prospecting and exploration for sulfide deposits, polymetallic nodules and cobalt-rich crusts in the Area.[197] In these regulations, the ISA has distinguished between prospecting and exploration. Prospecting is the search for deposits of minerals, including estimation of the size and distribution of mineral deposits and their economic values, without any exclusive rights. Exploration is broadly defined as searching for deposits with exclusive rights, undertaking steps towards developing technology and conducting studies to determine the commercial viability of the operation.[198] The regulations establish requirements for environmental assessment, monitoring and protection, which may be of interest to coastal States considering how to regulate mining in their jurisdiction.[199]

Considerable interest has developed in the prospect of seabed mining in light of the developments in Papua New Guinea. In the South Pacific, a number of States are updating their national legislation to provide for seabed mining. They are being assisted by a regional programme that is a collaboration between the Secretariat of

---

[192] Van Dover, 'Mining Seafloor Massive Sulphides and Biodiversity', p. 6. However, the recolonization may not always be by the same species comprising the original vent community. See Boschen et al., 'Mining of Deep-sea Seafloor Massive Sulfides', p. 59.

[193] Collins et al., 'A Biological Survey Method Applied to Seafloor Massive Sulphides', p. 91; Kaul Gena, 'Deep Sea Mining of Submarine Hydrothermal Deposits and Its Possible Environmental Impact in Manus Basin, Papua New Guinea', *Procedia Earth and Planetary Science* 6 (2013): pp. 226–33, p. 231.

[194] Koslow, *The Silent Deep*, p. 172.

[195] Clark and Smith, 'Environmental Management Considerations', p. 35.

[196] De Ronde et al., 'Intra-oceanic Subduction-related Hydrothermal Venting'.

[197] Regulations on Prospecting and Exploration for Polymetallic Sulphides in the Area, Doc. ISBA/16/A/12/Rev.1, 7 May 2010.

[198] Michael Lodge, 'The Deep Seabed' in Donald R Rothwell et al., (eds), *Oxford Handbook of the Law of the Sea* (Oxford: Oxford University Press, 2015): pp. 226–53, p. 242.

[199] Boschen et al., 'Mining of Deep-sea Seafloor Massive Sulfides', p. 61.

the Pacific Community and the European Union.²⁰⁰ However, the likelihood of mining operations taking place on the extended continental shelf in the near future is very slim. Compared to the sophisticated technology available to the hydrocarbon industry for exploiting resources on the deep seabed, the technology for deep-sea mining is still in its relative infancy.

F. Other Activities with a Direct Impact on the Seafloor

There are a number of issues that affect the entire marine environment that could also impact on the living resources of the continental shelf. Pollution is a problem for all parts of the ocean, but it has been suggested that concentrations of persistent organic pollutants in deep-sea dwelling fish may be an order of magnitude higher than in fish living at the surface.²⁰¹ Such pollutants may end up in the sediments on the sea floor or accumulated in benthic and demersal (close to the sea floor) organisms.²⁰²

Climate change has many reported impacts on ocean species and these will be felt on the continental shelf. For example, it can be expected that oxygen concentrations will decline in the deep ocean.²⁰³ Changes in circulation patterns can affect larval dispersal.²⁰⁴ The degree of impact of climate change is expected to be much greater than most direct human activities. However, the following discussion focuses on activities that are deliberately targeted at the sea floor.

(1) Waste disposal

As long as humans have been going to sea, dumping of waste has occurred, both as a way of disposing of ship-generated waste, but also in order to remove waste generated on land that is toxic, inconvenient or expensive to dispose of safely. In some cases, debris may enter the ocean from land-based sources. Some of this waste, particularly plastics, will remain at the surface of the ocean and may be ingested by marine animals.²⁰⁵ However, a large amount of debris finds its way to the sea floor, even when it is not the result of deliberate dumping. For example, a recent study found a considerable amount of litter at a variety of deep-sea sites, including a large amount of lost or abandoned fishing gear, food packaging, glass bottles and other

---

²⁰⁰ www.sopac.org/dsm/.
²⁰¹ Koslow et al., 'Biological Communities on Seamounts and Other Submarine Features Potentially Threatened by Disturbance', p. 11.
²⁰² Oliver Froescheis et al., 'The Deep-sea as a Final Global Sink of Semivolitile Persistent Organic Pollutants? Part I: PCBs in Surface and Deep-sea Dwelling Fish of the North and South Atlantic and the Monterey Bay Canyon (California)' *Chemosphere* 40 (2000): pp. 651–60, p. 659.
²⁰³ Koslow et al., 'Biological Communities on Seamounts and Other Submarine Features Potentially Threatened by Disturbance', p. 11.
²⁰⁴ Le Bris et al., 'Chapter 45: Hydrothermal Vents and Cold Seeps', p. 6.
²⁰⁵ Jose GB Derraik, 'The pollution of the marine environment by plastic debris: a review' *Marine Pollution Bulletin* 44 (2002): pp. 842–52.

general plastic and glass items.²⁰⁶ In one survey off the Californian coast at depths between 1000 and 2000 metres, it was observed that litter such as torpedo wire, plastic bags and miscellaneous items was the primary source of solid substrata in very deep areas.²⁰⁷

Deliberate dumping of waste has a long history. During the age of steam-powered shipping, the residue of burnt coal, known as clinker, was thrown overboard as a matter of course.²⁰⁸ This substance is found in a range of benthic environments, where it is toxic to many organisms. Following the Second World War, more than one million tonnes of defective, obsolete and surplus conventional and chemical weapons were dumped into the ocean off Europe.²⁰⁹ Radioactive waste has been disposed of in the deep sea since the 1940s, often in iron drums that corrode over time.²¹⁰ Other types of waste that have been deliberately dumped at sea include industrial and municipal wastes, sewage, chemical waste, chemical weapons, pharmaceuticals, and military and civilian ships.²¹¹ Much of this sort of waste dumping has been limited or stopped as a result of the London Dumping Convention.²¹²

The impact of the waste on the benthic environment varies according to the type of waste, the location and the ocean currents. Waste can produce a variety of substances that are detrimental to the local environment, including persistent organic pollutants, bacteria, heavy metals (such as mercury and cadmium), chlorinated hydrocarbons (such as DDT and PCBs) and radiation. These can have long-term, serious impacts on benthic organisms.

(2) Climate change mitigation measures

As States try to find a solution for climate change, scientists have considered the possibility of removing carbon from the atmosphere by injecting it into geological structures beneath the seabed or in the deep ocean. Because these activities could

---

²⁰⁶ Lucy C Woodall et al., 'Deep-sea Litter: a Comparison of Seamounts, Banks and a Ridge in the Atlantic and Indian Ocean Reveals Both Environmental and Anthropogenic Factors Impact Accumulation and Composition' *Frontiers in Marine Science* 2 (2015): pp. 3–10.
²⁰⁷ Ramirez-Llodra et al., 'Man and the Last Great Wilderness', p. 5.
²⁰⁸ Hjalmar Thiel, 'Anthropogenic Impacts on the Deep Sea' in PA Tyler (ed) *Ecosystems of the Deep Oceans* (Elsevier, 2003): pp. 427–72, p. 432.
²⁰⁹ Koslow reports that mustard gas has been caught in fishing trawls, causing injuries and contamination of fish catch. Koslow, *The Silent Deep*, p. 138. Glover and Smith suggest that several million tonnes of munitions were dumped in European waters between 1945 and 1976. Glover and Smith, 'The Deep-sea Floor Ecosystem', p. 223. See also Hans Sanderson et al., 'Environmental Hazards of Sea-Dumped Chemical Weapons' *Environmental Science and Technology* 44 (2010): pp. 4389–94.
²¹⁰ Thiel, 'Anthropogenic Impacts on the Deep Sea', p. 437; Koslow, *The Silent Deep*, p. 139; James Waczewski, 'Legal, Political and Scientific Response to Ocean Dumping and Sub-seabed Disposal of Nuclear Waste' *Journal of Transnational Law and Policy* 7 (1997–1998): pp. 97–118. Countries that have dumped radioactive waste at sea at various times include the United States, France, the United Kingdom, Germany, Italy, the Netherlands, Sweden, Japan, South Korea, New Zealand and Russia.
²¹¹ Ramirez-Llodra et al., 'Man and the Last Great Wilderness', pp. 6–8.
²¹² Convention on the Prevention of Marine Pollution by Dumping of Wastes and Other Matter 1972 (opened for signature 29 December 1972, entered into force 30 August 1975) 1046 UNTS 138 and the 1996 Protocol (opened for signature 7 November 1996, entered into force 24 March 2006).

have environmental consequences if they take place in or above the continental shelf, coastal States may need to regulate them.

It has been estimated that the world will need to sequester as much as 10 billion tonnes of carbon dioxide (CO_2) every year as a climate change mitigation method.[213] The ocean is already a carbon sink, absorbing approximately a quarter of the anthropogenic carbon dioxide emissions that are put into the atmosphere annually.[214] One high-profile example to increase absorption of carbon dioxide is the effort to explore the potential for iron fertilization of the oceans to increase the absorption of carbon dioxide into the depths of the ocean.[215] Some of the other proposals have significance for the seabed.

(a) Carbon sequestration in the seabed

This process injects compressed CO_2 into reservoirs in the sediments of the sea floor. The idea is that these geological formations can capture carbon dioxide for long periods, in a similar way to how oil and gas are kept in reservoirs in sediments. In fact, disused oil and gas fields are one possible location for this activity. Another option is to store the CO_2 in unmined coal seams.[216] This process is not new to the industry. Carbon dioxide is injected into operating oil and gas reservoirs to increase the pressure inside and extract residual hydrocarbons, known as enhanced oil recovery.[217] Recent research has suggested that basalt (a common rock found on the sea floor) can absorb CO_2 by producing stable, non-toxic minerals.[218] The most likely location for offshore carbon sequestration is on the continental shelf and in some adjacent sedimentary basins, because sediments on the abyssal deep sea floor are too thin and permeable to be suitable for storage.[219]

Technology exists today to undertake carbon sequestration under the sea floor. As an example, a long-term project in the North Sea has seen one million tonnes of CO_2 injected into a sandstone reservoir 1000 metres below the surface of the seabed.[220] The CO_2 is a by-product of oil and gas extraction in the Sleipner field, in Norwegian waters, and amounts to approximately 3 per cent of Norway's annual

---

[213] Daniel P Schrag, 'Storage of Carbon Dioxide in Offshore Sediments' *Science* 325 (2009): pp. 1658–9.
[214] Naomi E Vaughan and Timothy M Lenton, 'A Review of Climate Geoengineering Proposals' *Climate Change* 109 (2011): pp. 745–90, p. 753.
[215] Ibid, p. 755.
[216] Dennis YC Leung, Giorgio Caramanna and M Mercedes Maroto-Valer, 'An Overview of Current Status of Carbon Dioxide Capture and Storage' *Renewable and Sustainable Energy Reviews* 39 (2014): pp. 426–43, p. 433; IPCC, *Special Report on Carbon Dioxide Capture and Storage* (Cambridge: Cambridge University Press, 2005) p. 6.
[217] Leung et al., 'An Overview of Current Status of Carbon Dioxide Capture and Storage', p. 433.
[218] Sigurdur R Gislason and Eric H Oelkers, 'Carbon Storage in Basalt' *Science* 344 (2014): pp. 373–4; David S Goldberg, Dennis C Kent and Paul E Olsen, 'Potential on-shore and off-shore reservoirs for CO_2 sequestration in Central Atlantic magmatic province basalts' *PNAS* 107 (2010): pp. 1327–32.
[219] IPCC, *Special Report on Carbon Dioxide Capture and Storage*, p. 200.
[220] Schrag, 'Storage of Carbon Dioxide in Offshore Sediments', p. 1659.

carbon dioxide emissions.[221] Considerable work is being done around the world on testing carbon storage, usually in association with oil and gas activity.[222] There are still a number of questions associated with such storage, including whether the CO_2 remains captured in the longer term and under what conditions it is likely to leak. Some locations are less likely to leak than others, depending on the geological integrity of the sedimentary rocks and the likelihood of tectonic movement.[223]

(b) Release of carbon dioxide into the deep ocean

A second proposal has suggested the release of carbon dioxide into the deep ocean, with the expectation that the density of the CO_2 at such depths will cause it to form a 'lake' on the deeper parts of the seabed, below 3000 metres.[224] It is expected that owing to the slow exchange of seawater from the depths of the ocean to the surface, the CO_2 will remain isolated from the atmosphere.[225] The technology for achieving this result is not yet developed, and there is considerable uncertainty about how long the CO_2 would remain in the deep ocean before dissolving and being released into the atmosphere. Although the CO_2 would be released into the water column, the result is of potential interest to coastal States with continental shelves underlying the CO_2 lake.

The environmental impact of sea-floor release of CO_2 is greater than when the CO_2 is injected beneath the sea floor because of the effect of the concentrated CO_2 present in the seawater above the sea floor. The lakes of CO_2 are likely to be at depths where species have adapted to low-energy environments and are less able to cope with changes to the environment than species found nearer the sea surface.[226] The rapid increase in CO_2 is likely to lower the pH of the seawater, increasing its acidity, which may affect a range of already vulnerable deep-sea organisms, some of which will be sedentary species and less able to relocate.[227] In addition, a rapid increase in CO_2 may be absorbed by organisms leading to metabolic suppression, respiratory distress, narcosis and mortality in some.[228] The effect on the local ecosystems may differ depending on the depth, location, method of injection and the volume of CO_2 injected.[229]

---

[221] Karen N Scott, 'The Day After Tomorrow: Ocean CO_2 Sequestration and the Future of Climate Change' *Georgetown International Environmental Law Review* 18 (2005–2006): pp. 57–108, p. 64.
[222] IPCC, *Special Report on Carbon Dioxide Capture and Storage*, p. 201; Leung, 'An Overview of Current Status of Carbon Dioxide Capture and Storage', p. 434.
[223] IPCC, *Special Report on Carbon Dioxide Capture and Storage*, p. 213.
[224] Scott, 'The Day After Tomorrow', p. 86.
[225] IPCC, *Special Report on Carbon Dioxide Capture and Storage*, p. 283.
[226] Ibid., p. 299.
[227] Brad A Seibel and Patrick J Walsh, 'Potential Impacts of CO_2 Injection on Deep-sea Biota' *Science* 294 (2001): pp. 319–20; IPCC, *Special Report on Carbon Dioxide Capture and Storage*, pp. 298, 301.
[228] Seibel and Walsh, 'Potential Impacts of CO2 Injection on Deep-sea Biota'; IPCC, *Special Report on Carbon Dioxide Capture and Storage*, p. 301.
[229] James P Barry et al., 'Effects of Direct Ocean CO_2 Injection on Deep-sea Meiofauna' *Journal of Oceanography* 60 (2004): pp. 759–66, p. 764.

(3) Marine scientific research

Another activity that could have a negative impact on the marine environment is marine scientific research.[230] The impact of this activity is usually minimal and far less than the impact of natural events such as volcanic activity, or other human-source disturbances such as climate change, fishing or mining. A variety of scientific disciplines can be included in marine scientific research, including biology, geology, chemistry, physics, and chemical and physical hydrography. The means by which scientists conduct research will determine the environmental impact of the activity. Research that has an impact on the continental shelf includes removal of benthic organisms or habitat, drilling and coring, dredging, and trawling.

The environmental effects of marine scientific research vary. Depending on the nature of the research project, it is possible that ecological niches may be disturbed to some extent simply by studying them.[231] One example is that researchers investigating the mineral content of chimneys formed by hydrothermal vents may remove them, which can impact on the surrounding organisms.[232] In other places, the introduction of light, noise and heat may cause stress to organisms and the movement of sediment could result in the smothering of some creatures.[233] There is also the possibility that some scientific projects could interfere with each other.[234] However, in general, the impact of marine research will be very small when compared with other human activities that affect the sea floor. For example, the collection of one chimney out of hundreds at a vent field would not have a measurable impact on the overall ecosystem.

Seismic surveys used by scientists are more frequently criticized for their potential impact on the marine environment as they cover large areas and potentially impact on organisms including marine mammals. However, these would rarely have a significant impact on the benthic environment.

Scientists are aware of the potential for environmental impact as a result of their work. The International Oceanographic Commission has adopted a code of conduct for marine scientific research vessels calling for use of marine environmental management plans to minimize the environmental impact of research activities.[235] Where possible, researchers are encouraged to conduct a pre-site survey to determine possible

---

[230] Marine scientific research is discussed in ch 6.
[231] Philomène A Verlaan, 'Experimental Activities That Intentionally Perturb the Marine Environment: Implications for the Marine Environmental Protection and Marine Scientific Research Provisions of the 1982 Convention on the Law of the Sea' *Marine Policy* 31 (2007): pp. 210–16.
[232] L Glowka, 'Putting Marine Scientific Research on a Sustainable Footing at Hydrothermal Vents' *Marine Policy* 27 (2003): pp. 303–12, p. 304. At most hydrothermal vent fields there are multiple chimneys, meaning that the removal of one chimney has a limited effect on the overall ecosystem.
[233] 'Oceans and the Law of the Sea: Report of the Secretary General', Addendum (15 July 2005) UN Doc. A/60/63/Add.1, para. 174.
[234] Ibid.
[235] Patricio Bernal and Alan Simcock, 'Chapter 30: Marine Scientific Research' *First Global Integrated Marine Assessment* (New York: United Nations, 2015) www.un.org/Depts/los/global_reporting/WOA_RegProcess.htm, p. 15; Twenty-first International Research Ship Operators Meeting (2007) www.irso.info/wp-content/uploads/IRSO-2007-Meeting-Minutes-FINAL.pdf.

impacts and suitable mitigation measures. InterRidge, an organization coordinating research into oceanic ridges, issued a statement of commitment to responsible research practices at deep-sea hydrothermal vents in 2006.[236] The statement calls on researchers to avoid activities that might have an impact on the sustainability of vent organisms or might lead to significant alteration to the vent site. Cross-contamination of sites and unnecessary collections should be avoided. The International Marine Minerals Society has adopted a code for environmental management in consultation with the International Seabed Authority.[237] The Code includes calls to consider environmental implications of activities, observe the precautionary approach, consult with stakeholders and report publicly on the implementation of the Code.

(4) Ecotourism

Tourism is increasingly offering underwater-based experiences and, in the future, there is a remote possibility that this could include activities that will have an impact on the continental shelf beyond 200 nm. It was estimated in 2007 that as many as forty-nine submarines were operating in the tourism industry, although these were primarily operating in shallow water.[238] One company, Deep Ocean Expeditions, has offered tourist trips to sites 3000–6000 metres below the surface.[239] Their destinations included hydrothermal vents in the Mid-Atlantic ridge area near the Azores and shipwrecks such as the *Titanic* and the *Bismarck*. The company used Russian research submersibles and often included the tourists in a broader scientific research expedition.[240]

Another growing form of tourism is the use of permanent underwater facilities on the seabed, although again this type of activity is always located near the shore. Hotels and restaurants are seeking to create niche experiences for tourists underwater.[241] It is believed that such developments could have major impacts on the local ecosystems through introduction of lighting, using food to influence fish patterns and the destruction of habitats for the construction of the facilities.[242] However, it is highly unlikely that such facilities will be constructed on any continental shelf beyond 200 nm in the foreseeable future.

G. Conclusion

The physical area encompassed by the extended continental shelf (as established by article 76 of the LOSC) covers a huge variety of geological and geomorphological

---

[236] www.interridge.org/IRStatement.
[237] www.immsoc.org/IMMS_code.htm. The first code was adopted in 2001 and a revised code was adopted in 2011.
[238] Carl Cater and Erlet Cater, *Marine Ecotourism: Between the Devil and the Deep Blue Sea* (CABI, 2007): p. 99.
[239] Eoghan Macguire, 'Deep Sea Tourism: Voyage to the Bottom of the Sea' CNN (20 March 2012) http://edition.cnn.com/2012/03/19/travel/deep-sea-tourism/.
[240] Cater and Cater, *Marine Ecotourism*, p. 100.   [241] Ibid., p. 101.   [242] Ibid.

areas, with greatly differing resources and ecosystems. The extended shelf may host large mineral deposits on the sea floor, and oil and gas reserves may be located at depths that are increasingly exploitable with new technology. Seamounts on some extended continental shelves may support rich fisheries and diverse benthic communities. The possibility of discovering sources of economic wealth has been a factor in so many States seeking to establish their entitlements to the continental shelf beyond 200 nm.

One clear message emerges from the literature: there is simply not sufficient information available at present to understand fully the nature of the living and non-living resources that may be present on those parts of the extended continental shelf that are both distant and deep. Although this chapter has attempted to summarize the resources that may be found, it is likely that new resources and uses may be discovered in the future. Coastal States will need to adapt constantly to these new discoveries through their regulation of activities on the continental shelf.

Similarly, the human uses of the oceans will continue to develop. Humanity already impacts on the deep ocean floor, and the impacts will grow as activities increase in number and scope. Information about the environment of the continental shelf and the effects of human activity is important for coastal States seeking to sustainably manage economic activity on the continental shelf. Ideally, this also includes information on the deep pelagic communities in the water column above the seabed, as benthic and pelagic biodiversity are linked components of deep-sea ecosystems. However, regulation of ocean-based activities will always have to deal with a degree of uncertainty relating to the effects of human activity. Successful development will be based on agreement among stakeholders as to the acceptable level of effects, monitoring of the operations and adapting regulatory approaches in light of new information.

# 3

The Development of Sovereign Rights to Continental Shelf Resources

The rights and obligations of coastal and non-coastal States in respect of the continental shelf are grounded in rights that have existed in international law since before the 1958 Continental Shelf Convention.[1] Although the Law of the Sea Convention (LOSC)[2] dramatically altered the provisions relating to the delineation of the outer limits of the extended continental shelf, Part VI of the LOSC largely replicates the provisions of the Continental Shelf Convention when setting out States' rights and obligations. Therefore, in order to understand the LOSC provisions on the continental shelf, it is useful to explore the customary and treaty law that ultimately led to Part VI of the LOSC dealing with the continental shelf. Early law and State practice relating to the intersection between the high seas and the continental shelf regimes provides useful parallels to the LOSC situation, where the extended continental shelf lies under the high seas.

During the negotiations for the Continental Shelf Convention, rights over sedentary species were included in addition to rights over non-living resources. The justification for this was that species that were essentially 'fixed' to the seabed should be treated in the same way as minerals or hydrocarbons. The LOSC did not alter coastal State rights over these resources. The definition of 'sedentary species' is less than ideal and can be difficult to apply. Over time, there has been controversy about the species that should be included in the sedentary species category. Despite this uncertainty, the inclusion of sedentary species in article 77 of the LOSC provides the basis for coastal State jurisdiction over organisms found on the continental shelf that meet the definition. Because States have rights over the living resources of the continental shelf, they also have obligations in relation to the protection of biodiversity of sedentary species. Therefore, this chapter outlines the historical development of the rights of coastal States over sedentary species and illustrates some of the difficulties that may be associated with applying the definition in a modern context.

The process for establishing the outer limits of the continental shelf under article 76 of the LOSC is discussed briefly in this chapter. Although article 76 creates a

---

[1] Convention on the Continental Shelf (opened for signature 29 April 1958, entered into force 10 June 1964) 499 UNTS 312 (Continental Shelf Convention).
[2] United Nations Convention on the Law of the Sea (opened for signature 10 December 1982, entered into force 16 November 1994) 1834 UNTS 397 (LOSC).

process by which coastal States can establish the outer limits of the extended continental shelf, this process is not a precondition for the existence of State rights in the extended shelf. Coastal States are permitted to exercise their sovereign rights to the resources of the extended continental shelf before the Commission on the Limits of the Continental Shelf (CLCS) issues its recommendations, owing to the inherent character of their rights.

Finally, this chapter discusses the customary international law status of Part VI of the LOSC, with a particular focus on articles 76 and 82. These are the articles in Part VI that departed substantially from the Continental Shelf Convention, most of which are considered to reflect customary international law. Therefore, there are real questions as to their applicability to States that have not signed the LOSC.

A. Overview of the Development of Continental Shelf Law

(1) The origins of the continental shelf doctrine

For hundreds of years, coastal States have made various claims to the exclusive use of the oceans near their coasts. Initial interest was in protecting the fisheries. In 1758, Vattel argued that States should claim marine areas adjacent to their coasts to 'convert to their own profit, an advantage which nature has so placed within their reach as to enable them conveniently to take possession of it, in the same manner as they possessed themselves of the dominion of the land they inhabit'.[3] This reflected the move to establish a territorial sea adjacent to the coast over which the coastal State exercised sovereignty.

A key aspect of the territorial sea was the claim to exclusive control over the resources found within it. This included fisheries on the seabed as well as in the water column and, as early as 1926, was recognized as including rights over everything found above or below the subsoil of the territorial sea, including 'coral-reefs, oil-wells, tin-mines'.[4] States considered the waters beyond the territorial sea to be the high seas. However, there was debate about whether States could appropriate areas of the seabed—thus removing them from the high seas category.[5] This perspective of the seabed reflected a categorization of the seabed under the high seas as *res nullius* and capable of appropriation. Under this approach, a claim to the resources of the seabed beyond the territorial sea would depend on occupation by

---

[3] Emer de Vattel and Joseph Chitty (eds), *The Law of Nations*, 6th American edn (Philadelphia: T & JW Johnson, 1844): p. 126.
[4] League of Nations, 'Second Session of the Committee of Experts for the Progressive Codification of International Law: Territorial Waters' *American Journal of International Law Special Supplement* 20 (1926): pp. 62–147, p. 107. See also Dominic Roughton and Colin Trehearne, 'The Continental Shelf' in DJ Attard, M Fitzmaurice and NA Martinez Gutierrez (eds), *The IMLI Manual on International Maritime Law: Volume I: The Law of the Sea* (Oxford: Oxford University Press, 2014): pp. 137–76, p. 143.
[5] DP O'Connell, *The International Law of the Sea: Volume I*, edited by Ivan Shearer (Oxford: Clarendon Press, 1982): p. 449.

a history of fishing or mining.[6] A number of States claimed rights to areas beyond the territorial sea, including over oyster beds and sponge fisheries on the basis of prolonged usage. These States included Ceylon (Great Britain had legislation on the matter since 1811), Venezuela and Panama.[7] In other cases, coastal States asserted the right to build mines into the subsoil beyond the territorial sea.[8] These included Great Britain from 1858, as well as Australia, Canada, Chile and Japan.

The Treaty of Paria, signed between the United Kingdom (on behalf of Trinidad) and Venezuela in 1942, broke ground as the first treaty to essentially claim the seabed beyond the territorial sea by dividing the seabed between the States.[9] The Treaty defined the area subject to each State's interests, but also made it clear that it did not affect the status of the waters of the Gulf of Paria or the freedom of navigation on the high seas.[10]

A growing scarcity of onshore hydrocarbon reserves, coupled with increasing energy demands and technological developments, meant that offshore oil and gas resources became increasingly important after the Second World War.[11] This interest in the resources of the seabed led to the Truman Proclamation, issued by the United States in 1945.[12] In that Proclamation, President Truman asserted that the resources of the United States' continental shelf were subject to its jurisdiction and control, at the same time clarifying that the legal status of the waters above the shelf remained unaffected.[13] The basis for the claim was that the continental shelf was 'an extension of the land-mass of the coastal nation and thus naturally appurtenant to it'.[14] The resources under the shelf were considered to be a seaward extension of

---

[6] RR Churchill and AV Lowe, *The Law of the Sea*, 3rd edn (Manchester: Manchester University Press, 1999): p. 143; O'Connell, *The International Law of the Sea: Volume I*, p. 449. This interpretation was by no means accepted by all legal scholars. O'Connell outlines the jurisprudential debates about the legal basis for claiming the continental shelf: see pp. 467–75.

[7] Suzette V Suarez, *The Outer Limits of the Continental Shelf: Legal Aspects of their Establishment* (Berlin: Springer, 2008): pp. 21–2; O'Connell, *The International Law of the Sea: Volume I*, pp. 450–2. On the other hand, Judge Shigeru Oda denied that the isolated practice of controlling fisheries on the seabed amounted to ownership of the seabed. See his dissenting opinion in Continental Shelf *(Tunisia v Libyan Arab Jamahiriya)* [1982] ICJ Reports 18, p. 200 (*Tunisia v Libya* case).

[8] Churchill and Lowe, *The Law of the Sea*, p. 143; Jean-François Pulvenis, 'The Continental Shelf Definition and Rules Applicable to Resources' in René-Jean Dupuy and Daniel Vignes (eds), *A Handbook on the New Law of the Sea: Volume 1* (Dordrecht: Martinus Nijhoff, 1991): pp. 315–81, p. 323; Øystein Jensen, *Commission on the Limits of the Continental Shelf: Law and Legitimacy* (Leiden: Martinus Nijhoff Publishers, 2014): p. 9.

[9] Roughton and Trehearne, 'The Continental Shelf', p. 144; David Leary, *International Law and the Genetic Resources of the Deep Sea* (Leiden: Martinus Nijhoff, 2007): p. 85.

[10] 1942 Treaty between Great Britain and Northern Ireland and Venezuela relating to the Submarine Areas of the Gulf of Paria (26 February 1942) 205 LNTS 121. See articles 2 and 6.

[11] Yoshifumi Tanaka, *The International Law of the Sea* (Cambridge: Cambridge University Press, 2012): p. 132; Suarez, *The Outer Limits of the Continental Shelf*, pp. 26–7.

[12] Donald Cameron Watt, 'First Steps in the Enclosure of the Oceans: The Origins of Truman's Proclamation on the Resources of the Continental Shelf, 28 September 1945' *Marine Policy* 3 (1979): pp. 211–24.

[13] Policy of the United States with Respect to the Natural Resources of the Subsoil and Sea Bed of the Continental Shelf, Proclamation 2667 of 28 September 1945, 10 Fed. Reg. 12,305 (1945) (Truman Proclamation).

[14] Preamble, Truman Proclamation.

deposits within the territory and it was reasonable to claim them because no other nation could exploit the resources without the cooperation of the coastal State. Where a continental shelf was shared with an adjacent State, the boundary would be negotiated between them. A press release issued at the time of the Proclamation announced that the United States considered the continental shelf to be the submerged land that was covered by no more than 100 fathoms (600 feet) of water.[15] This Proclamation reflected a move away from the concept of occupation as the basis for sovereign rights and towards the concept of natural prolongation of the territory instead. The latter concept was to become the prevailing theory underpinning the development of the regime of the continental shelf.

State practice following the Truman Proclamation was varied.[16] Some States followed suit in very similar terms.[17] Other States claimed to exercise rights to the seabed contiguous to their coasts in terms that differed from the Truman Proclamation. For example, in 1949, the Kingdom of Saudi Arabia claimed the subsoil and seabed outside its territorial sea, despite the fact that it possessed no physical continental shelf beyond the territorial sea.[18] In 1953, Australia's declaration claimed 'sovereign rights' rather than 'jurisdiction and control' over the shelf resources. Chile, Peru, Costa Rica and El Salvador stretched the concept to its limits by declaring full sovereignty not only over the seabed, but also over the water column above.[19] In 1952, Chile, Ecuador and Peru issued the Santiago Declaration, which stated that 'all countries which had no continental shelf … claimed sole jurisdiction and sovereignty over the area of sea extending 200 nm from their coast including sovereignty and jurisdiction over the sea floor and subsoil thereof'.[20] As Lauterpacht noted, the various declarations had two features in common. First, all disclaimed any intention of interfering with the freedom of navigation on the high seas. Secondly, only the most extreme declarations prompted any protest by other States.[21]

The question of whether the right to the resources of the continental shelf required a claim or not was discussed rigorously among academics and by States at the time.[22] Various theories were proffered to explain the ability of a coastal

---

[15] Robert W Smith and George Taft, 'Legal Aspects of the Continental Shelf' in Peter J Cook and Chris M Carleton (eds), *Continental Shelf Limits: The Scientific and Legal Interface* (Oxford: Oxford University Press, 2000): pp. 17–24, p. 18.

[16] H Lauterpacht, 'Sovereignty over Submarine Areas' *British Yearbook of International Law* 27 (1950): pp. 376–433.

[17] Churchill and Lowe, *The Law of the Sea*, p. 144.

[18] Royal Pronouncement Concerning the Policy of the Kingdom of Saudi Arabia with Respect to the Subsoil and Seabed of Areas of the Persian Gulf Contiguous to the Coasts of the Kingdom of Saudi Arabia, *American Journal of International Law Supplement* 43 (1949): pp. 156–7; Richard Young, 'Saudi Arabian Offshore Legislation' *American Journal of International Law* 43 (1949): pp. 530–2.

[19] Lauterpacht, 'Sovereignty over Submarine Areas', pp. 381–433; Richard Young, 'Recent Developments with Respect to the Continental Shelf' *American Journal of International Law* 42 (1948): pp. 849–57.

[20] Suarez, *The Outer Limits of the Continental Shelf*, pp. 28–9.

[21] Lauterpacht, 'Sovereignty over Submarine Areas', p. 383.

[22] The United Kingdom Foreign Office's view in 1945 was that the seabed could be acquired by occupation. This contrasted with the Truman Proclamation's use of natural prolongation as the justification for the claim to the continental shelf. See O'Connell, *The International Law of the Sea: Volume I*, p. 470.

State to exercise control over seabed resources.[23] Such theories included historical use, occupation of the seabed or proclamation accompanied by effective control.[24] This debate was related to questions about the legal character of the continental shelf: was it *res nullius* and therefore open to acquisition by States; or was it the coastal State's by virtue of prolongation of land territory? By the time of the Continental Shelf Convention, consensus was emerging in favour of the idea that the rights of the coastal State amounted to an inherent, exclusive and permanent title due to the prolongation of its land territory.[25]

(2) The International Law Commission and the 1958 Continental Shelf Convention

The International Law Commission (ILC) was created by the United Nations General Assembly in 1947 to work on the codification of international law. One of the first topics selected for study by the ILC was the high seas. The continental shelf was initially considered to be a subset of the high seas topic but the vastly different State practice on the continental shelf following the Truman Proclamation prompted the ILC to attempt to codify the doctrine.[26] The early work of the ILC resulted in a series of draft articles on the law of the sea.

Several issues were raised during the ILC deliberations. One important issue was the identification of the outer limits of the continental shelf.[27] Rather than focusing on a geological description of the continental shelf, the ILC's first attempt to define the shelf referred to 'the seabed and subsoil of the submarine areas contiguous to the coast, but outside the area of territorial waters, where the depth of the superjacent waters admits of the exploitation of the natural resources …'.[28] No outer limit was fixed in terms of distance from the coast because the Commission found 'no practical reason' for doing so.[29]

A later version of the draft articles in 1953 removed the reference in article 1 to the exploitability criterion, instead referring to a depth of 200 metres.[30] This

---

[23] The theoretical debate over the status of the continental shelf prior to the Continental Shelf Convention was eloquently summarized in Judge Shigeru Oda's dissenting opinion in the *Tunisia v Libya* case, pp. 174–8.

[24] See Shigeru Oda, 'The Reconsideration of the Continental Shelf Doctrine' *Tulane Law Review* 32 (1957–1958): pp. 21–36; O'Connell, 'Sedentary Fisheries and the Australian Continental Shelf' *American Journal of International Law* 49 (1955): pp. 185–209, pp. 188–90; Suarez, *The Outer Limits of the Continental Shelf*, pp. 34–8.

[25] Richard Young, 'The Geneva Convention on the Continental Shelf: A First Impression' *American Journal of International Law* 52 (1958): pp. 733–8, p. 734; O'Connell, *The International Law of the Sea: Volume I*, p. 472.

[26] Suarez, *The Outer Limits of the Continental Shelf*, p. 30.

[27] See Bernard Oxman, 'The Preparation of Article 1 of the Convention on the Continental Shelf' *Journal of Maritime Law and Commerce* 3 (1971–1972): pp. 245–305.

[28] Article 1, 1951 ILC Draft Articles on the Continental Shelf and Related Subjects *American Journal of International Law Supplement* 45 (1951): pp. 139–47.

[29] Ibid., p. 140.

[30] Article 1, 1953 Draft Articles on the Continental Shelf *American Journal of International Law Supplement* 48 (1954): pp. 27–43, p. 28.

approach was preferred for reasons of certainty and because that was considered the depth at which the continental shelf generally came to an end.[31] However, disagreement among the members meant that, by 1956, the ILC decided to leave the determination of the outer limits of the continental shelf to coastal States and chose to combine the two criteria of distance and exploitability.[32]

An overriding concern for the ILC during its discussions was to ensure that the continental shelf regime did not affect the legal status of the waters above. Any suggestion that the coastal State had sovereignty over the shelf was rejected for fear of expanding the uses of the high seas that could come under the jurisdiction of the coastal State. The ILC considered that the approach taken in the Truman Proclamation of assigning 'jurisdiction and control' was preferable. The ILC recognized that coastal States could not exploit the shelf without having an impact on the high seas. However, the draft articles sought to minimize the impact of the coastal State's activities on the high seas. The freedom of the sea was the paramount principle that guided the negotiations.[33] The concept of 'sovereign rights' was seen as drawing the necessary balance between freedoms and the coastal State's rights to the seabed.

Another significant issue for the ILC was the nature of the resources over which the coastal State had rights. Many preferred that the resources of the shelf referred only to mineral or hydrocarbon resources. However, in 1953, it was suggested that the draft articles should include reference to sedentary species that were considered to belong to the shelf more than the high seas. This was ultimately adopted by the ILC.[34]

The Continental Shelf Convention[35] was one of four draft conventions considered and subsequently adopted by the international community at the first United Nations Law of the Sea Conference in Geneva in 1958.[36] Article 1 of the Continental Shelf Convention defined the continental shelf as:

[t]he seabed and subsoil of the submarine areas adjacent to the coast but outside the area of the territorial sea, to a depth of 200 metres or, beyond that limit, to where the

---

[31] Ibid., p. 30. See also Ernest Katin, *The Legal Status of the Continental Shelf as Determined by the Conventions Adopted at the 1958 United Nations Conference on the Law of the Sea: An Analytical Study of an Instance of International Law Making* (Minneapolis: University of Minnesota Press, 1962): pp. 61–71.
[32] Suarez, *The Outer Limits of the Continental Shelf*, p. 32.
[33] DR Rothwell and T Stephens, *The International Law of the Sea* (Oxford: Hart Publishing, 2010): p. 102.
[34] See below Section B(1) for further discussion of the inclusion of sedentary species in the continental shelf regime.
[35] The Continental Shelf Convention entered into force 10 June 1964 and ultimately had fifty-eight State parties.
[36] See also the Convention on the Territorial Sea and the Contiguous Zone (opened for signature 29 April 1958, entered into force 10 September 1964) 516 UNTS 206; Convention on the High Seas (opened for signature 29 April 1958, entered into force 30 September 1962) 450 UNTS 82; and Convention on Fishing and the Conservation of the Living Resources of the High Seas (opened for signature 29 April 1958, entered into force 20 March 1966) 559 UNTS 286. The conventions on the continental shelf, the territorial sea and the high seas were considered to be largely based on customary international law, and so gained widespread acceptance. Churchill and Lowe, *The Law of the Sea*, p. 15.

depth of the superjacent waters admits of the exploitation of the natural resources of the said areas ...

Thus, the Continental Shelf Convention adopted the approach recommended by the ILC in relation to the outer limits of the shelf. This adoption was not made without considerable debate but the opponents of the 'exploitability' criteria were unable to agree on an alternative.[37]

According to the Continental Shelf Convention, the rights to the continental shelf were exclusive and inherent—meaning that the rights did not depend on occupation or express proclamation by the coastal State.[38] The coastal State was given 'sovereign rights for the purpose of exploring ... and exploiting [the] natural resources' of the continental shelf.[39] Natural resources consisted of 'mineral and other non-living resources of the seabed and subsoil together with living organisms belonging to sedentary species'.[40] The rights of other States in the water above the shelf were unaffected by the Convention and coastal States were instructed not to unjustifiably interfere with navigation, fishing, conservation of the living resources or scientific research.[41] The relationship between the continental shelf and the high seas was clearly defined and the idea that the continental shelf regime did not affect the status of the waters above was 'beyond dispute'.[42]

The Continental Shelf Convention also gave coastal States jurisdiction over installations for the exploration and exploitation of the resources of the shelf.[43] States were entitled to establish safety zones of up to 500 metres, which had to be respected by foreign vessels.[44] The object and character of the safety zones were not discussed at any great length at the Geneva Conference.[45]

One question that troubled commentators at the time was the status of the continental shelf doctrine under customary international law. In 1950, Lauterpacht raised the prospect that a customary principle was developing, based on the declarations of key 'law-abiding' States and an absence of protest by other States.[46] However, this was not accepted by all scholars.[47] Although State practice gave considerable support to the creation of a customary international law norm consistent with the Proclamation's terms, opinion is divided as to whether such a customary norm existed in 1958.[48]

---

[37] Jensen, *Commission on the Limits of the Continental Shelf*, pp. 16–17.
[38] Article 2, Continental Shelf Convention. [39] Ibid. [40] Ibid., article 2(4).
[41] Ibid., articles 3 and 5(1).
[42] Shigeru Oda, 'Proposals for Revising the Convention on the Continental Shelf', *Columbia Journal of Transnational Law* 7 (1968): pp. 1–31. For further discussion of these concepts, see ch 7.
[43] Articles 5(2) and 5(3), Continental Shelf Convention.
[44] UN Conference Official Records, p. 90, UN Doc. A/CONF.13/42 (1958).
[45] Oda, 'Proposals for Revising the Convention on the Continental Shelf', p. 22. For a summary of the history of article 5, see Marjorie M Whiteman, 'Conference on the Law of the Sea: Convention on the Continental Shelf' *American Journal of International Law* 52 (1958): pp. 629–59.
[46] Lauterpacht, 'Sovereignty over Submarine Areas', p. 395.
[47] See, e.g. Josef L Kunz, 'Continental Shelf and International Law: Confusion and Abuse' *American Journal of International Law* 50 (1956): pp. 828–53, p. 829.
[48] For a discussion of the status of the continental shelf in customary international law, see LFE Goldie, 'Australia's Continental Shelf: Legislation and Proclamations' *International and Comparative*

In 1969, the International Court of Justice (ICJ) in the *North Sea Continental Shelf* case considered that the first three articles of the Geneva Convention were 'regarded as reflecting, or as crystallizing, received or at least emergent rules of customary international law'.[49] These related to the question of the seaward extent of the shelf, the juridical character of the coastal State's entitlement, the nature of the rights exercisable, the kind of natural resources to which these relate and the preservation intact of the legal status of the waters over the shelf as high seas, as well as the legal status of the superjacent airspace.[50] However, the Court rejected an assertion that the provisions of the Convention relating to delimitation were reflective of customary international law. This was a significant development and reinforced the international community's general acceptance of the Continental Shelf Convention's core provisions.

The modern development of the continental shelf doctrine has been heavily based on the concept that the shelf is the natural prolongation of the territory of the State.[51] The *North Sea Continental Shelf* case was notable for the approach the ICJ took regarding the nature of the shelf itself. The ICJ confirmed that the continental shelf was the 'natural prolongation of [a State's] land territory into and under the sea' that existed '*ipso facto* and *ab initio* by virtue of its sovereignty over the land …'.[52] The Court found that if a submarine area did not constitute a natural extension of the land territory, then it could not be regarded as appertaining to that State, even if the State was the closest to the submarine area.[53] The concept of natural prolongation was influential in the negotiation of the LOSC and in delimitation cases up to the 1985 *Libya v Malta* case.[54]

(3) The continental shelf in the Law of the Sea Convention

The fundamental concept of the continental shelf and the coastal State's rights over it was not seriously changed as a result of the LOSC. However, the negotiations in relation to the outer limits of the shelf were hard fought.[55] Agreement that the

---

Law Quarterly 3 (1954): pp. 535–75, pp. 554 ff; Shigeru Oda, *International Control of Sea Resources* (Leyden: AW Sythoff, 1963), p. 153; ZJ Slouka, *International Custom and the Continental Shelf* (The Hague: Martinus Nijhoff, 1968), p. 170; Juraj Andrassy, *International Law and the Resources of the Sea* (New York: Columbia University Press, 1970), pp. 56–66; ED Brown, *The Legal Regime of Hydrospace* (London: Stevens & Sons, 1971), p. 100; MS McDougal and WT Burke, *The Public Order of the Oceans: A Contemporary International Law of the Sea* (New Haven: New Haven Press, 1986); Suarez, *Outer Limits of the Continental Shelf*, pp. 35–7.

[49] North Sea Continental Shelf *(Federal Republic of Germany v Denmark and the Netherlands)* [1969] ICJ Reports 3, p. 39 (*North Sea Continental Shelf* case).
[50] Ibid.
[51] RY Jennings, 'The Limits of Continental Shelf Jurisdiction: Some Possible Implications of the North Sea Case Judgment' *International and Comparative Law Quarterly* 18 (1969): pp. 819–32.
[52] *North Sea Continental Shelf* case, p. 22.   [53] Ibid, p. 31.
[54] Continental Shelf *(Libyan Arab Jamahiriya v Malta)* [1985] ICJ Reports 13; Bjarni Már Magnússon, *The Continental Shelf Beyond 200 Nautical Miles* (Leiden: Brill Nijhoff, 2015): p. 17.
[55] For a comprehensive discussion of the negotiations under the LOSC, see Suarez, *The Outer Limits of the Continental Shelf*; Jensen, *Commission on the Limits of the Continental Shelf*; and the dissenting opinion of Judge Shigeru Oda, *Tunisia v Libya* case, pp. 211–22.

coastal State's sovereign rights over the seabed should extend to the 200 nm mark was achieved early in the negotiations. This reflected the swift consensus that was achieved on the concept of the exclusive economic zone (EEZ).[56] Early proposals to join the continental shelf and EEZ regime conceptually were rejected. Therefore, in order for a coherent regime to be established, it made sense for the continental shelf and the EEZ to coexist spatially.

More problematic was the issue of States (known as broad margin states) whose physical continental shelves extended beyond 200 nm from their baselines. The outer boundary of the continental shelf was particularly significant under the LOSC because the seabed beyond national jurisdiction would be treated as the common heritage of mankind and managed by the International Seabed Authority (ISA). States such as the United Kingdom, Canada, Brazil, India, Venezuela, Australia and New Zealand with large continental shelves argued that they should have sovereign rights over the entire shelf, not just that within 200 nm.[57] A number of arguments were put forward to support this position. First, these coastal States relied on the concept of the continental shelf as the natural prolongation of the territory as set out in the *North Sea Continental Shelf* case.[58] There was a clear distinction, they argued, between the abyssal plains that largely characterized the Area and the continental shelves, which were a geological extension of the coastal State. Therefore, there was no justification to take rights over the shelves away from States and include them in the Area. Secondly, under the 1958 Convention, coastal States would have had jurisdiction over their shelves to the limit of exploitability. Therefore, as the entire shelf would eventually be exploitable, why should the coastal States give up potentially valuable rights?

The significant opposition to this position was based upon the fact that, if some parts of the continental shelves were beyond national jurisdiction, then the resources on the outer parts of the shelves could be utilized for the benefit of all mankind.[59] Because continental shelves are often the location of oil, gas and mineral reserves, including the shelf beyond 200 nm in the Area[60] would enable more benefits to flow to the international community. The development of the concept of the common heritage of mankind was a significant advance in the law of the sea.[61] Developing

---

[56] By 1975 the basic concept of a 200 nm exclusive zone in which coastal States had rights over the resources of the zone, marine scientific research and the preservation of the marine environment had emerged. Myron H Nordquist (ed), *United Nations Convention on the Law of the Sea 1982: A Commentary Volume II* (The Hague: Martinus Nijhoff Publishers, 1993): p. 531.

[57] These positions were held from the beginning of negotiations in the Sea-bed Committee. Nordquist (ed), *United Nations Convention on the Law of the Sea 1982: A Commentary Volume II*, p. 842.

[58] See Jensen, *Commission on the Limits of the Continental Shelf*, pp. 24–5; Alex Oude Elferink, 'Article 76 of the Law of the Sea Convention on the Definition of the Continental Shelf: Questions Concerning its Interpretation from a Legal Perspective' *International Journal of Marine and Coastal Law* 21 (2006): pp. 269–85, p. 272.

[59] See SC Vasciannie, *Land-locked and Geographically Disadvantaged States in the International Law of the Sea* (Oxford: Clarendon Press, 1990): p. 105; articles 136 and 140, LOSC.

[60] The Area is the deep seabed beyond national jurisdiction to which the concept of the common heritage of mankind applies. See articles 1, 136 and 140, LOSC.

[61] See generally Kemal Baslar, *The Concept of the Common Heritage of Mankind in International Law* (The Hague: Martinus Nijhoff Publishers, 1998); Alexandre Kiss, 'The Common Heritage of

countries were generally keen to ensure that as much of the seabed as possible would be included within the Area and its resources. However, many developing countries were also coastal States who saw that they stood to benefit from the expansion of coastal State jurisdiction. Therefore, it was difficult to present a united front to insist on the limitation of rights over the continental shelf.[62]

Ultimately, broad margin States won their argument, but some restrictions were placed on the exercise of sovereign rights beyond 200 nm. Two significant compromises were made. First, coastal States are required to make payments or contributions in kind to the ISA in relation to the exploitation of non-living resources on the continental shelf beyond 200 nm. The payments are to begin after the first five years of production with the rate rising from one per cent of production to seven per cent by the twelfth year.[63] This payment is in recognition of the fact that coastal States were permitted to extend their sovereign rights over an area that might otherwise have been subject to the common heritage of mankind.[64]

The second compromise was in relation to the rights of the coastal State over marine scientific research on the extended continental shelf. In relation to the EEZ and continental shelf within 200 nm, the coastal State has the right to regulate marine scientific research by foreign researchers. Although it is expected to give consent for marine scientific research in normal circumstances, there are a number of reasons why the coastal State may refuse consent, including where the research is of direct significance for the exploration and exploitation of marine resources.[65] However, this right to refuse consent is limited where the research relates to the continental shelf beyond 200 nm. Article 246(6) provides that coastal States may only refuse consent on the basis that the research is of significance for marine resources where the coastal State has publicly designated areas in which exploration or exploitation is occurring, or will occur, within a reasonable period of time. Outside those areas, the coastal State may not refuse consent on the basis that the research is of significance for the exploitation of marine resources.[66]

This latter compromise is not as significant as it may first appear. It must be remembered that there is a distinction between exploratory/exploitative activities (such as fishing or mining) and marine scientific research. Where the activity amounts to exploration or exploitation, the coastal State retains the right to prohibit that activity in any location on the continental shelf. Nevertheless, the

---

Mankind: Utopia or Reality?' *International Journal* 40 (1985): pp. 423–41; Rüdiger Wolfrum, 'The Principle of the Common Heritage of Mankind' *ZaöRV* 44 (1983): pp. 312–37; AO Adede, 'The System for Exploitation of the "Common Heritage of Mankind" at the Caracas Conference' *American Journal of International Law* 69 (1975): pp. 31–49.

[62] Carl August Fleischer, 'The Continental Shelf beyond 200 Nautical Miles—a Crucial Element in the "Package Deal": Historic Background and Implications for Today' in Davor Vidas (ed), *Law, Technology and Science for Oceans in Globalisation* (Leiden: Brill, 2010): pp. 429–48.

[63] Article 82, LOSC.

[64] Ted L McDorman, 'The New Definition of "Canada Lands" and the Determination of the Outer Limit of the Continental Shelf' *Journal of Maritime Law and Commerce* 14 (1983): pp. 195–223, p. 202. See also ch 5.

[65] Article 246, LOSC.   [66] See ch 6.

paragraph does impose a restriction in relation to coastal State authority on the extended continental shelf. The reason for the compromise was to reflect the fact that, had the shelf been included in the Area, the freedom of scientific research would have been retained.[67]

The majority of the debate during the negotiations for the Convention was related to the definition in article 76 of the outer limits of the continental shelf. Ultimately, States concluded a complicated formula for the calculation of the limits, and established the CLCS to assist with the finalization of the outer limits.

B. Sedentary Species

The issue of the rights of coastal States over living resources located on the continental shelf was generally of minor importance to most States. There is no doubt that the primary impetus for the development of the continental shelf regime was the growing interest in exploiting offshore oil reserves. The initial interest in including living resources in the continental shelf regime came from a few States concerned about protecting specific valuable fisheries from foreign fishermen. Over time, the economic contribution of sedentary species has been considerable—for example lobsters, scallops, oysters and crabs are important and lucrative resources for coastal communities. Once again, when it was time to consider the issue of the extended continental shelf beyond 200 nm in the LOSC, mineral and hydrocarbon resources were the primary interest for States. It seemed very unlikely that fishing of sedentary species would be conducted at the depths likely to be involved. However, more recent developments in the exploitation of genetic resources means that rights over sedentary species may become important for coastal States on the extended continental shelf.

(1) The origins of the sedentary species doctrine

The inclusion of sedentary species in the Continental Shelf Convention is often attributed to Australia, which was very concerned about the impacts on its oyster fishery by Japanese fishers who exploited the resource in the areas beyond Australia's territorial sea.[68] Unlike other countries,[69] Australia was considered unable to claim control of the fishery on the basis of effective occupation because the history of exploitation by Malaysian, Indonesian and Japanese fishermen in the area

---

[67] Article 143, LOSC.
[68] Sir Kenneth Bailey, 'Australia and the Law of the Sea' *Adelaide Law Review* 1 (1960–1962): pp. 1–22, p. 9.
[69] For example, the pearl fisheries of Ceylon and Bahrain. Such States were considered to have acquired the right to exclusive control of the fishery on the basis of a historical exercise of jurisdiction or occupation. DP O'Connell, 'Sedentary Fisheries and the Australian Continental Shelf' *American Journal of International Law* 49 (1955): pp. 185–209, p. 189; Oda, 'Proposals for Revising the Convention on the Continental Shelf', p. 11; Richard Young, 'Sedentary Fisheries and the Convention on the Continental Shelf' *American Journal of International Law* 55 (1961): pp. 359–73, p. 360.

undermined any claim to exclusive control of the area.[70] Following the failure of bilateral negotiations in 1953, Australia proclaimed its rights over the living and non-living resources of the seabed and purported to require foreign nationals to obtain a licence to harvest the oysters.[71] Consequently, Australia took its argument for the inclusion of sedentary species in the continental shelf doctrine to the ILC.

The ILC had originally decided to exclude the issue of sedentary species from the question of the continental shelf.[72] However, at the urging of Australia, the ILC included all sedentary fisheries in the concept of the continental shelf in its draft articles.[73] The emphasis was on those species that, when harvested, lived in constant contact with the sea floor. Despite some opposition, the ILC approach was accepted and included in article 2(4) of the Continental Shelf Convention.[74]

One justification for the inclusion of sedentary species was that these species were analogous to crops in a field, and so they were more properly considered as belonging to the seabed rather than the high seas.[75] Not all commentators accepted this analogy. Goldie argued that the graft of sedentary species onto the continental shelf regime 'was camouflaged by a legal fiction, namely the pretense of an untrue state of facts (sessile sea animals are growths of the soil as crops are) to reach a legal conclusion whose propriety depends on the acceptability of the erroneous premise (sessile animals are legally classifiable as crops)'.[76] However, other arguments focused on the illogic of granting rights over the seabed and not over 'the coral sponges and the living organisms which never moved more than a few inches or a few feet on the floor of the sea'.[77]

Oda was also sharply critical of the inclusion of sedentary fisheries in the Continental Shelf Convention, for several reasons.[78] He did not consider that the situation of States controlling parts of the seabed owing to long practice required the inclusion of living resources in the continental shelf doctrine. He also pointed out that the fishing activity itself would be carried out in the high seas, using similar methods to fishing for non-sedentary species. He saw no reason to treat one type

---

[70] SV Scott, 'The Inclusion of Sedentary Fisheries within the Continental Shelf Doctrine' *International and Comparative Law Quarterly* 41 (1992): pp. 788–807, p. 794.
[71] Ibid., p. 799. Australia amended its acceptance of the jurisdiction of the ICJ in 1954 to exclude disputes arising out of rights claimed by Australia in respect of the continental shelf. Shabtai Rosenne, *Documents on the International Court of Justice* (Alphen aan den Rijn: Sijthoff & Noordhoff, 1979): p. 346.
[72] O'Connell, *International Law of the Sea: Volume I*, p. 499.
[73] Yearbook of the International Law Commission [1953] II, p. 214.
[74] O'Connell, *International Law of the Sea: Volume I*, pp. 500–1; Oda, *International Control of Sea Resources*, p. 192. The issue was controversial because many States still wanted the regime to cover only mineral resources, while other States argued that the regime should include all bottom fisheries. See Scott, 'The Inclusion of Sedentary Fisheries', pp. 804–5.
[75] For the use of such analogies in US law, see Charles E Curtis, 'Alaska's Regulation of King Crab on the Outer Continental Shelf' *UCLA Alaska Law Review* 6 (1976–1977): pp. 375–410, p. 384.
[76] LFE Goldie, 'Sedentary Species and Article 2(4) of the Convention on the Continental Shelf—a Plea for a Separate Regime' *American Journal of International Law* 63 (1969): pp. 86–97, p. 90. See also Young, 'Sedentary Fisheries and the Continental Shelf', p. 361.
[77] Scott, 'The Inclusion of Sedentary Fisheries', p. 806.
[78] Oda, 'Proposals for Revising the Convention on the Continental Shelf', pp. 11–12.

of fishing differently from another. On the other hand, in 1982 O'Connell argued that the objections to the inclusion of the sedentary species 'seem dated'.[79]

Young has observed that the proponents of coastal State control over sedentary species were able to profit from the fortuitous development of the continental shelf concept in 1945. 'No longer was it necessary to rely on dubious factual arguments or on debatable applications of the principles of prescription, usage, occupation, and acquiescence'.[80] Another commentator stated at the time that 'economic necessity is the generating impulse of the doctrine, and contiguity is relied upon as the test of its validity'.[80] Of course, economic reality drove the development of the continental shelf regime as a whole. However, the inclusion of sedentary species was subject to particular criticism.

Regardless of the dissatisfaction that remained in relation to the rights over sedentary species, the concept became entrenched in international law. In 1969, the ICJ stated in the *North Sea Continental Shelf* case that the first three articles of the Continental Shelf Convention were 'regarded as reflecting, or as crystallizing, received or at least emergent rules of customary international law'.[81] As this included article 2(4), it would have been difficult to revisit the sedentary species issue. Indeed, the LOSC adopted the Continental Shelf Convention approach to sedentary species with seemingly little debate.[82]

(2) The definition of sedentary species

The definition of sedentary species in article 2(4) of the Continental Shelf Convention and article 77(4) of the LOSC are identical in substance. The articles read:

> The natural resources referred to in [these articles/this Part] consist of the mineral and other non-living resources of the seabed and subsoil together with living organisms belonging to sedentary species, that is to say, organisms which, at the harvestable stage, either are immobile on or under the seabed or are unable to move except in constant physical contact with the seabed or the subsoil.

During the negotiations for the Continental Shelf Convention several proposals failed to gain acceptance. These included a proposal to limit the Convention's application to mineral resources only and a proposal that the article should end with the words 'but crustacea and swimming species are not included'. This latter proposal almost gained acceptance, but ultimately failed.[83] Clearly, States had varying perspectives on whether crustacea—including crabs and lobsters—should be included in the category of sedentary species. The sponsors of the move to include sedentary

---

[79] O'Connell, *International Law of the Sea: Volume I*, p. 500.
[80] Young, 'Sedentary Fisheries and the Continental Shelf', p. 369.
[80] DP O'Connell, 'Sedentary Fisheries and the Australian Continental Shelf' *American Journal of International Law* 49 (1955): pp. 185–209, p. 196.
[81] *North Sea Continental Shelf* case, p. 39.
[82] Nordquist, *United Nations Convention on the Law of the Sea 1982: A Commentary Volume II*, p. 897; Dissenting opinion of Judge Shigeru Oda, *Tunisia v Libya* case, p. 209.
[83] Scott, 'The Inclusion of Sedentary Fisheries', pp. 805–6.

species, at least, considered that crustacea would not be sedentary species under the definition.[84]

Gutteridge considered that the definition supposed a scale of organisms. At one end were sedentary species which, once the brief larval phase is over, have absolute physical association with the seabed. These species include sponges, coral and pearl oysters. At the other end of the scale are demersal species, which swim much of the time in close proximity to the seabed, such as halibut, cod and hake. In between the extremes were crustacea, which although closely associated with the seabed make 'limited seasonal migrations', such as lobsters, crabs, shrimp and prawns.[85] In his view, swimming crustacea had to be excluded. The question remained—does the definition apply to species that 'walk' on the sea floor but are also capable of swimming in certain situations?

One of the problems is that the definition of sedentary species 'has little or no relationship to biological taxonomy'.[86] Scallops, often considered to be associated with the seabed, can swim by means of jet propulsion.[87] Even within species, there can be considerable degrees of variation in the ability of the organisms to move. For example, different types of crabs may have substantial variability in the level of movement and ability to swim. Sea cucumbers can vary from those that are unable to swim and simply 'walk' to those that swim regularly.[88]

Inevitably, disputes have arisen about which species are included in the definition of sedentary species. The disputes have usually involved crustacea and resulted in the coastal State retaining some control over the resource. In all cases, the species in question were capable of some sort of movement independent of the sea floor. The disputes arose under the Continental Shelf Convention or were resolved on the basis of customary international law.

The United States had to resolve a dispute with Japan as to the status of king crab on the United States continental shelf under the Continental Shelf Convention.

---

[84] ILC, 8th Plenary Meeting 22 April 1958, Official Records Vol II, A/Conf.13/38, pp. 11–15; Scott, 'The Inclusion of Sedentary Fisheries', p. 806; O'Connell *International Law of the Sea: Volume 1*, p. 501. Australia, in arguing at the ILC for the inclusion of sedentary species, stated that crustacea should be excluded on the basis that they could move considerable distances. See Whiteman 'Conference on the Law of the Sea', p. 639 (quoting extensively from the Australian submission) and Oda, *International Control of Sea Resources*, p. 192. Australia's subsequent legislation appeared to be consistent with this view. See RD Lumb, 'Australian Legislation on Sedentary Resources of the Continental Shelf' *University of Queensland Law Journal* 7 (1970–1971): pp. 111–14, p. 113.

[85] JAC Gutteridge, 'The 1958 Geneva Convention on the Continental Shelf' *British Yearbook of International Law* 35 (1959): pp. 117–18. See also Young, who stated that: 'in nature there is no simple line of demarcation between sedentary and other fish, but only a long series of gradations from the unquestionably fixed at one extreme to the unquestionably free at the other'. Young, 'Sedentary Fisheries and the Continental Shelf', p. 365.

[86] Craig H Allen, 'Protecting the Oceanic Gardens of Eden: International Law Issues in Deep-sea Vent Resource Conservation and Management' *Georgetown International Environmental Law Review* 13 (2001): pp. 563–660, p. 621.

[87] JD Moore and ER Trueman, 'Swimming of the Scallop, Chlamys Operculis' *Journal of Exp. Mar. Biol. Ecol.* 6 (1971): pp. 179–85.

[88] Andrey Gerbruck, 'Locomotory Organs in the Elasipodid Holothurians: Functional-Morphological and Evolutionary Approaches' in Roland Emerson, Andrew Smith and Andrew Campbell (eds), *Echinoderm Research 1995* (Rotterdam: AA Balkema, 1995): pp. 95–101.

Japan considered that king crabs were a high seas fishery and the United States considered them to be a sedentary species. Without resolving the status of the crabs, the two countries reached an agreement in relation to king crab that allowed Japanese vessels to continue their historical catch of these crabs, subject to conservation measures.[89] The United States also signed an Agreement with the USSR in 1965 in relation to king crab and tanner crab fisheries on the United States' continental shelf.[90] In that Agreement, the States accepted that king crabs were sedentary. It is likely that this was an attempt to strengthen the argument against Japan in relation to king crabs.[91]

In 1963, Brazil arrested three French vessels that were fishing for lobsters on the Brazilian continental shelf. Considerable tensions arose between the two countries, which disagreed about whether lobsters were sedentary species. Neither was a party to the Continental Shelf Convention, but France relied on the negotiating history and statements of State parties to argue that crustacea were not sedentary species.[92] Brazil argued the opposite. It was not until December 1964 that the two States reached a settlement in which twenty-six French vessels were permitted to catch lobsters on Brazil's continental shelf in return for a 'tribute in lobsters and fish'.[93]

Another dispute involving the United States arose with Canada over the classification of scallops. In July 1994, Canada seized two United States fishing vessels that were dredging for Icelandic scallops on the extended continental shelf on the Grand Banks.[94] The United States protested against the arrest, arguing that because molluscs can move above the seafloor by clapping their fluted bivalve shells together, they were not sedentary species.[95] However, the United States eventually accepted that the scallops were sedentary species and Canada released the vessels.[96]

---

[89] Exchange of Notes Constituting an Agreement between the United States of America and Japan Relating to the King Crab Fishery in the Eastern Bering Sea (25 November 1964) 1965 UNTS 32; O'Connell, *International Law of the Sea: Volume 1*, p. 502.

[90] Agreement Between the Government of the United States of America and the Government of the Union of Soviet Socialist Republics Relating to Fishing for King Crab, concluded and entered into force on 5 February 1965, *International Legal Materials* 4 (1965): pp. 359–61; LFE Goldie, 'The Ocean's Resources and International Law: Possible Developments in Regional Fisheries Management' *Columbia Journal of Transnational Law* 8 (1969): pp. 1–53, p. 14.

[91] Curtis, 'Regulation of King Crab', p. 385.

[92] Issam Azzam, 'The Dispute between France and Brazil over Lobster Fishing in the Atlantic' *International and Comparative Law Quarterly* 13 (1964): pp. 1453–9.

[93] Goldie, 'The Ocean's Resources and International Law', p. 13.

[94] At the time of the dispute, both Canada and the United States were party to the Continental Shelf Convention but not the LOSC. See James H Andrews, 'Behind US-Canada Flap' *The Christian Science Monitor* (8 August 1994): p. 8.

[95] Jon M Van Dyke, 'Modifying the 1982 Law of the Sea Convention: New Initiatives on Governance of High Seas Fisheries Resources: The Straddling Stocks Negotiations' *International Journal of Marine and Coastal Law* 10 (1995): pp. 219–27, p. 222. The issue of the classification of scallops as sedentary species was also raised in the context of 1968 Australian legislation by Lumb, who speculated that the reason for exclusion of scallops from the legislation was attributable to the fact that they were capable of movement. Lumb, 'Australian Legislation on Sedentary Resources of the Continental Shelf', p. 112.

[96] J Ashley Roach, 'Dispute Settlement in Specific Situations' *Georgetown International Law Review* 7 (1994–1995): pp. 775–89, p. 784.

According to United States fisheries law, sedentary species are now considered to include a wide range of crustacea, many of which are capable of independent movement. These include various species of tanner crab, king crab, lobster, Dungeness crab, stone crab and red crab.[97] It has been observed that this deviates from the strict meaning of article 77(4) by including organisms capable of movement above the seabed.[98] However, State practice indicates that coastal States that wish to assert jurisdiction over such species as sedentary will be successful.

C. The Outer Limits of the Continental Shelf

Although the juridical nature of the shelf and its attendant rights has attracted significant attention over the years, arguably the most difficult issue has been the determination of the outer limits of the continental shelf. Under the Continental Shelf Convention, the issue was purposefully left fluid, based on the exploitability of the shelf. This made sense when it was considered that there were fairly strict constraints on the depths to which the shelf could be exploited. On this understanding, other States had little interest in the seabed beyond the continental shelf. However, as technology improved, this fluid definition began to cause problems.

The issue of the outer limits was more acute under the LOSC. As already explained, the boundary between the continental shelf and the Area had significance for whether the benefits of any resource on the shelf accrued to the coastal state, or to the international community. Therefore, the conclusion of article 76 became a contentious topic during the negotiations. Because of the importance of the issue and its connection with the existence of sovereign rights to the continental shelf, it is useful to outline the history and content of article 76.

(1) Outer limits under the Continental Shelf Convention

It will be recalled that the Continental Shelf Convention defined the outer limit of the shelf as extending to 'a depth of 200 metres, or beyond that limit, to where the depth of the superjacent waters admits of the exploitation of the natural resources of the said areas'. One of the main criticisms directed at the Continental Shelf Convention was the fact that the outer limit of the shelf was not defined and reliant on the depth to which exploitation could occur. This issue was hotly debated in the ILC proceedings and again at the 1958 Conference.[99] The uncertainty of the definition was challenged by many of the negotiators.[100]

---

[97] Magnuson-Stevens Fishery Conservation and Management Act, 16 U.S.C. § 1802(7).
[98] Allen, 'Protecting the Oceanic Gardens of Eden', pp. 563–660, p. 625; Curtis, 'Regulation of King Crab', p. 385.
[99] See, e.g. Gutteridge, 'The 1958 Geneva Convention on the Continental Shelf', pp. 102–23; Oda, 'Proposals for Revising the Convention on the Continental Shelf', p. 6; Oxman, 'Preparation of Article 1'.
[100] Gutteridge, 'The 1958 Geneva Convention on the Continental Shelf', p. 110.

As commentators pointed out, the definition could potentially result in ever-expanding areas of the seabed coming under coastal State control.[101] One possible outcome was that the Geneva Convention theoretically divided the seabed among the coastal states, leaving no area potentially out of their reach.[102] Oxman suggested that the inability of the Geneva Conventions to identify precisely where the regime applied was 'a symptom of the re-emergence of the territorial temptation at sea'.[103]

Another problem surrounded the question as to which standard of exploitability should be used. As noted by Oda, States have greatly varying technological capability, so 'the concept of exploitability must be interpreted each time in terms of the most advanced standards of technology in the world'.[104] However, there remained some doubt as to whether the precise boundary might be connected to the technical capability of the individual coastal State.[105] Brown, in examining exploration licences granted by the United States for areas at considerable depths, argued that the depths at which exploration might take place did not necessarily mean that exploitation would be possible and so exploration activity would not be counted in the exploitability criteria.[106] Young decried the 'confusion and controversy which must arise in the course of ascertaining, verifying, and publishing the latest data on such a maximum depth'.[107]

During the later debate in the United Nations General Assembly regarding the status of the deep seabed, it became obvious that the Continental Shelf Convention definition would not be consistent with the idea that the deep seabed would be exploited for the benefit of mankind.[108] It seemed clear that there would need to be a division between the resources of the continental shelf under coastal State control, and the resources of the deep seabed.[109] Thus, the issue of the delineation of the outer limits of the continental shelf became an important topic for the third United Nations Conference on the Law of the Sea.

---

[101] Oda, 'Proposals for Revising the Convention on the Continental Shelf', p. 9; O'Connell, *The International Law of the Sea: Volume I*, p. 493.

[102] Shigeru Oda, 'Proposals for Revising the Convention on the Continental Shelf', p. 9. McDorman argues that the expansion of coastal State jurisdiction should be limited by the concept of 'adjacency' found in article 1. McDorman, 'The New Definition of "Canada Lands"', p. 199. See also CL Rozakis 'Compromises of States Interests and their Repercussions upon the Rules on the Delimitation of the Continental Shelf: From the Truman Proclamation to the 1982 Convention on the Law of the Sea' in CL Rozadkis and CA Stephanou (eds), *The New Law of the Sea* (Amsterdam: Elsevier Science Publishers, 1983): pp. 155–84, p. 166.

[103] Bernard Oxman, 'The Territorial Temptation: A Siren Song at Sea' *American Journal of International Law* 100 (2006): pp. 830–51, p. 833.

[104] Oda, 'Proposals for Revising the Convention on the Continental Shelf', p. 9. See also Richard Young, 'The Geneva Convention on the Continental Shelf: A First Impression' *American Journal of International Law* 52 (1958): pp. 733–8, p. 735.

[105] Young, 'The Geneva Convention on the Continental Shelf', p. 735.

[106] ED Brown, *The Legal Regime of Hydrospace* (London: Stevens & Sons, 1971): p. 20.

[107] Young, 'The Geneva Convention on the Continental Shelf', p. 735.

[108] Oda, 'Proposals for Revising the Convention on the Continental Shelf', pp. 5, 10.

[109] Churchill and Lowe, *The Law of the Sea*, p. 147.

(2) Outer limits under the Law of the Sea Convention

It is not the purpose of this book to set out the numerous legal questions that have arisen in the interpretation and implementation of article 76. A comprehensive literature has been developed on this topic.[110] The following section will outline the general approach taken in article 76 for the purposes of determining the outer limits of the continental shelf.

Article 76 of the LOSC defines the continental shelf and the delineation of its outer limits in the following terms:

1. The continental shelf of a coastal State comprises the seabed and subsoil of the submarine areas that extend beyond its territorial sea throughout the natural prolongation of its land territory to the outer edge of the continental margin, or to a distance of 200 nautical miles from the baselines from which the breadth of the territorial sea is measured where the outer edge of the continental margin does not extend up to that distance.

...

3. The continental margin comprises the submerged prolongation of the land mass of the coastal State, and consists of the seabed and subsoil of the shelf, the slope and the rise. It does not include the deep ocean floor with its oceanic ridges or the subsoil thereof.

4. (a) For the purposes of this Convention, the coastal State shall establish the outer edge of the continental margin wherever the margin extends beyond 200 nm from the baselines from which the breadth of the territorial sea is measured, by either:
   (i) a line delineated in accordance with paragraph 7 by reference to the outermost fixed points at each of which the thickness of sedimentary rocks is at least 1 per cent of the shortest distance from such point to the foot of the continental slope; or
   (ii) a line delineated in accordance with paragraph 7 by reference to fixed points not more than 60 nautical miles from the foot of the continental slope.
   (b) In the absence of evidence to the contrary, the foot of the continental slope shall be determined as the point of maximum change in the gradient at the base.

5. The fixed points comprising the line of the outer limits of the continental shelf on the seabed, drawn in accordance with paragraph 4(a)(i) and (ii), either shall not exceed 350 nautical miles from the baselines ... or shall not exceed 100 nautical miles from the 2,500 metre isobath, which is a line connecting a depth of 2,500 metres.

6. Notwithstanding the provisions of paragraph 5, on submarine ridges, the outer limit of the continental shelf shall not exceed 350 nautical miles from the baselines ... This paragraph does not apply to submarine elevations that are natural components of the continental margin, such as its plateaux, rises, caps, banks and spurs.

---

[110] In particular, Magnússon, *The Continental Shelf Beyond 200 Nautical Miles*; Suarez, *The Outer Limits of the Continental Shelf*; Jensen, *Commission on the Limits of the Continental Shelf*; PJ Cook and CM Carleton, *Continental Shelf Limits: The Scientific and Legal Interface* (Oxford: Oxford University Press, 2000).

7. The coastal State shall delineate the outer limits of its continental shelf, where that shelf extends beyond 200 nm from the baselines ... by straight lines not exceeding 60 nautical miles in length, connecting fixed points, defined by coordinates of longitude and latitude.

...

10. The provisions of this article are without prejudice to the question of delimitation of the continental shelf between States with opposite or adjacent coasts.

The definition above contains a range of geological and geomorphological terms.[111] States may establish a continental shelf either based on distance (up to 200 nm) or natural prolongation. The concept of prolongation recognizes the reasoning of the *North Sea Continental Shelf* case.[112] However, the concept of prolongation can be used in both a geologic and a geomorphological sense. The former is concerned with the composition of the rocks and sediments that lie beneath the seabed that reflect the formation of the shelf. The latter focuses on the contours of the seabed.[113] One geological view of the edge of the continental margin tends to focus on the point at which continental crust changes to oceanic crust. This may have been behind the use of prolongation by the ICJ and in the LOSC. However, the geological transition can be difficult to determine, and so the LOSC uses primarily geomorphological characteristics to help draw a political boundary.[114]

It was important to the negotiators that the definition of the outer continental shelf not be left open-ended in the same way that it was in the Continental Shelf Convention. Therefore, paragraph 4 creates two alternative methods for States to use to determine the outer limits and States are permitted to use a combination of the two to arrive at the most advantageous outcome for them. The first, known as the 'Irish formula' after the nationality of its proponent, focuses on the sediment thickness. This is based on a geological concept and is designed to ensure that sovereign rights would cover areas where hydrocarbon resources were expected to exist. The second, known as the 'Hedburg formula', focuses on the distance from the foot of the continental slope and is easier to

---

[111] See PA Symonds et al., 'Characteristics of Continental Margins' in PJ Cook and CM Carleton (eds), *Continental Shelf Limits: The Scientific and Legal Interface* (Oxford: Oxford University Press, 2000): pp. 25–63, pp. 25–7.

[112] United Nations, *Law of the Sea: Definition of the Continental Shelf* (New York: United Nations, 1993): p. 2.

[113] See Symonds et al., 'Characteristics of Continental Margins', pp. 27–8.

[114] Ibid., p.29. The CLCS has pointed to the fact that the LOSC does not in fact refer to the types of crust. The type of crust is therefore not considered relevant to the determination of the outer limits. The ITLOS rejected an argument that geological discontinuity was a relevant factor in determining entitlement to an extended continental shelf. See Delimitation of the Maritime Boundary between Bangladesh and Myanmar in the Bay of Bengal *(Bangladesh v Myanmar)* (2012) 51 ILM 844, para. 438 (*Bay of Bengal* case).

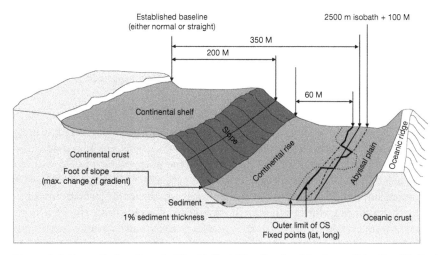

Figure 3.1 From Rothwell et al., The Oxford Handbook of the Law of the Sea (Oxford: Oxford University Press, 2015): p. 191.

determine.[115] An illustration of the operation of these paragraphs is contained in Figure 3.1.

There is one important way in which article 76 of the LOSC does not reflect geological concepts. In geological terms, the 'continental shelf' is usually applied only to the relatively flat part of the overall continental margin.[116] However, article 76 uses the term to apply to the entire area, including the continental slope and rise. Thus, the legal and geological definitions of the 'continental shelf' are inconsistent as they refer to different parts of the seafloor.[117]

(3) The Commission on the Limits of the Continental Shelf

The negotiators of article 76 recognized that the definition was complex and relied on a mixture of legal, geological and geomorphological concepts that could be open to debate. In order to forestall serious conflict over the interpretation of the article, paragraph 8 established a new institution, the CLCS.[118]

---

[115] Robert W Smith and George Taft, 'Legal Aspects of the Continental Shelf' in PJ Cook and CM Carleton (eds), *Continental Shelf Limits: The Scientific and Legal Interface* (Oxford: Oxford University Press, 2000): pp. 17–24, p. 19; Andrew Serdy, 'The Commission on the Limits of the Continental Shelf and its Disturbing Propensity to Legislate' *International Journal of Marine and Coastal Law* 16 (2011): pp. 355–83, p. 358.
[116] Smith and Taft, 'Legal Aspects of the Continental Shelf', p. 18.
[117] United Nations, *Law of the Sea: Definition of the Continental Shelf*, p. 10.
[118] For discussion of the role of the CLCS, see Roughton and Trehearne, 'The Continental Shelf'.

8. Information on the limits of the continental shelf beyond 200 nautical miles from the baselines ... shall be submitted by the coastal State to the Commission on the Limits of the Continental Shelf ... The Commission shall make recommendations to coastal States on matters related to the establishment of the outer limits of their continental shelf. The limits of the shelf established by a coastal State on the basis of these recommendations shall be final and binding.

Annex II of the LOSC further articulates the purpose and structure of the CLCS. The CLCS has two primary aims: to consider the information submitted by States relating to their extended shelves and make recommendations, and to provide scientific and technical advice to coastal states.[119] The members of the CLCS are experts in the fields of geology, geophysics or hydrography, elected by State parties.[120] In making recommendations, the CLCS must assess whether the information submitted by States for the delineation of their extended shelves is in accordance with the LOSC. Although this was originally thought to be a technical matter, it has become clear that the CLCS has had to engage in legal interpretation of article 76.[121] This has not always met with approval by commentators.[122] The CLCS has issued Rules of Procedure and Scientific and Technical Guidelines to establish the procedures of the Commission, and also to provide guidance on how the CLCS will interpret article 76.

Coastal States were originally expected to submit information relating to their extended continental shelf limits within ten years of the entry into force of the LOSC for each State.[123] This was intended to ensure that the boundary between the Area and coastal State jurisdiction would be settled promptly. For 129 States, this deadline expired on 16 November 2004, ten years after the LOSC came into effect. Unfortunately, it quickly became clear that most States would be unable to meet this target owing to the fact that compiling the required information was both onerous and costly. The initial solution was to fix the starting point for the ten years at 1999, the date of adoption of the CLCS's Scientific and Technical Guidelines. Subsequently it was decided by the meeting of the State parties to the LOSC to allow coastal States (who now had a 2009 deadline) to file preliminary information

---

[119] Article 2, Annex II, LOSC. On the latter function, see Suzette Suarez, 'The Commission on the Limits of the Continental Shelf and its Function to Provide Scientific and Technical Advice' *Chinese Journal of International Law* 12 (2013): pp. 339–62.

[120] Article 1, Annex II, LOSC. See Ted McDorman, 'The Role of the Commission on the Limits of the Continental Shelf: A Technical Body in a Political World' *International Journal of Marine and Coastal Law* 17 (2002): pp. 301–24.

[121] For a discussion of the legal implications of the CLCS interpretation of article 76, see Jensen, *Commission on the Limits of the Continental Shelf*, pp. 71–92.

[122] See, e.g. Magnússon, *The Continental Shelf Beyond 200 Nautical Miles*, p. 49; Serdy, 'Propensity to Legislate'; Øystein Jensen, 'The Commission on the Limits of the Continental Shelf: An Administrative, Scientific, or Judicial Institution?' *Ocean Development and International Law* 45 (2014): pp. 171–85; Kunoy Bjørn, 'The Terms of Reference of the Commission on the Limits of the Continental Shelf: A Creeping Legal Mandate' *Leiden Journal of International Law* 25 (2012): pp. 109–30.

[123] Article 4, Annex II, LOSC.

instead of a full submission.[124] By 31 May 2016, sixty-six coastal States had filed full or partial submissions with the CLCS and a further fourteen States had submitted preliminary information but not a submission.[125] The Commission had issued twenty-four recommendations.

Any failure by a coastal State to meet the deadline for making submissions to the CLCS does not deprive that State of its rights under the LOSC. The rights of a coastal State over the continental shelf, including that part beyond 200 nm, are inherent, exclusive and do not depend on occupation or express proclamation.[126] Annex II does not indicate the consequences of not meeting the deadline but, at worst, the coastal State may lose the right to set the outer limits according to the CLCS recommendations.[127]

It should be noted that the CLCS is not the final determinant of the outer limits of the continental shelf. This is a function that can only be fulfilled by the coastal State.[128] The importance of a submission to the CLCS is that if the final limits are established in accordance with the recommendations of the Commission, then the limits are final and binding under international law.[129] Magnússon has pointed out the negative consequences of failing to establish the outer limits of the continental shelf in accordance with article 76(8). These primarily relate to the potential uncertainties surrounding the area over which the coastal State has jurisdiction, which

---

[124] Surya P Subedi, 'Problems and Prospects for the Commission on the Limits of the Continental Shelf in Dealing with Submissions by Coastal States in Relation to the Ocean Territory Beyond 200 Nautical Miles' *International Journal of Marine and Coastal Law* 26 (2011): pp. 413–31, p. 419; Erik Franckx, 'The International Seabed Authority and the Common Heritage of Mankind: The Need for States to Establish the Outer Limits of their Continental Shelf' *International Journal of Marine and Coastal Law* 25 (2010): pp. 543–67, p. 554.

[125] See www.un.org/depts/los. In total, forty-one States filed preliminary information with the Commission but subsequently twenty-seven of these have made full or partial submissions. It can be expected that further submissions will be made. Many States have not yet filed full submissions in respect of their entire claimed extended continental shelf. In addition, some States with extended shelves such as the United States and Venezuela are not party to the LOSC but may become parties in the future.

[126] Article 77(2) and (3), LOSC.

[127] See ILA Committee on the Legal Issues of the Outer Continental Shelf, Second Report (2006) (ILA Second Report), Conclusion 16.

[128] ILA Second Report, ibid., Conclusion 1; Ted L McDorman, 'The Role of the Commission on the Limits of the Continental Shelf: A Technical Body in a Political World' *International Journal of Marine and Coastal Law* 17 (2002): pp. 301–24, p. 306.

[129] McDorman has argued that this should not deprive other States of the right to reject or disagree with the coastal State boundaries. Rather, it is a requirement that the coastal State not change its outer limits once they have been set in accordance with the recommendations of the CLCS and no State has challenged them. See McDorman, 'The Role of the CLCS', p. 315. The International Law Association agrees that the coastal State may not change the limits once set, but that the limits are also binding on other States as long as the limits are set 'on the basis of' the CLCS's recommendations. ILA Second Report, Conclusion 11. In the *Bay of Bengal* case, the ITLOS described the 'final and binding' as 'opposable' in relation to other States. *Bay of Bengal* case, para. 407. See also Alex G Oude Elferink, 'Causes, Consequences and Solutions Relating to the Absence of Final and Binding Outer Limits of the Continental Shelf' in Clive R Symmons (ed), *Selected Contemporary Issues in the Law of the Sea* (Leiden: Martinus Nijhoff, 2011): pp. 253–72.

may hinder activities on both the continental shelf itself and also in the Area.[130] Additionally, the lack of clarity may lead to an inadvertent infringement of rights to resources in either the Area or on the shelf if the final boundary is not determined, or is disputed.

The practice of the CLCS has come under intense scrutiny by States and commentators. Among the issues discussed are: the question of the definition and role of oceanic ridges, submarine ridges and elevations;[131] the CLCS's interpretation of the LOSC and its own Scientific and Technical Guidelines;[132] the application of article 76 to particular geographical areas, including the Arctic;[133] the resource constraints on the CLCS;[134] the refusal of the CLCS to deal with submissions where there are disputed areas of continental shelf;[135] and whether and how coastal States can establish their outer limits if they disagree with the CLCS.[136] It is beyond the scope of this volume to enter into these questions.

(4) The relationship between CLCS recommendations and the rights of the coastal State

There will be a period of time, possibly decades in some cases, in which CLCS recommendations are not available to coastal States.[137] Its workload means that only a few submissions can be addressed each year. In addition, the CLCS has refused to consider any submissions in respect of areas under dispute. This raises the question as to the existence of coastal State rights on the extended continental shelf in circumstances where the CLCS has not issued recommendations. Can, for example, a coastal State authorize oil and gas exploration prior to the issuance of CLCS recommendations? This matter has not been directly addressed by any international court or tribunal but three delimitation cases have made comments

---

[130] Magnússon, *The Continental Shelf Beyond 200 Nautical Miles*, p. 82.
[131] Harald Brekke and Philip Symonds, 'Submarine Ridges and Elevations of Article 76 in Light of Published Summaries of Recommendations of the Commission on the Limits of the Continental Shelf' *Ocean Development and International Law* 42 (2011): pp. 289–306.
[132] Serdy, 'Propensity to Legislate', p. 372.
[133] Betsy Baker, 'Law, Science, and the Continental Shelf: The Russian Federation and the Promise of Arctic Cooperation' *American University International Law Review* 25 (2010): pp. 251–81; Ted L McDorman, 'The Continental Shelf Beyond 200 NM: Law and Politics in the Arctic Ocean' *Journal of Transnational Law and Policy* 89 (2008): pp. 155–93; Vladimir Jares, 'Continental Shelf Beyond 200 Nautical Miles: The Work of the Commission on the Limits of the Continental Shelf and the Arctic' *Vanderbilt Journal of Transnational Law* 42 (2009): pp. 1265–305.
[134] Chris Carleton, 'Article 76 of the UN Convention on the Law of the Sea: Implementation Problems from the Technical Perspective' *International Journal of Marine and Coastal Law* 21 (2006): pp. 287–308, p. 306.
[135] Serdy, 'Propensity to Legislate', p. 362; Alex Oude Elferink, 'The Establishment of Outer Limits of the Continental Shelf Beyond 200 Nautical Miles by the Coastal State: The Possibilities of Other States to Have an Impact on the Process' *International Journal of Marine and Coastal Law* 24 (2009): pp. 535–56.
[136] Oude Elferink, 'Article 76 of the LOSC', p. 281.
[137] Clive Schofield, 'Securing the Resources of the Deep: Dividing and Governing the Extended Continental Shelf' *Berkeley Journal of International Law* 33 (2015): pp. 274–93, p. 286.

The Outer Limits of the Continental Shelf 75

about the relationship between delineation (of the outer limits) and delimitation (of maritime boundaries). Some of these comments have raised the possibility that the existence of sovereign rights on the extended continental shelf are dependent on the CLCS process. Other judicial comments have denied that the CLCS process is linked to the exercise of sovereign rights. However, an examination of these inconsistent authorities reveals there does not appear to be any legal impediment to coastal States exercising their sovereign rights independently of the process in the CLCS.

Prior to any delimitation cases involving the continental shelf beyond 200 nm, the International Law Association established a committee to examine legal issues relating to the extended continental shelf. This committee acknowledged the uncertainty that arises for the exercise of sovereign rights on the extended continental shelf where no outer limits have been declared. In 2006, the report of the committee found, first, that the entitlement to the continental shelf is not dependent on establishment of outer limits.[138] Secondly, the report found that the exact extent of the continental shelf is dependent on establishing the outer limit. Therefore, 'the absence of outer limit lines of the continental shelf beyond 200 nm is bound to raise doubts over the exact extent of the continental shelf, with the attendant difficulties for the coastal State to exercise sovereign rights over such specific areas of the continental shelf'. This was not a denial that the rights existed, but pointed out the practical problems for coastal States where no limits were in place.

(a) *Nicaragua v Honduras*

The question of the existence of State rights prior to a recommendation by the CLCS was called into question by comments of the ICJ in the *Nicaragua v Honduras* case in 2007.[139] In that case, the request for a delimitation beyond 200 nm might possibly have had an effect on Jamaica's claim to a continental shelf. The Court acknowledged the possibility that third-party States might be affected. Therefore, the ICJ declined to exercise jurisdiction in relation to the continental shelf beyond 200 nm in that case. The Court stated that:[140]

[i]n no case may the [boundary] line be interpreted as extending more than 200 nautical miles from the baselines from which the breadth of the territorial sea is measured; any claim of continental shelf rights beyond 200 miles must be in accordance with Article 76 and reviewed by the Commission on the Limits of the Continental Shelf.

The possible implication of this statement in a maritime boundary delimitation context was that the ICJ saw the review by the CLCS as a vital step in the

---

[138] International Law Association, *Legal Issues of the Outer Continental Shelf*, Second Report, Toronto Conference (2006).
[139] Territorial and Maritime Dispute in the Caribbean Sea *(Nicaragua v Honduras)* [2007] ICJ Reports 659 *(Nicaragua v Honduras)*.
[140] *Nicaragua v Honduras*, para. 319.

determination of coastal State rights to the extended continental shelf. It might have been interpreted as saying that the rights to the extended continental shelf depended on the recommendations of the CLCS.[141] However, in light of the fact that the ICJ was discussing the fact that third parties might have a claim to a continental shelf in the part of the shelf, this statement could also be seen as recognition of the fact that the issues would be better resolved through the CLCS process.

(b) *Bay of Bengal* case

The ITLOS was much clearer about this issue in the *Bay of Bengal* case between Bangladesh and Myanmar in 2012.[142] The parties had requested the Tribunal to delimit a maritime boundary for the territorial sea, the EEZ and the continental shelf. Myanmar relied in part on the *Nicaragua v Honduras* case and the existence of possibly affected third parties to argue that it was inappropriate for the Tribunal to delimit the boundary beyond 200 nm.[143] Bangladesh argued that, because the CLCS will not consider claims subject to disputes (such as the one in that case), requiring the Tribunal to wait for the outcome of the CLCS would mean no action could ever be taken in relation to the extended shelf.[144]

In response to the parties' arguments, the Tribunal set out several important principles. First, the Tribunal emphasized that articles 76 and 77 of the LOSC do not distinguish between the areas of continental shelf within and outside 200 nm. The rights of the coastal State are in relation to the continental shelf in its 'entirety'.[145] Secondly, there is a distinction between the delimitation of the continental shelf in article 83 and the delineation of the outer limits under article 76.[146] The exercise by the Tribunal of its power to delimit does not interfere with the CLCS's functions as it is possible to create a delimitation line without fixed outer edges.[147] Thirdly, in relation to the *Nicaragua v Honduras* decision, the Tribunal noted that: 'the determination of whether an international court or tribunal should exercise its jurisdiction depends on the procedural and substantive circumstances of each case'.[148] Fourthly, and most importantly, the entitlement of coastal States to an extended shelf is a separate issue to the outer limits of the shelf.[149] At paragraph 409, the Tribunal stated:[150]

A coastal State's entitlement to the continental shelf exists by the sole fact that the basis of entitlement, namely, sovereignty over the land territory, is present. It does not require the establishment of outer limits. Article 77, paragraph 3, of the Convention confirms that the existence of entitlement does not depend on the establishment of the outer limits of the continental shelf by the coastal State.

---

[141] See Roughton and Trehearne, 'The Continental Shelf', pp. 170–1 for a discussion of the consequences of this decision.
[142] *Bay of Bengal* case. [143] Ibid., para. 349. [144] Ibid., para. 358.
[145] Ibid., para. 361. [146] Ibid., para. 376. [147] Ibid., para. 393.
[148] Ibid., para. 384. [149] Ibid., para. 406.
[150] Article 77(3) provides that the 'rights of the coastal state over the continental shelf do not depend on occupation, effective or notional, or on any express proclamation'.

The Tribunal then confirmed that it is up to the coastal State to determine the outer limits of its continental shelf:[151]

Although this is a unilateral act, the opposability with regard to other States of the limits thus established depends upon satisfaction of the requirements specified in article 76, in particular compliance by the coastal State with the obligation to submit to the [CLCS] and issuance by the Commission of relevant recommendations in this regard.

The significance of these findings is that, as long as a coastal State is able to demonstrate that it has an entitlement to a continental shelf beyond 200 nm in terms of article 76, the final determination of the outer limits following recommendations from the CLCS is not connected to the fact that it may still exercise sovereign rights over the continental shelf.

(c) Nicaragua v Colombia

In the dispute between Nicaragua and Colombia over maritime boundaries, Nicaragua asked the ICJ to define a continental shelf boundary by dividing the overlapping entitlements equally.[152] The parties agreed that coastal States have *ipso facto* and *ab initio* rights to the continental shelf but disagreed about the nature and content of the 'rules governing the entitlements of coastal States to a continental shelf beyond 200 nm'.[153] In response to Nicaragua's argument that it had information demonstrating its entitlement to an extended continental shelf, the ICJ noted that Nicaragua had only filed preliminary information with the CLCS but not full submissions.[154] Colombia argued that preliminary information does not allow the CLCS to make recommendations 'and therefore Nicaragua has not established any entitlements to an extended continental shelf'.[155] Colombia also noted the fact that the CLCS would not make recommendations in a disputed matter.[156] Nicaragua relied on the *Bay of Bengal* case to argue that its entitlements existed prior to determination by the CLCS.

The ICJ observed that the Bay of Bengal was a unique situation that had been acknowledged in the course of negotiations at the United Nations Conference on the Law of the Sea.[157] It also noted that, unlike the *Nicaragua v Colombia* case, both countries were party to the LOSC and had made full submissions to the CLCS. The Court reiterated the statement quoted above from the *Nicaragua v Honduras* case.[158] The ICJ then decided that, because Nicaragua had not complied with

---

[151] *Bay of Bengal* case, para. 407.
[152] Territorial and Maritime Dispute *(Nicaragua v Colombia)* [2012] ICJ Reports 624, para. 106.
[153] Ibid., para. 115.
[154] Ibid., para. 120. Interestingly, Nicaragua's argument was that its continental shelf extended to an area within 200 nm of Colombia's coast. It argued that the right to a continental shelf based on prolongation overrode Colombia's entitlement to jurisdiction over the seabed to 200 nm. Ibid., paras. 121–22.
[155] Ibid., para. 122.   [156] Ibid., para. 124.
[157] Ibid., para. 125. The Bay of Bengal is covered in thick layers of sedimentary rocks.
[158] Ibid., para. 126.

article 76(8) of the LOSC and had 'not established that it has a continental margin that extends far enough to overlap' with Colombia's EEZ, the Court could not delimit the extended continental shelf boundary.[159]

The Court's decision in relation to article 76(8) was criticized by Judge Donoghue and Judge ad hoc Mensah. Both judges found the ICJ's reference to the *Nicaragua v Honduras* case to be misleading.[160] Judge Donoghue believed the case failed because Nicaragua did not provide enough evidence that the continental shelf existed beyond 200 nm and the delimitation method proposed by Nicaragua would have required delineation of the outer limits as the first step in the negotiation.[161] In relation to the *Nicaragua v Honduras* quote she stated:[162]

I regret that the Court reaffirms that statement today without acknowledging that delimitation is not precluded in every case in which an UNCLOS State party seeks delimitation of continental shelf beyond 200 nautical miles before having established the outer limits of such continental shelf. ... The Bangladesh/Myanmar case illustrates that where the existence of continental shelf in the relevant area is not in dispute and the methodology and geography do not require a court or tribunal to make any factual finding regarding the outer limits of the continental shelf, the 'distinct' exercises of delimitation and delineation of the outer limits of the continental shelf may proceed in parallel, regardless of whether a State has established the outer limits of its continental shelf.

It is unfortunate that the Court did not reconcile the potential inconsistency between the statements in the *Nicaragua v Honduras* case and the *Bay of Bengal* case. It would have been possible for the ICJ to state that the *Nicaragua v Colombia* case failed for the reasons given by Judge Donoghue. In addition, the ICJ seemed to confine the effect of the Bay of Bengal to the 'unique situation' reflected by its geology and geography. The position was left very ambiguous as a result.

The ICJ has an unusual opportunity to revisit its finding in this case because Nicaragua issued a new application to the ICJ in 2013 for a delimitation of the extended continental shelf. The ICJ issued a decision in March 2016 responding to preliminary objections raised by Colombia.[163] The Court refuted arguments that recommendations from the CLCS were a prerequisite for delimitation. Colombia claimed that the Court could not consider Nicaragua's application because the CLCS had not 'ascertained that the conditions for determining the extension of the outer edge of Nicaragua's continental shelf ... are satisfied and, consequently,

---

[159] Ibid., para. 129.
[160] Separate Opinion of Judge Donoghue, ibid., para. 24; Declaration of Judge ad hoc Mensah, ibid., para. 2.
[161] Separate Opinion of Judge Donoghue, ibid., paras 9 and 18.
[162] Separate Opinion of Judge Donoghue, ibid., para. 25.
[163] Question of the Delimitation of the Continental Shelf between Nicaragua and Colombia beyond 200 nautical miles from the Nicaraguan Coast (*Nicaragua v Colombia*) (Preliminary Objections) ICJ, 17 March 2016(*Continental Shelf beyond 200 nm* case).

has not made a recommendation'.[164] Colombia argued that there was a distinction between a coastal State's entitlement to a continental shelf up to 200 nm and its entitlement to the shelf beyond 200 nm, and the latter was dependent on the CLCS process to confirm that the State had complied with article 76(4)–(6).[165] If this argument were taken to its logical conclusion, the possible implication is that a coastal State would not have any entitlement to exercise rights in respect of the extended continental shelf until the CLCS had issued recommendations.

The ICJ rejected Colombia's argument. It found that Nicaragua had an obligation to submit information on the limits of the continental shelf to the CLCS and this was a prerequisite for the delimitation of the continental shelf beyond 200 nm.[166] The CLCS's main role is to ensure that coastal States do not extend beyond the limits in article 76 and encroach on the Area.[167] However, this process is distinct from delimitation of the shelf, which is governed by article 83.[168] Because the delimitation of the extended shelf can be undertaken independently of a recommendation from the CLCS, the ICJ found that 'the latter is not a prerequisite that needs to be satisfied by a State party to UNCLOS before it can ask the Court to settle a dispute with another State over such a delimitation'. The merits stage of this case may well return to the point as it will be necessary for Nicaragua to establish an entitlement to the shelf beyond 200 nm—and, in the absence of the CLCS's recommendations, this judgment may have to be made by the ICJ.

In each of the above cases, the ICJ and the ITLOS were responding to a request to delimit the boundaries of the extended continental shelf between two coastal States. Therefore, the jurisprudence is not directed squarely at the question of when the entitlement to rights over a continental shelf beyond 200 nm comes into existence. It could be argued that delimitation is only an issue that arises when a State has an entitlement to (and rights over) an extended shelf. In the *Bay of Bengal* case the ITLOS was prepared to find the entitlement existed, based on the evidence presented by the parties and not on the basis of a recommendation from the CLCS. In contrast, the ICJ jurisprudence might suggest that a full submission to the CLCS is a prerequisite for such entitlement. That is, that the coastal State has no rights over the shelf until the CLCS has made recommendations as to the outer limit. The ITLOS decision in the *Bay of Bengal* case did not turn on this point because, by 2012, both countries had made full submissions to the CLCS.

On the other hand, it is persuasive that the ICJ and ITLOS cases were focused on the process of delimitation rather than the existence, or otherwise, of sovereign rights. If the issue had been, instead, whether Nicaragua or Colombia could authorize oil and gas exploration and exploitation on a part of their extended shelf that was not in dispute, it is harder to maintain the argument that the CLCS recommendations are a prerequisite for the exercise of sovereign rights. In this context, the statements by the ITLOS in the *Bay of Bengal* case about the undivided nature of the continental shelf regime should logically prevail. Overall, nothing in the ICJ

[164] Ibid., para. 97. [165] Ibid., paras. 98 and 99. [166] Ibid., para. 105.
[167] Ibid., para. 109. [168] Ibid., para. 112.

cases should be taken as rejecting the argument that a coastal State exercises rights over the resources of the continental shelf beyond 200 nm *ipso facto* and *de jure*. The CLCS process is relevant only to ensuring that any delineation of the outer limits is consistent with article 76.

The ITLOS's findings in the *Bay of Bengal* case are consistent with the development of the continental shelf regime. The distinction between entitlement (as established in article 76(1)) and the establishment of the outer limits (as set out in article 76(5)–(7) of the LOSC) is clear. The existence of rights to the continental shelf has never depended on clarity as to its outer limits—as is demonstrated by the confusion over the outer limits of the shelf under the Continental Shelf Convention. Rights over the continental shelf are inherent and not dependent on any declaration.[169]

One example where these rights have been recognized in practice was in the case of Portugal's extended continental shelf. In 2006, Portugal proposed that a hydrothermal vent field located 235 nm from its coast should be made a marine protected area under the OSPAR Convention.[170] None of the members of OSPAR objected on the basis that Portugal did not have jurisdiction over the field.[171] At the time, Portugal had not filed submissions with the CLCS.[172]

The problem for coastal States where no recommendation has yet been received from the CLCS, or a maritime boundary has not been negotiated, relates not to entitlement to sovereign rights but to preventing disputes over infringement of the sovereign rights of other States. If there are overlapping claims to the continental shelf beyond 200 nm, there is the potential for disputes to arise about the ownership of resources found in the overlapping area. Where the line between the extended continental shelf and the Area is unclear, the risk is that the coastal State will exploit resources rightly belonging to the Area, or vice versa. The LOSC does not provide any guidance in dealing with the latter situation. However, in relation to the delimitation of maritime boundaries for the EEZ and continental shelf, articles 74(3) and 83(3) require States, pending agreement on delimitation, to make every effort to enter into provisional arrangements of a practical nature and not to jeopardize or hamper the reaching of a final agreement.[173] It is arguable that these principles could apply to any indeterminacy created by the lack of final outer limits and the coastal State's relationship with the ISA. It is highly advisable for States to act with restraint in any area that could later be found to belong to another State or the Area, as it may engage international responsibility if the State exploits resources that later turn out not to be within its jurisdiction.

---

[169] Article 77, LOSC. See also Frida M Armas-Pfirter, 'Submissions on the Outer Limit of the Continental Shelf: Practice to Date and Some Issues of Debate' in Davor Vidas (ed), *Law, Technology and Science for Oceans in Globalisation* (Leiden: Brill, 2010): pp. 477–98, p. 493.

[170] Established by the Convention for the Protection of the Marine Environment of the North-East Atlantic (opened for signature 22 September 1992, entered into force 25 March 1998) 32 ILM 1069.

[171] Marta Chantal Ribeiro, 'The "Rainbow": The First National Marine Protected Area Proposed under the High Seas' *International Journal of Marine and Coastal Law* 25 (2010): pp. 183–207, p. 196. This example is discussed further in ch 9.

[172] Portugal's submissions were made on 11 May 2009.   [173] See ch 5.

D. The Status of Part VI of the Law of the Sea Convention in Customary International Law

Although the vast majority of States are now party to the LOSC, there are still some that are not.[174] The United States, which has exercised jurisdiction in respect of its extended continental shelf by delimiting the boundary with Mexico in the Gulf of Mexico and tendering for exploration licences, is a key example. This raises an important question about the status of the provisions of the LOSC relating to the continental shelf. Can all the provisions of Part VI of the LOSC be regarded as customary international law, with the effect that they will bind non-party States so long as they are not persistent objectors? If they are not custom, then the principle of *pacta tertiis* would mean that the non-parties are not bound by certain provisions of Part VI of the LOSC.

According to article 38(1)(b) of the Statute of the ICJ, international custom is 'evidence of a general practice accepted as law'. This is generally said to require evidence of two elements: State practice in conformity with a rule, and *opinio juris*—the belief that the principle is law.

The provisions of a treaty can be part of customary international law if they either codify existing customary international law, lead to the crystallization of a rule that may be emerging, or lead to a practice accepted as law.[175] The ICJ in the *North Sea Continental Shelf* case identified conditions for the transformation of a treaty provision into customary international law.[176] First, the provision must be of a fundamentally norm-creating character that could be regarded as forming the basis of a general rule of law.[177] Secondly, State practice in respect of the provision should be extensive and virtually uniform and include specially affected States.

At the conclusion of the third United Nations Conference on the Law of the Sea, there was recognition that the LOSC both incorporated existing rules of customary international law and represented progressive development.[178] Ambassador Tommy Koh acknowledged the progressive development at the closing of the conference, where he referred to the regimes of transit passage through straits, archipelagic sea lanes passage and article 76 on the continental shelf as representing 'new concepts'.[179]

Many provisions of the LOSC have been found by international tribunals to represent customary international law. This is because, in some cases, the principles

---

[174] At the time of writing, 168 States were parties to the LOSC. Non-parties that are also coastal States include Eritrea, Israel, Peru, Syria, Turkey, the United States and Venezuela.
[175] Michael Wood, Rapporteur, *Third Report on the Identification of Customary International Law*, 2015, International Law Commission, A/CN.4/682, para. 34.
[176] *North Sea Continental Shelf* case, paras 70–4.
[177] Wood suggests that provisions providing a role for institutions are unlikely to be of a norm-creating character. *Third Report on the Identification of Customary International Law*, para. 39.
[178] See the Preamble, LOSC.
[179] Official Records Vol XVII, 193rd Meeting, 19 December 1982, pp. 132–6, para. 48.

reflected custom at the time that they were drafted.[180] Others have developed into customary international law through universal adoption by non-parties as well as parties to the Convention.[181]

Much of Part VI of the LOSC can be seen as reflecting or codifying customary international law. The substantive rights over the shelf were adopted in a manner largely consistent with the Continental Shelf Convention, most of which was regarded as custom. In the *North Sea Continental Shelf* case, the ICJ found that the provisions of the Continental Shelf Convention relating to the outer limits of the shelf, the nature of the coastal States' entitlement and rights and the preservation of the status of the superjacent water were all regarded as reflecting or crystallizing customary international law.[182] Judge Shigeru Oda, referring to article 77, noted in the *Tunisia v Libya* case that 'the actual régime of the continental shelf is represented as remaining in 1982 exactly the same as in 1958', despite the change in relation to the outer limits of the shelf.[183] Therefore, it is safe to assume that articles 77 and 78 are customary international law.

For the purposes of this study, the key question arises about the extension of outer limits to the continental margin in article 76, the creation of the CLCS as a method for checking claims to the outer limits are according to article 76, and the substance of article 82. These articles were the result of negotiations during the United Nations Conference on the Law of the Sea and were considered by some not to reflect customary international law at the conclusion of the conference.[184] The question is whether some or all of the provisions have crystallized into custom since that time.

(1) Customary international law status of article 76

An examination of State practice by non-parties to the LOSC must consider the United States which, as a maritime nation, is 'specially affected' by the law of the sea. The United States takes the position that the delineation provisions of article 76 are customary international law and that the United States will apply articles 76(1)–(7) in establishing the outer limits of the continental shelf.[185] It has been gathering

---

[180] Maritime Delimitation and Territorial Questions between Qatar and Bahrain *(Qatar v Bahrain)* [2001] ICJ Reports 40, paras 184, 195 *(Qatar v Bahrain)*. The ICJ found that the status of islands in article 121 was customary law.
[181] See, e.g. *Libya v Malta*, para. 34. The ICJ found that the 'institution of the exclusive economic zone' had become part of customary international law prior to the entry into force of the LOSC.
[182] *North Sea Continental Shelf* case, para. 63. Many of these provisions are echoed in Part VI of the LOSC. See also O'Connell, *International Law of the Sea: Volume I*, pp. 475–6.
[183] Dissenting opinion, *Tunisia v Libya* case, p. 222.
[184] Statement by Ambassador Tommy Koh at the closing session of the conference. Official Records Vol XVII, 193rd Meeting, 19 December 1982, pp. 132–6, para. 48. Of course, some new provisions in the LOSC were considered to have entered into customary international law by, or soon after, 1982, especially the EEZ regime.
[185] Memorandum from Assistant Secretary John D Negroponte to Deputy Legal Adviser Elizabeth Verville, 17 November 1987, State Department File No. P89 0140-428, II *Cumulative Digest* 1878, as

information in support of this process since 2001.[186] Another non-party, Venezuela, has also claimed rights to an extended continental shelf on the basis of customary international law.[187] Coastal States have asserted their rights to the continental shelf beyond 200 nm against non-parties to the Convention, which indicates that they consider the rights to be based on more than the LOSC. Third-party States cannot be bound by treaties they are not party to unless the provisions reflect customary international law.[188]

Some States and academics have taken the position that the Continental Shelf Convention authorized the exercise of jurisdiction beyond 200 nm, and it is not necessary to rely on article 76 of the LOSC.[189] This argument is based first on the fact that article 1 of the Continental Shelf Convention did not impose an outer limit to the continental shelf, and secondly, on State practice by parties to the Convention.[190] The ICJ in the *North Sea Continental Shelf* case based its decision on the idea that the continental shelf was based on natural prolongation of the land territory and this also might provide a basis for claiming the continental shelf to the outer edges of the continental rise.[191] O'Connell concluded that: 'except for technicalities of delineation, and the granting of seabed rights to States with continental shelves narrower than 200 nm, the [LOSC] provisions may be taken to represent the present position in customary international law'.[192] As an illustration of this position, in the letter of submittal of the delimitation treaty between the United States and Mexico to the Senate, it was suggested that the basis for exercising jurisdiction beyond 200 nm lay *both* in the Continental Shelf Convention and in the LOSC.[193] Kwiatkowska conducted an examination of the practice of States in 1991 and noted that at least

---

cited in J Ashley Roach and Robert W Smith, *Excessive Maritime Claims*, 3rd edn (Leiden: Martinus Nijhoff, 2012): p. 188.

[186] http://www.state.gov/e/oes/continentalshelf/.

[187] Letter, Permanent Mission of the Bolivarian Republic of Venezuela to the United Nations, 17 September 2008, http://www.un.org/depts/los/clcs_new/submissions_files/brb08/ven_sept_2008.pdf.

[188] Articles 34 and 38, Vienna Convention on the Law of Treaties (concluded 23 May 1969, entered into force 27 January 1980) 1155 UNTS 331.

[189] For example, this argument was made by Libya in the *Tunisia v Libya* case (1982); see the dissenting opinion of Judge Oda, p. 221. David Joseph Attard, *The Exclusive Economic Zone in International Law* (Oxford: Clarendon Press, 1987): p. 134.

[190] Ted McDorman, 'The Entry into Force of the 1982 LOS Convention and the Article 76 Outer Continental Shelf Regime' *International Journal of Marine and Coastal Law* 10 (1995): pp. 165–87, p. 167.

[191] See the statement by Mr Beesley (Canada) introducing a working paper to the conference in 1974, justifying the extension of the continental shelf beyond 200 nm on the basis that existing practice of States is to exercise jurisdiction to the edge of the continental margin. A/Conf.62/SR.46, p. 203, http://legal.un.org/diplomaticconferences/lawofthesea-1982/docs/vol_I/a_conf-62_sr-46.pdf.

[192] O'Connell, *International Law of the Sea: Volume 1*, p. 497. O'Connell argued that the *North Sea Continental Shelf* case created a connection between the legal and physical extent of the continental shelf, so that the rights of the coastal State extended to the entire continental margin. See pp. 492–5.

[193] See Letter of Submittal, Treaty between the United States and Mexico on the Delimitation of the Continental Shelf in the Western Gulf of Mexico Beyond 200 Nautical Miles, as cited in *Digest of United States Practice in International Law 2000*, p. 699 http://www.state.gov/documents/organization/139599.pdf.

twenty-three States had claimed a continental shelf to the outer edge of the continental margin or to a distance of 200 nm where the outer edge did not reach that far. Of the twenty-three, only eight had ratified the LOSC.[194]

Many commentators did not accept that the Continental Shelf Convention created an 'ambulatory limit that moved forward as marine technology progressed',[195] or that the LOSC provisions of article 76 reflected customary international law. Burke strongly criticized the United States' position that the delineation provisions of article 76 were customary international law. He argued that there was 'zero state practice' to insist that article 76 was already customary international law.[196] An important factor in his argument was the fact that article 82 could not be customary international law, so acceptance of article 76 as custom undermined the package deal on which the LOSC was negotiated. Burke suggested that the claim arose 'more from the general political usefulness of the reliance than from the soundness of the specific legal proposition at stake'.[197] Judge Shigeru Oda, in his dissenting opinion in the *Tunisia v Libya* case, expounded the history of the negotiations of article 76 and concluded that:[198]

[t]here is no comparison between the degree to which the actual régime of the continental shelf has found acceptance, not to mention its endorsement by the Court in 1969 as customary law, and the status of the latest definition of the shelf's expanse, as it has hitherto emerged from UNCLOS III. It should be crystal clear that Article 76 of the draft convention is essentially a product of compromise—not consensus—between the conflicting positions of various groups which have different, and sometimes opposite, interests in the use of sea-bed areas. Well may the draft convention be expected eventually to become binding upon many nations, once it has become widely accepted and received a sufficient number of ratifications. Until that time, there is no doubt that Article 76 is not a provision of any worldwide multilateral convention, and can hardly be considered as enshrining established rules of international law.

Despite this opposition, support for the idea that parts of article 76 have become customary law has also come from the ICJ.[199] In the *Nicaragua v Colombia* case, the ICJ was

---

[194] Barbara Kwiatkowska, 'Creeping Jurisdiction Beyond 200 Miles in the Light of the 1982 Law of the Sea Convention and State Practice' *Ocean Development and International Law* 22 (1991): pp. 153–87, p. 154. This does not necessarily lead to the conclusion that States were relying on customary international law rather than article 76. Many States made claims to an EEZ prior to the conclusion of the negotiations and well before they ratified the LOSC. Certainly, States that claimed continental shelves to 200 nm in the absence of a physical shelf would not have been able to base their entitlement on the Continental Shelf Convention because there was no natural prolongation.

[195] Attard, *The Exclusive Economic Zone in International Law*, p. 133, citing Brownlie and Bowett.

[196] William T Burke, '"Customary Law as Reflected in the LOS Convention": A Slippery Formula' in John P Craven, Jan Schneider and Carol Stimson (eds), *The International Implications of Extended Maritime Jurisdiction in the Pacific* (Honolulu: Law of the Sea Institute, 1987): pp. 402–9, p. 404.

[197] Ibid., p. 403.

[198] Dissenting opinion of Judge Shigeru Oda, *Tunisia v Libya* case, p. 220, para. 104. See also the separate opinion of Judge Jiménez de Aréchaga, who found that the 200 nm criterion in article 76 destroyed the concept of natural prolongation as the determinant of the outer limits of the continental shelf. *Tunisia v Libya* case, p. 114, para. 51.

[199] See *Libya v Malta* case, paras 26, 34 and 39. In that case, the ICJ considered that the extension of the continental shelf based on a distance criterion of 200 nm was customary international law prior to the entry into force of the Convention, in part as a consequence of its connection with the EEZ, which had been widely accepted.

asked to determine maritime boundaries, including a request by Nicaragua to delimit the extended continental shelf boundary. Colombia is not party to the LOSC. Without devoting any significant analysis to the question, the Court found that article 76(1) formed part of customary international law.[200] Colombia agreed, but argued that there was no evidence of State practice indicating that paragraphs (4) to (9) of article 76 were customary international law.[201] The ICJ found that its role was to decide whether it could carry out a delimitation of the continental shelf and, in the circumstances, it decided that it was not necessary to consider the status of other paragraphs in article 76.[202]

Academic commentators have been divided on whether the rest of article 76 is customary law. If a State is relying solely on the Continental Shelf Convention or the *North Sea Continental Shelf* case as the source of a customary rule, then it would be inconsistent with that position that articles 76(3)–(5) are also custom because their detail can be found neither in the Convention nor in that case. They were purely the result of negotiation at the third UN Conference on the Law of the Sea. However, if it is accepted that article 76(1) has independently become customary law through consistent State practice, there is a strong argument that article 76 establishes an integrated regime which sets out how the outer limits in paragraph 1 are to be determined.[203] Therefore, all parts of article 76 relevant to the establishment of the outer limits will also be custom.

There is one particular difficulty in considering that *all* paragraphs of article 76 reflect customary international law. Article 76(8) provides that information on the outer limits shall be submitted to the CLCS for its recommendations. It can be argued that because the CLCS was established by the LOSC and is a technical process, it is not possible for this provision to reflect customary international law.[204] This would be an example of a provision that does not have a norm-creating character.

Judge Donoghue and Judges ad hoc Mensah and Cot were highly critical of the ICJ decision in *Nicaragua v Colombia* that Nicaragua had to complete its submission to the CLCS prior to delimitation. In their view, this position was erroneous because Colombia was not a party to the LOSC. Judge Donoghue pointed out that, where two non-parties to the LOSC are in a dispute, 'it goes without saying that such States have no duty to make submissions to the Commission, so the Court's observations regarding Nicaragua's obligations to States parties to UNCLOS cannot be extended to them'.[205] Judge ad hoc Mensah argued that:[206]

[t]here is nothing in the Preamble or any provision of UNCLOS that can legitimately be interpreted to mean that the obligations under the Convention are owed also to States that

---

[200] *Nicaragua v Colombia*, para. 118.   [201] Ibid., para. 117.
[202] Ibid., para. 118.
[203] Fleischer, 'The Continental Shelf beyond 200 Nautical Miles—a Crucial Element in the "Package Deal"', p. 436. See the Dissenting Opinion of Judge Weil, Delimitation of Maritime Areas between Canada and France (St Pierre and Miquelon), 10 June 1992, 31 ILM 1145, p. 1215, para. 42.
[204] McDorman, 'The Entry into Force of the 1982 LOS Convention', p. 168.
[205] Separate Opinion of Judge Donoghue, *Nicaragua v Colombia*, para. 28. Judge Donohue also found that Nicaragua had not adequately proved the existence of a shelf beyond 200 nm.
[206] Declaration of Judge ad hoc Mensah, *Nicaragua v Colombia*, para. 8. See also Declaration of Judge ad hoc Cot, *Nicaragua v Colombia*, para. 19.

are not parties thereto. In my opinion, the obligations under Article 76, paragraphs 8 and 9, are 'treaty obligations' that apply only as between States that have expressed their consent to be bound by the UNCLOS treaty. Those provisions cannot be considered as imposing mandatory obligations under customary international law.

This position is undoubtedly correct. It may be possible for the States parties to accept in the future that a non-party State might avail itself of the CLCS process, but no customary international law obliges non-party States to use the CLCS processes.

As a final matter, it is interesting to note that non-party States have responded to submissions by coastal States to the CLCS. Venezuela notified the CLCS through the United Nations of the fact that Barbados had not consulted it on its submission, which should not prejudice the delimitation of the maritime boundary between the two countries.[207] In 2009, the United States drew the CLCS's attention to the existence of a potential maritime boundary issue with Cuba.[208] No objections to these statements appear to have been received from other States, indicating that they accept the propriety of non-parties interacting with the CLCS. This does not, however, amount to State practice that would establish article 76(8) as a principle of customary international law because the non-party States have not indicated that they consider themselves bound by article 76(8).

(2) Customary international law status of article 82

A second question is whether article 82, requiring coastal States to make payments in respect of production on the extended shelf, is customary international law. This will be a key issue in the future if non-party States exploit the resources of their continental shelves beyond 200 nm. The primary objection to an argument that article 82 is customary law is that the entitlement to an extended continental shelf in article 76 is the result of a carefully negotiated compromise that is linked to the obligation in article 82. To rely on the entitlement to a continental shelf in article 76 without making payments under article 82 would be seen by many States as a violation of the 'deal' on which articles 76 and 82 are based.

The key debate in the negotiations relating to the continental shelf was the outer limit of the continental shelf.[209] Because it was clear that the seabed beyond national jurisdiction was to be managed according to the common heritage of mankind, many States (especially land-locked and geographically disadvantaged states) wished to limit coastal State jurisdiction over the continental shelf.[210] There was strong resistance to the extension of coastal State jurisdiction beyond 200 nm. Some

---

[207] Letter, Permanent Mission of the Bolivarian Republic of Venezuela to the United Nations, 17 September 2008, http://www.un.org/depts/los/clcs_new/submissions_files/brb08/ven_sept_2008.pdf.
[208] Letter, United States Mission to the United Nations, 30 June 2009, http://www.un.org/depts/los/clcs_new/submissions_files/cub51_09/usa_re_cuba_2009.pdf.
[209] The history of the negotiations for Article 82 is discussed in more detail in ch 5.
[210] See Vasciannie, 'Land-locked and Geographically Disadvantaged States'.

broad margin States argued that an application of the concept of natural prolongation meant that, in principle, they already had rights to the edge of the continental margin. In order to break this deadlock, article 82 was developed, requiring coastal States to make payments or contributions in kind for a portion of the production of non-living resources beyond 200 nm. This concession was seen as a 'quid pro quo' for agreement that coastal States be permitted to extend their jurisdiction where the physical shelf went beyond 200 nm.[211]

The view that article 82 was a quid pro quo for the extension of the continental shelf in article 76 was highlighted by Ambassador Koh in his closing statement to the United Nations Conference on the Law of the Sea. For Ambassador Koh, the 'integral' package of the LOSC meant that 'it is not possible for a State to pick what it likes and to disregard what it does not like. It was also said that rights and obligations go hand in hand and it is not permissible to claim rights under the Convention without being willing to shoulder the corresponding obligations'.[212]

Given that article 76(1)—and possibly article 76(2)–(7)—are now considered to be customary international law, what is the situation in relation to article 82?

Neither the United States nor any other non-party States have made any statements about their view of the status of article 82. In a summary document on the LOSC prepared for the United States Senate, it was acknowledged by officials that article 82 was an important aspect of Part VI:[213]

Revenue sharing for exploitation of the continental shelf beyond 200 miles from the coast is part of a package that establishes with clarity and legal certainty the control of coastal States over the full extent of their geological continental margins. At this time, the United States is engaged in limited exploration and no exploitation of its continental shelf beyond 200 miles from the coast. At the same time, the United States is a broad margin State, with significant resource potential in those areas and with commercial firms that operate on the continental shelves of other States. On balance, the package set forth in the Convention, including revenue sharing at the modest rate set forth in article 82, clearly serves United States interests.

Notably, this comment is silent about the United States' view on the status of article 82 under customary international law, and to date the United States does not appear to have taken an official position on this issue. The comment does indicate a clear awareness that article 82 was an integral aspect of the bargain that allowed broad margin States to claim a shelf beyond 200 nm.

The United States Minerals Management Service has advised potential lease holders that additional royalty payments may apply 'If the United States (US) Government becomes a party to the 1982 Law of the Sea Convention … prior to

---

[211] Churchill and Lowe, *The Law of the Sea*, p. 157.
[212] Official Records Vol XVII, 193rd Meeting, 19 December 1982, pp. 132–6, para. 47. This position was also taken by the Group of 77, and by delegates at the signing of the LOSC. See Luke T Lee, 'The Law of the Sea Convention and Third States' *American Journal of International Law* 77 (1983): pp. 541–68, p. 547; Hugo Caminos and Michael R Molitor, 'Progressive Development of International Law' *American Journal of International Law* 79 (1985): pp. 871–90, p. 877.
[213] Commentary—The 1982 United Nations Convention on the Law of the Sea and the Agreement on Implementation of Part XI, Appendix 8, Senate Treaty Doc. 103-39, cited in Roach and Smith, *Excessive Maritime Claims*, p. 192.

or during the life of a lease issued by the US Government on a block or portion of a block located beyond the US Exclusive Economic Zone'.[214] This stipulation is contingent on the United States becoming a party to the LOSC, and does not indicate that it considers article 82 to apply as a result of customary international law.

There are a number of cogent reasons why article 82 is not currently customary international law. It must be remembered that not every treaty provision is capable of becoming customary international law. In the *North Sea Continental Shelf* case, the ICJ rejected the argument that article 6 of the Continental Shelf Convention was customary international law because of 'doubts as to the norm-creating character' of the rule. The ICJ in the *Gulf of Maine* case referred to the need to draw a distinction between 'what are principles and rules of international law ... and what could better be described as the various equitable criteria and practical methods that may be used to ensure *in concreto* that a particular situation is dealt with in accordance with the principles and rules in question'.[215] Given that both article 76(8) and article 82 contain references to institutions created by the LOSC and to processes embedded within the LOSC regime, this may be an obstacle to them ever becoming customary international law.[216]

The second obvious problem is that there is no State practice by non-parties in relation to either article 76(8) or article 82. State practice can be based on a range of sources including administrative acts, decisions of courts and statements in international meetings.[217] Although the United States and Venezuela have submitted observations on the submissions of other States to the CLCS, they have not yet declared a formal position on this issue. Importantly, no exploitation has yet occurred on any non-party's extended continental shelf.

Some might suggest that the status of article 82 is tied to the customary international law status of article 76. This relies on the proposition that one article or paragraph of the LOSC is so dependent on another that the customary status of one confers customary status on another. In the *Nicaragua v Colombia* case, the ICJ was concerned with paragraph 3 of article 121 of the LOSC on islands. The first two paragraphs of article 121 were treated as customary law by the ICJ in the *Qatar v Bahrain* case.[218] In *Nicaragua v Colombia*, the ICJ observed that the entitlement to maritime rights accorded to an island by the provisions of article 121(2) is expressly limited by reference to the provisions of article 121(3).[219] It considered that the

---

[214] See, e.g. MMS, Department of the Interior, Lease Stipulations, Western Planning Area, Oil and Gas Lease Sale 207, Final Notice of Sale, Stipulation 4; Gulf of Mexico OCS Oil and Gas Lease Sale: 2012, Central Planning Area Lease Sale 216/222. BOEM 2012-058. See also International Seabed Authority, *Issues Associated with the Implementation of Article 82 of the United Nations Convention on the Law of the Sea*, Technical Study No. 4 (2009): p. 5.

[215] Delimitation in the Maritime Boundary in the Gulf of Maine Area (*Canada v United States*) [1984] ICJ Reports 246, para. 80. See also para. 82.

[216] See Wood, *Third Report on the Identification of Customary International Law*, para. 34; Jonathan I Charney, 'The United States and the Law of the Sea after UNCLOS III—The Impact of General International Law' *Law and Contemporary Problems* 46 (1983): pp. 37–54, p. 49.

[217] Malcolm M Shaw, *International Law*, 6th edn (Cambridge: Cambridge University Press, 2008): p. 82.

[218] *Qatar v Bahrain*, paras 167, 185 and 195.

[219] *Nicaragua v Colombia*, para. 139.

legal regime of islands in article 121 forms an 'indivisible regime', all of which has the status of customary international law.

Could article 82 be considered an 'indivisible regime' together with article 76 along the lines that, if article 76 is customary law, then so is article 82? The arguments set out above about the quid pro quo deal that underpinned the agreement on article 76 might support such an approach. The argument would be that the 'package deal' was so important to the conclusion of the LOSC that new negotiating practices were created to facilitate it.[220] However, this would be extending the concept beyond the approach of the ICJ in *Nicaragua v Colombia*. There is no express connection between the provisions on the face of article 76 and article 82—unlike article 121(2), which is limited by the words 'except as provided for in paragraph 3'. In the alternative, McDorman has argued that the manner in which the LOSC was negotiated and the use of 'coastal State' in article 82 means that it creates obligations that could only be avoided by an explicit rejection, even by non-parties.[221]

If a traditional view of the formation of customary international law is adopted according to the *North Sea Continental Shelf* cases, it is hard to argue that article 82 is custom in the absence of either State practice or unilateral State declarations. It would require a tribunal or court to take an original approach in order to confer customary law status on article 82.[222] In all previous cases involving the discussion of the status of the LOSC in customary international law, the judges have looked at individual articles without considering the 'package deal' approach to the treaty negotiations.[223]

Another argument is that article 82 is customary international law because it includes, to a partial extent, the principle of the common heritage of mankind. Ong has suggested that the compromise between the proponents of the Area and coastal states, and the fact that payments are collected by the ISA, means that article 82 is a limited application of the principle of the common heritage of mankind.[224] This

---

[220] Caminos and Molitor, 'Progressive Development of International Law and the "Package Deal"', p. 884. The authors argue that an approach based on a package deal concept was applied to the Vienna Convention on the Law of Treaties, where procedural safeguards in the Convention were considered so integral to the bargain that they could also be considered part of customary international law.

[221] McDorman, 'The Entry into Force of the 1982 LOS Convention', p. 180.

[222] McDorman argues that it is hard to pigeonhole the LOSC into the constraints imposed by a customary and treaty law paradigm. Thus, he suggests that the negotiation process and the character of the treaty may have created obligations especially where the issue has not been rejected by States. McDorman, 'The Entry into Force of the 1982 LOS Convention', p. 181.

[223] The closest the Court has come to acknowledging the package deal was in the *Gulf of Maine* case. The judgment put emphasis on the consensus reached at UNCLOS III on large portions of the instrument, and that some provisions including those on the EEZ and continental shelf were adopted without question, in order to determine whether the delimitation provisions were part of customary international law. The Court considered that: 'these provisions, even if in some respects they bear the mark of the compromise surrounding their adoption may nevertheless be regarded as consonant at present with customary international law'. *Gulf of Maine*, para. 94.

[224] David M Ong, 'A Legal Regime for the Outer Continental Shelf? An Inquiry as to the Rights and Duties of Coastal States within the Outer Continental Shelf', Paper Presented to the Third ABLOS Conference, 2003, p. 3, http://www.iho.int/mtg_docs/com_wg/ABLOS/ABLOS_Conf3/PAPER7-4.PDF.

assertion is discussed in more detail in Chapter 5, where it is argued that article 82 does not represent an application of the common heritage of mankind to the continental shelf, but merely reflects the compromise insisted on by States who wished to limit the extent of the continental shelf to maximize the Area. Ong suggests that the common heritage of mankind is customary international law, and therefore article 82 binds non-parties to the LOSC. However, even if Ong's assertion about the status of the common heritage of mankind is correct, there is some question about its application in this context.

Many other commentators have argued that the common heritage of mankind principle is customary international law. Wolfrum argued that the content of the principle is customary only in its general respects: no agreement exists as to the specific obligations it imposes.[225] Noyes has suggested that it may be possible to argue that the principle is customary international law but also questions what the exact content of such a norm may be.[226] Joyner argued in 1986 that there was no evidence of State practice or *opinio juris* that the principle should be a mandatory legal norm in common space areas such as the high seas.[227] One of the reasons that the LOSC took some time to come into force was the opposition of some States to the seabed mining provisions in Part XI, which applies the principle.[228] These issues were in large part resolved with the adoption of the 1994 Implementation Agreement.[229] However, the ongoing failure of some States, including the United States, to ratify the LOSC raises some questions whether it is safe to say that the common heritage of mankind is customary international law. Even if the principle in general is custom, it is a large step to conclude that non-party States must, as a matter of legal obligation, apply article 82 as a result.

Assuming article 82 is not customary international law, is it possible for non-parties to the LOSC to completely ignore article 82? At the very least, there will be a strong expectation on the part of the international community that a non-party State will comply with article 82. The ICJ in the *Nuclear Tests* case held that 'one of the basic principles governing the creation and performance of legal obligations, whatever their source, is the principle of good faith'.[230] It is possible to imagine an argument that a non-party to the LOSC that exercises rights over the continental shelf beyond 200 nm in conformity with the provisions of article 76 without

---

[225] Rüdiger Wolfrum, 'The Principle of the Common Heritage of Mankind' *Zeitschrift für Ausländisches Öffentliches Recht und Voelkerrecht* 43 (1983): pp. 312–37, p. 335.
[226] John E Noyes, 'The Common Heritage of Mankind: Past, Present and Future' *Denver Journal of International Law and Policy* 40 (2011–2012): pp. 447–71, p. 456. Noyes also said that the possibility that some States were persistent objectors needed to be considered. See p. 455.
[227] Christopher C Joyner, 'Legal Implications of the Concept of the Common Heritage of Mankind' *The International and Comparative Law Quarterly* 35 (1986): pp. 190–9, p. 198.
[228] Michael W Lodge, 'The Common Heritage of Mankind' *International Journal of Marine and Coastal Law* 27 (2012): pp. 733–42, p. 736.
[229] Agreement relating to the Implementation of Part XI of the United Nations Convention on the Law of the Sea of 10 December 1982 (1994) 33 ILM 1309.
[230] Nuclear Tests *(Australia v France)* (Merits) [1974] ICJ Reports 253, para. 46; *(New Zealand v France)* (Merits) [1974] ICJ Reports 457, para. 49.

complying with article 82 is acting in bad faith by avoiding the very obligation that made article 76 possible in the first place. Of course, the problem with this argument is that the obligation of good faith 'is not in itself a source of obligation where none would otherwise exist'.[231] Nevertheless, any State that tried to argue that article 82 was voluntary would be inviting significant resistance from the rest of the international community. It would be sensible for a non-party State to consider complying with the obligation in article 82 as a matter of policy, even if it does not accept that the obligation is part of customary international law.

E. Conclusion

The law of the sea developed very quickly and dramatically through the twentieth century prior to the adoption of the LOSC. Technological advances that allowed States to exploit resources at greater depths and distances challenged the prevailing paradigm of the freedom of the high seas. The sea's resources were no longer seen as inexhaustible and, increasingly, States began to view the waters and sea floor off their coasts as within their sphere of interest. The idea that the continental shelf is the natural prolongation of the land territory played into the increasingly popular narrative that coastal States should have exclusive rights to the resources found off their shores. At the same time, an increasing demand for more resources for food and energy motivated coastal States to assert greater control over areas close to the coast. The continental shelf was an early manifestation of 'creeping jurisdiction'. In the Continental Shelf Convention, States deliberately left the outer limits of the shelf indeterminate to allow for changing circumstances. There was resistance by some States to the increase of coastal State jurisdiction over the shelf, but because most States benefited from the extension to some degree the concept rapidly became accepted by the international community. The retention of the freedoms of navigation and fishing meant that the Continental Shelf Convention satisfied the interests of the majority of States.

The aspiration for greater coastal State control over the ocean's resources was once again on display during the third United Nations Conference on the Law of the Sea. At this conference, however, the international community pushed back against the expansion of coastal State jurisdiction. Coastal States were able to achieve a significant advantage in the creation of the EEZ and the expansion of the continental shelf where it extended beyond 200 nm. But the extension of the continental shelf was only achieved in return for some concessions, most notably in article 82 of the LOSC. The process for establishing the outer limits of the extended continental shelf was another important feature of the LOSC, which was intended to provide certainty as to the boundary between areas under coastal State jurisdiction and the deep seabed.

---

[231] Border and Transborder Armed Actions *(Nicaragua v Honduras)* [1988] ICJ Reports 69, para. 94.

There is no indication in Part VI of the LOSC that the negotiators intended that the rights to the shelf beyond 200 nm could not be exercised until the establishment of the outer limits according to the CLCS recommendations. Instead, the rights of the coastal State are inherent and exclusive. Any doubts about the outer limits or boundaries with other States will need to be resolved by methods that are in the LOSC or otherwise permitted under international law. However, so long as a coastal State's continental shelf physically extends beyond 200 nm, the coastal State has sovereign rights to those resources that can be exercised immediately.

4

Living Resources and Protection of the Environment on the Continental Shelf Beyond 200 Nautical Miles

The coastal State has jurisdiction over sedentary species found on the continental shelf. As set out in Chapter 2, sedentary species can include corals, sponges, tubeworms, molluscs, crustaceans and even microbes. They may be found on seamounts, at hydrothermal vents or any part of the continental margin. To the extent that the individual species within the ecosystem are sedentary, they come within the coastal State's jurisdiction. However, other species that are part of the same ecosystem will be non-sedentary. Within the exclusive economic zone (EEZ), the coastal State has control over conservation and exploitation of all aspects of the ecosystem in the water column and on the sea floor. Beyond 200 nm, the State has sovereign rights only in respect of the sedentary species. Of course, focusing on 'sedentary species' tends to gloss over the fact that these species are part of an ecosystem.

This distinction between sedentary and non-sedentary species is likely to cause some difficulties in practice. The nature and source of the environmental obligations on coastal States in relation to sedentary species will be different to the obligations in relation to the non-sedentary species above the continental shelf beyond 200 nm. For example, the coastal State has responsibility to protect biodiversity and the marine environment within its jurisdiction under the Convention on Biological Diversity (CBD)[1] and the Law of the Sea Convention (LOSC).[2] Conservation of the non-sedentary species under these conventions is possible and recommended, but States will need to cooperate and work through international agreements in relation to species found in the high seas. The dual legal character of the ecosystem means that a coastal State will be advised to work closely with other States and international organizations in order to protect the ecosystem as a whole.

---

[1] Convention on Biological Diversity (opened for signature 5 June 1992, entered into force 29 December 1993) 760 UNTS 79.
[2] United Nations Convention on the Law of the Sea (opened for signature 10 December 1982, entered into force 16 November 1994) 1834 UNTS 397.

The definition of sedentary species is difficult to apply in a modern context. It made sense when States were negotiating the Continental Shelf Convention,[3] when the main focus was on commercial fisheries such as oysters. It seemed a practical solution when the intention was to allocate rights to major fish species on the sea floor as opposed to in the water column. However, the development of scientific understanding about shelf ecosystems, the broadening of commercial interest to include genetic resources and modern obligations to protect marine biodiversity mean that the artificial distinction between sedentary and non-sedentary species makes management of the living resources of the continental shelf beyond 200 nm complex.

One purpose of this chapter is to highlight the international environmental obligations that exist in relation to sedentary species on the extended continental shelf. In most respects, these obligations are the same for the continental shelf within 200 nm. However, it is important that coastal States consider their obligations in relation to the extended shelf. Also, the coastal State's jurisdiction is much more limited on the extended continental shelf where it cannot infringe on the exercise of high seas freedoms such as fishing. The obligation not to cause transboundary harm or harm to areas beyond national jurisdiction will be particularly important beyond 200 nm because activities directed at the continental shelf will be taking place in the high seas—a commons area.

The second part of the chapter explores the legal regime for the exploitation of living resources in the context of biotechnology and genetic resources. The fact that coastal State rights relate to sedentary species is a significant limitation on the coastal State's ability to regulate bioprospecting beyond 200 nm because sessile species generally spend a part of their life cycle in the water before settling onto the sea floor as adults. This chapter discusses some of the important considerations for coastal States in regulating the uses of genetic resources on their extended continental shelves.

The third part of the chapter deals with fisheries. It is fairly unlikely that commercially exploitable sedentary species will be located on most States' extended continental shelves. There will be some exceptions: for example, there have been scallop and crab fisheries operating on the Canadian extended shelf, which led to disputes with the United States in 1994 and 2001.[4] However, the issue that is more likely to arise in relation to fishing on the extended shelf is where foreign vessels target high seas species using bottom-fishing techniques that have a destructive impact on benthic communities containing sedentary species.

---

[3] Convention on the Continental Shelf (opened for signature 29 April 1958, entered into force 10 June 1964) 499 UNTS 312.

[4] JM Van Dyke, 'Modifying the 1982 Law of the Sea Convention: New Initiatives on Governance of High Seas Fisheries Resources: the Straddling Stocks Negotiations' *International Journal of Marine & Coastal Law* 10 (1995): pp. 219–27, pp. 221–2; *R v Perry*, 2003 Carswell Nfld 23, 222 Nfld & PEIR 313, 663 APR 313, 59 WCB (2d) 92.

A. Environmental Obligations in Relation to the Extended Continental Shelf

There are several sources of obligations on coastal States to protect the marine environment. Despite the fact that Part VI of the LOSC contains no explicit obligation to conserve the resources of the sea floor, coastal States do have obligations to protect the marine environment on the continental shelf beyond 200 nm. These obligations can be found in the LOSC, in multilateral conventions, customary international law and other instruments.

(1) Law of the Sea Convention

The LOSC contains a number of obligations for States to conserve the marine environment and living resources. Part VI of the LOSC—which deals with the continental shelf—is comparatively silent about environmental principles compared to other Parts dealing with other maritime zones. However, there are obligations that can be found in the LOSC that apply to sedentary species on the continental shelf.

(a) Part VI of the Law of the Sea Convention: The continental shelf

Compared to the legal regime for the EEZ, Part VI of the LOSC does not directly oblige coastal States to conserve the living resources of the continental shelf. In this regard, two aspects of the EEZ regime are not reflected in Part VI. First, article 61 requires coastal States to ensure that the living resources of the EEZ are not endangered by over-exploitation. Coastal States must maintain or restore harvested species to levels that produce the maximum sustainable yield and consider the effects on associated or dependent species. There is no equivalent requirement in relation to the sedentary species of the continental shelf in Part VI. Secondly, the sovereign rights of the coastal States in the EEZ are 'for the purposes of exploring and exploiting, conserving and managing the natural resources'. The rights in relation to the resources of the continental shelf are 'for the purpose of exploring it and exploiting its natural resources'. No reference is made to conservation and management.

The EEZ gives rights to coastal States over the seabed and subsoil, as well as the water column.[5] It might be tempting to argue that this means that the sustainability provisions of the EEZ also apply to the continental shelf. However, this argument is ruled out by Part V itself. Article 56(3) states that the rights set out in respect of the seabed and subsoil shall be exercised in accordance with Part VI. Article 68 makes the position even clearer by stating that Part V 'does not apply to sedentary species as defined in article 77, paragraph 4'. Therefore, sedentary species are not covered by the sustainability obligations in relation to the EEZ.

---

[5] Article 56(1)(a), LOSC.

The explanation for this narrower focus for continental shelf resources is not spelt out in the LOSC but is related to the history of Part VI, which was heavily based on the Continental Shelf Convention. In 1958, environmental concerns were not high on the international agenda and instead the focus was on the exclusive right of the coastal State to exploit the resources of the shelf. Rothwell and Stephens suggest the continental shelf regime does not contain a general duty to conserve resources because the concept of the continental shelf is based on a natural prolongation of land territory.[6] In contrast to the sedentary species, the fisheries resources in the water column of the EEZ often migrate across boundaries and into the high seas. In some circumstances, coastal States are required to provide access to the fisheries resources of the EEZ to other States.[7] Therefore, the international community has more interest in the sustainable use of fish in the water column than in sedentary species.[8]

Another question is the significance of the fact that rights over continental shelf resources are 'for the purposes of exploring it and exploiting its natural resources'. There is no reference to conservation and management as set out in relation to the sovereign rights in the EEZ.[9] Although it may seem unusual, these factors do raise a question whether the coastal State can exercise rights to *conserve* the living resources of the continental shelf.

It is doubtful that any reasonable argument could be made that coastal States are forbidden from exercising their sovereign rights to conserve the living resources of the continental shelf.[10] First, the exclusive right to exploit a resource implicitly involves the right not to exploit it. A coastal State that chooses not to exploit its oil and gas reserves, for whatever reason, does not lose exclusive rights to those reserves.[11] Penick asserts that the rights to resources on the continental shelf are similar to property rights, and so the limitation of rights to those of exploration and exploitation is not to be interpreted strictly.[12] He argues that the words 'for the purpose of exploring and exploiting' were intended to preserve the freedoms of the high seas: 'so long as these traditional freedoms are preserved, the coastal State may exercise any right ... in the natural resources of the continental shelf which will facilitate their development'. Although his argument relates to whether States

---

[6] Donald R Rothwell and Tim Stephens, *The International Law of the Sea* (Oxford: Hart, 2010): p. 119.

[7] Article 62(2), LOSC.

[8] Attard suggested that the omission of 'conservation and management' reflects the exclusive nature of continental shelf rights compared to the EEZ, where States are under an obligation to share the surplus total allowable catch with other States. David Attard, *The Exclusive Economic Zone in International Law* (Oxford: Clarendon Press, 1987): p. 142.

[9] Article 56(1)(a), LOSC.

[10] This argument first appeared in Joanna Mossop, 'Protecting Marine Biodiversity on the Continental Shelf Beyond 200 Nautical Miles' *Ocean Development and International Law* 38 (2007): pp. 283–304, p. 289.

[11] Article 77(2), LOSC.

[12] FVW Penick, 'The Legal Character of the Right to Explore and Exploit the Natural Resources of the Continental Shelf' *San Diego Law Review* 22 (1985): pp. 765–78, p. 771.

could assign property rights in shelf resources, it could also be extended to a right to conserve those resources.

The focus on exploration and exploitation dates back to the 1958 Continental Shelf Convention.[13] O'Connell argued that the purpose of the use of these terms was to limit the power of the coastal State to the seabed and leave the waters untouched.[14] The use of 'for the purpose of' was not intended to qualify the rights of the coastal State, but to indicate their location.[15] It is fair to assume that the intention in relation to Part VI of the LOSC was similar and therefore not intended to limit the way in which the State dealt with its resources. A State that has sovereign rights over resources must have the right to manage those resources consistently with its policy and international law obligations.

It must also be remembered that coastal States have environmental obligations that derive from other parts of the LOSC, as well as multilateral environmental treaties and customary international law. These obligations require a coastal State to protect the marine environment, despite the absence of an express requirement in Part VI.

(b) Part XII of the Law of the Sea Convention: Protection and preservation of the marine environment

A number of environmental obligations are contained in Part XII of the LOSC, which deals with the protection and preservation of the marine environment. It is important to note that Part XII binds all States, including coastal States and flag States. Foremost amongst the obligations in this Part is the general obligation to protect and preserve the marine environment.[16] This has been described as a positive obligation to take active measures to protect and preserve the marine environment and a negative obligation not to degrade the marine environment.[17] In addition, article 193 requires States to exercise their sovereign rights to exploit their natural resources 'in accordance with' their duty to protect and preserve the marine environment. Article 193 links the obligations in Part XII to the sovereign rights of coastal States, including Part VI relating to the continental shelf. This means that all sovereign rights in the LOSC must be exercised consistently with the obligations in Part XII.[18] At the very least, coastal States must consider their obligations to protect and preserve the marine environment when exercising their sovereign rights. Where a State fails to exercise appropriate control over its nationals to prevent them

---

[13] Article 2, 1958 Convention.
[14] DP O'Connell, *The International Law of the Sea: Volume I*, edited by Ivan Shearer (Oxford: Clarendon Press, 1982): p. 477.
[15] Ibid., p. 478.   [16] Article 192, LOSC.
[17] South China Sea (*Philippines v China*) (Award), PCA Case No. 2013-19 (12 July 2016), para. 941 (*South China Sea* case).
[18] Rothwell and Stephens, *The International Law of the Sea*, p. 342. See also the comments of the Arbitral Tribunal in the *South China Sea* case, paras. 950–75.

from causing significant harm to the marine environment, it will have breached the LOSC.[19]

Although Part XII is primarily aimed at preventing and mitigating pollution, it contains obligations that are relevant to the living resources on the extended continental shelf. Coastal States are required to prevent, reduce and control pollution of the environment from any source and coastal States will be conscious of the effect of pollution on the resources of the continental shelf.[20] Article 206 imposes a requirement on all States to conduct assessments where they 'have reasonable grounds for believing that planned activities under their jurisdiction or control may cause substantial pollution of or significant and harmful changes to the marine environment'. Although the article does not refer to an 'environmental impact assessment' as used in modern environmental management practice, it is the basis for an obligation to conduct environmental impact assessments where the threshold in article 206 is met.[21] The obligation applies to areas within and beyond national jurisdiction and extends to activities that not only cause pollution, but also activities that cause significant and harmful changes to the marine environment.[22] In the *South China Sea* case, the Arbitral Tribunal found that a State must not only prepare an environmental impact assessment, but also communicate it to the relevant international organizations.[23] The obligation to conduct environmental impact assessments has been developed as a principle of customary international law by the International Court of Justice (ICJ) and the International Tribunal for the Law of the Sea (ITLOS), and is discussed in further detail later in this chapter.

States are also required, under article 204 of the LOSC, to monitor the effects of activities they undertake. This is also reflected in customary international law in that the obligation of due diligence also requires ongoing monitoring in relation to potential transboundary harm.

As mentioned above, the obligations in Part XII apply to flag States. In the Advisory Opinion on IUU fishing, the ITLOS found that article 192 imposed obligations on flag States to ensure their vessels comply 'with the relevant conservation measures concerning living resources enacted by the coastal State for its exclusive economic zone because ... they constitute an integral element in the protection and preservation of the marine environment'.[24] The Tribunal was quite clear that this

---

[19] *South China Sea* case, paras. 954–66. Among China's actions that the Tribunal found were breaches of Part XII included: allowing the harvesting of endangered species on a large scale and the infliction of significant damage on rare or fragile ecosystems and habitat; failing to enforce rules against fishermen; and the construction of artificial islands in a manner that caused long-lasting damage to the marine environment.
[20] Article 194, LOSC.
[21] *South China Sea* case, para. 948; Lingjie Kong, 'Environmental Impact Assessment under the United Nations Convention on the Law of the Sea' *Chinese Journal of International Law* 10 (2011): pp. 651–69, p. 658.
[22] Alex G Oude Elferink, 'Environmental Impact Assessment in Areas beyond National Jurisdiction' *International Journal of Marine and Coastal Law* 27 (2012): pp. 449–80, p. 455.
[23] *South China Sea* case, para. 991.
[24] Request for an Advisory Opinion Submitted by the Sub-Regional Fisheries Commission, 54 ILM 893 (2015), para. 120.

obligation is independent of the obligation to comply with coastal State regulation in articles 58(3) and 62(4). Therefore it is arguable that article 192 imposes an obligation on flag States to comply with coastal State regulation aimed at protecting the marine environment on the continental shelf beyond 200 nm even if this restricts in some way the exercise of high seas freedoms. Of course, this does not remove the requirement for coastal States not to infringe or unjustifiably interfere with high seas freedoms.

(2) Multilateral treaties

A range of multilateral environmental treaties impose legal obligations on States to take steps to preserve the marine environment. The most relevant in this context is the CBD.

The CBD requires States to develop national plans for the conservation and sustainable use of biodiversity.[25] States are encouraged to establish systems of protected areas and take other measures to seek compatibility between uses of biodiversity and the conservation and sustainable use of its components.[26] The CBD provides a much more detailed requirement to conduct impact assessments than article 206 of the LOSC.[27] Environmental impact assessments are required when projects 'are likely to have significant adverse effects on biological diversity'. The CBD applies to areas within the limits of national jurisdiction and to activities and processes under the jurisdiction or control of the State in areas beyond national jurisdiction.[28] States are instructed to implement the Convention with respect to the marine environment consistently with the rights and obligations of States under the law of the sea.[29] Therefore, the provisions of the CBD are applicable to coastal States and flag States, so long as they are compatible with the LOSC.[30] Coastal States will need to consider how the provisions of the CBD will apply to the extended continental shelf.

The CBD parties have been working on how to achieve the goal of establishing representative marine protected areas (MPAs) set at the Johannesburg World Summit on Sustainable Development in 2002.[31] In 2004, the Conference of the Parties (COP) committed to establishing 'marine and coastal protected areas that are effectively managed, ecologically based and contribute to a global network of marine and coastal protected areas'.[32] That decision endorsed the 2002 commitment

---

[25] Article 6, CBD. The CBD also contains provisions relating to the access of foreign researchers to genetic material, but that is discussed later in this chapter.
[26] Article 8(a) and (i), CBD.   [27] Article 14, CBD.   [28] Article 4, CBD.
[29] Article 22, LOSC.
[30] A Charlotte de Fontaubert, David R Downes and Rundi S Agardy, 'Biodiversity in the Seas: Implementing the Convention on Biological Diversity in Marine and Coastal Habitats' *Georgetown International Environmental Law Review* 10 (1997–1998): pp. 753–854, p. 756.
[31] Para. 32(c), Plan of Implementation of the World Summit on Sustainable Development, Johannesburg, 2002, A/CONF.199/20. The call was for 'the establishment of marine protected areas consistent with international law and based on scientific information, including representative networks by 2012'.
[32] Para. 18, Decision VII/5 *Marine and Coastal Biological Diversity* COP VII (2004).

to establishing representative networks of MPAs by 2012.[33] Target 11 of the Aichi Biodiversity Targets, agreed in 2010, reaffirmed the call that by 2020, at least '10 per cent of coastal and marine areas, especially areas of particular importance for biodiversity and ecosystem services, are conserved through effectively and equitably managed, ecologically representative and well-connected systems of protected areas and other effective area-based conservation measures'.[34]

During this time, the parties developed criteria to identify Ecologically and Biologically Significant Areas (EBSAs), which could be used to identify possible sites for MPAs. This work was approved in 2008 and State parties were urged to use the scientific criteria to identify EBSAs.[35] A number of regional expert workshops were held to assist States in identifying EBSAs and build capacity for this work.[36] EBSAs will be identified in a CBD EBSA repository and on the CBD website.[37] Designation of a site as an EBSA carries no legal consequences but can be used to promote conservation measures in domestic law or international organizations including, but not limited to, the establishment of MPAs. All of these moves require parties to the CBD to consider the option of establishing MPAs on the extended continental shelf as it is also an area within national jurisdiction.

Many regional fisheries management organizations (RFMOs) are taking measures to protect the benthic marine environment by creating areas closed to bottom fishing,[38] or 'move-on' rules that require a vessel to move on a certain distance when by-catch indicates a vessel is fishing above a vulnerable marine ecosystem.[39] There is potential for RFMOs to develop rules responding to coastal State concerns about bottom fishing above coastal States' extended continental shelves where vulnerable ecosystems are identified.

Other related treaties are directly targeted at reducing environmental harm, including MARPOL[40] and the London Dumping Convention.[41] Both Conventions are relevant to the extended continental shelf without containing specific provisions regarding this area. Additionally, in theory a coastal State could, through the International Maritime Organization (IMO), seek particular protection for the

---

[33] Ibid., para. 19.
[34] *Strategic Plan for Biodiversity 2011–2020 and the Aichi Biodiversity Targets*, Annex, Decision X/2, COP 10, 2010.
[35] Para. 21, Decision X/29, *Marine and Coastal Biodiversity*, COP 10, 2010.
[36] Daniel C Dunn et al., 'The Convention on Biological Diversity's Ecologically or Biologically Significant Areas: Origins, Development, and Current Status' *Marine Policy* 49 (2014): pp. 137–45; Nicholas J Bax et al., 'Results of Efforts by the Conservation on Biological Diversity to Describe Ecologically or Biologically Significant Marine Areas' *Conservation Biology* 30 (2015): pp. 571–81.
[37] www.cbd.int/ebsa.   [38] See, e.g. the North East Atlantic Fisheries Organization.
[39] See, e.g. the South Pacific Regional Fisheries Management Organization.
[40] International Convention for the Prevention of Pollution from Ships, as Modified by the Protocol of 1978 (opened for signature 2 November 1971 and 17 February 1978, entered into force 2 October 1982) 1340 UNTS 62 (MARPOL 73/78).
[41] Convention on the Prevention of Marine Pollution by Dumping of Wastes and Other Matter 1972 (opened for signature 29 December 1972, entered into force 30 August 1975) 1046 UNTS 138.

outer continental shelf, for example, through seeking the designation of a vulnerable area as a particularly sensitive sea area.[42]

(3) Customary international law

In addition to those obligations imposed by treaties, there are also two significant customary international law obligations that will apply to a coastal State's actions on the extended continental shelf. The first is the principle that a State should not allow its territory to be used to cause harm to another State or to areas beyond national jurisdiction. Associated with this principle is a general obligation to prevent harm to the environment and to reduce, limit or control harmful activities. Secondly, States have an obligation to undertake due diligence to meet their environmental obligations.

(a) Obligation not to cause transboundary harm and duty of prevention

A fundamental international environmental obligation in customary international law is that a State must not allow activities under its jurisdiction to cause serious harm to the environment of another State.[43] This principle is articulated in Principle 21 of the Stockholm Declaration[44] and Principle 2 of the Rio Declaration.[45] These two principles are almost identical in form. The Rio Declaration provides that:[46]

> States have, in accordance with the Charter of the United Nations and the principles of international law, the sovereign right to exploit their own resources pursuant to their own environmental policies *and developmental policies*, and the responsibility to ensure that activities within their jurisdiction or control do not cause damage to the environment of other States or of areas beyond the limits of national jurisdiction.

This principle (often referred to as the prohibition on transboundary harm) reflects two important concepts, which are echoed in the LOSC. First, States have the sovereign right to exploit their resources. For the continental shelf, this principle is reflected in article 77 of the LOSC, which confirms that coastal States have exclusive rights to the living and non-living resources of the continental shelf.[47] Secondly, States have an obligation not to cause damage to the environment of other States or of areas beyond national jurisdiction. Article 194(2) provides that coastal States must prevent, reduce and control pollution of the marine environment and

---

[42] See generally, Markus J Kachel, *Particularly Sensitive Sea Areas: The IMO's Role in Protecting Vulnerable Marine Areas* (Springer, 2008); Hélène Lefebvre-Chalain, 'Fifteen Years of Particularly Sensitive Sea Areas: A Concept in Development' *Ocean and Coastal Law Journal* 13 (2007): pp. 47–69. See ch 9.
[43] Trail Smelter *(United States v Canada)* 3 RIAA 1911.
[44] Declaration of the United Nations Conference on the Human Environment, UN Doc. A/Conf.48/14/Rev. 1(1973); 11 ILM 1416 (1972).
[45] Rio Declaration on Environment and Development, UN Doc. A/Conf.151/26 (vol. I); 31 ILM 874 (1992).
[46] The italicized words only appear in Principle 2 of the Rio Declaration. Emphasis added.
[47] See also article 193, LOSC.

ensure that activities under their jurisdiction do not cause pollution damage that spreads 'beyond the areas where they exercise sovereign rights in accordance with' the LOSC. Although the Stockholm and Rio Declarations were non-binding under international law, the prohibition on transboundary harm they expressed has been held to be customary international law. In the advisory opinion on the *Legality of the Threat or Use of Nuclear Weapons*, the ICJ confirmed that:[48]

The existence of the general obligation of States to ensure that activities within their jurisdiction and control respect the environment of other States *or of areas beyond national control* is now part of the corpus of international law relating to the environment.

It is important to emphasize that the obligation applies not only to areas in other States' jurisdiction, but also to the areas beyond such jurisdiction including the high seas and the Area.[49] This aspect becomes particularly significant for the coastal State in relation to the extended continental shelf that lies beneath the high seas. The interconnected nature of the ecosystems located on or near the seabed means that there is a high degree of probability that any activities on the extended continental shelf will affect areas beyond national jurisdiction. Almost any activity targeting the extended continental shelf takes place in the high seas. However, the principle will not be invoked for minor or temporary harm to the areas above or near the extended shelf. Most commentators agree that the threshold of the obligation is transboundary harm that is 'serious' or 'significant'.[50] Therefore, not all activities on the shelf will trigger this obligation and its consequences.[51]

The duty to prevent damage to the environment is closely related to the obligation not to cause transboundary harm; however, the duty of prevention can apply to areas within national control, as well as beyond national jurisdiction.[52] It has been suggested that the obligation to prevent harm arises once evidence demonstrates that harm to the environment will occur.[53] This obligation is often discussed in conjunction with the obligation not to cause transboundary harm.[54]

---

[48] Legality of the Threat or Use of Nuclear Weapons (Advisory Opinion) [1996] ICJ Reports 266, para. 29 (emphasis added). See also Gabčíkovo-Nagymaros Project *(Hungary v Slovakia)* [1997] ICJ Reports 7, para. 53.

[49] Responsibilities and Obligations of States Sponsoring Persons and Entities with Respect to Activities in the Area (Advisory Opinion) (2011) 50 ILM 458, para. 148.

[50] Günther Handl, 'Transboundary Impacts' in Daniel Bodansky, Jutta Brunnée and Ellen Hey (eds), *The Oxford Handbook of International Environmental Law* (Oxford: Oxford University Press, 2008): pp. 531–49, p. 535. See also the ILC Draft Articles on the Prevention of Transboundary Harm from Hazardous Activities.

[51] Contrast articles 192 and 194 of the LOSC, which do not mention a threshold of serious or significant harm.

[52] See, e.g. Philippe Sands and Jacqueline Peel, *Principles of International Environmental Law*, 3rd edn (Cambridge: Cambridge University Press, 2012): p. 201.

[53] Yoshifumi Tanaka, 'Obligations and Liability of Sponsoring States Concerning Activities in the Area: Reflections on the ITLOS Advisory Opinion of 1 February 2011' *Netherlands International Law Review* 60 (2013): pp. 205–30.

[54] See Pulp Mills on the River Uruguay *(Argentina v Uruguay)* (Judgment) [2010] ICJ Reports 14, para. 101 *(Pulp Mills* case).

(b) Obligation to exercise due diligence

Increasingly, international dispute settlement bodies are referring to the duty to prevent harm to the environment as an obligation on States to exercise due diligence in implementing their obligations. This does not impose a duty to avoid all harm, but a State must take measures within its legal system that are 'reasonably appropriate' to fulfil its obligations.[55] In the *Pulp Mills* case, the ICJ considered the obligation of States under a bilateral treaty for the use of a river on the border of the two countries.[56] The treaty stated that the parties were to adopt appropriate rules and measures to protect and preserve the aquatic environment of the river and to prevent pollution.[57] What is notable about the decision is that the ICJ based much of its reasoning on the existence of customary international law principles, giving the case significance for other environmental issues.[58] The Court affirmed the customary international law status of the preventative and transboundary harm principles, and stated that these were linked to the obligation of due diligence.[59] The ICJ stated that due diligence is:[60]

[a]n obligation which entails not only the adoption of appropriate rules and measures, but also a certain level of vigilance in their enforcement and the exercise of administrative control applicable to public and private operators, such as the monitoring of activities undertaken by such operators.

The due diligence obligation was applied by the ITLOS in its advisory opinion on the Responsibilities and Obligations of States Sponsoring Persons and Entities with Respect to Activities in the Area.[61] In that case, the Tribunal found that an obligation to ensure that activities in the Area are carried out in accordance with the LOSC[62] was not an obligation to make that occur in every instance. Instead, the Tribunal stated that due diligence was 'an obligation to deploy adequate means, to exercise best possible efforts, to do the utmost, to obtain the result'.[63] According to

---

[55] Advisory Opinion on Responsibility of States, para. 120.
[56] See Donald K Anton, 'Case Concerning Pulp Mills on the River Uruguay *(Argentina v Uruguay)* (Judgment)' *Australian International Law Journal* 17 (2010): pp. 213–23; Xiaopin Zhu and Jinlong He, 'International Court of Justice's Impact on International Environmental Law: Focusing on the Pulp Mills Case' *Yearbook of International Environmental Law* 23 (2012): pp. 106–30.
[57] *Pulp Mills* case, para. 61.
[58] See, e.g. Anton, 'Case Concerning Pulp Mills on the River Uruguay', p. 214.
[59] *Pulp Mills* case, para. 101.
[60] Ibid., para. 197. See also International Law Commission draft articles on the Prevention of Transboundary Harm from Hazardous Activities and associated commentary, particularly article 3. ILC, 'Report of the International Law Commission on the work of its fifty-third session' *Yearbook of the International Law Commission* Part 2 (2001–2).
[61] Advisory Opinion on Responsibilities of States. For commentary, see David Freestone, 'Responsibilities and Obligations of States Sponsoring Persons and Entities with Respect to Activities in the Area' *American Journal of International Law* 105 (2011): pp. 755–60; Hui Zhang, 'The Sponsoring State's "Obligation to Ensure" in the Development of the International Seabed Area' *International Journal of Marine and Coastal Law* 28 (2013): pp. 681–99; Tim Poisel, 'Implications of Seabed Disputes Chamber's Advisory Opinion' *Australian International Law Journal* 19 (2012): pp. 213–33.
[62] Article 139, LOSC.   [63] Advisory Opinion on Responsibilities of States, para. 110.

the ITLOS, the obligation of due diligence is variable and may change over time and in relation to the risks of the activity.[64] Therefore, the standard of due diligence is more severe for riskier activities.

Both the ICJ and ITLOS have emphasized that due diligence involves some specific procedural requirements. The most significant of these is the obligation to conduct an environmental impact assessment where there is a risk that a proposed activity may have a significant adverse impact in a transboundary context,[65] or on resources that are the common heritage of mankind.[66] Although in both cases the governing treaties did contain obligations to conduct environmental impact assessments, the judges expressly noted that the obligation also arose as a customary principle of international law.

The customary rule does not extend to prescribing the scope and content of the environmental impact assessment.[67] In *Pulp Mills*, the ICJ acknowledged that the United Nations Environment Programme Guidelines and Principles should be taken into account by the parties because the treaty required them to consider appropriate rules of international law.[68] However, the Court decided that it was up to each State to determine the content of the environmental impact assessment in light of the nature of the project.[69]

The obligation to conduct an environmental impact assessment needs to be more than a post hoc assessment. In the *Nicaragua v Costa Rica* case, the ICJ criticized the actions of Costa Rica in purporting to conduct an environmental assessment.[70] Costa Rica had undertaken studies evaluating the effects of a road that had already been built. The ICJ stated that the environmental impact assessment must be done prior to commencing the project and must evaluate the risk of future harm.[71]

In the Advisory Opinion, the ITLOS suggested that the precautionary approach is also an integral part of the obligation of due diligence.[72] It defined the precautionary approach in terms set out in Principle 15 of the Rio Declaration, which describes the approach as '[w]here there are threats of serious or irreversible damage, lack of full scientific certainty shall not be used as a reason for postponing cost effective measures to prevent environmental degradation'. The Tribunal stated that the incorporation of the precautionary approach into a range of treaties and instruments reflects a 'trend towards making this approach part of customary international law'.[73]

---

[64] Ibid., para. 117.   [65] *Pulp Mills* case, para. 204.
[66] Advisory Opinion on Responsibilities of States, para. 148.   [67] Ibid., para. 149.
[68] 1987 Goals and Principles of Environmental Impact Assessment of the United Nations Environment Programme, Adopted by the Governing Council of UNEP, Decision 14/25, 17 June 1987.
[69] *Pulp Mills* case, para. 205.
[70] Certain Activities Carried out by Nicaragua in the Border Area and Construction of a Road in Costa Rica along the San Juan River *(Costa Rica v Nicaragua)* (Merits) ICJ, 16 December 2015.
[71] *Costa Rica v Nicaragua*, para. 161.
[72] Advisory Opinion on Responsibilities of States, para. 131. Not all commentators share the Tribunal's views in relation to the connection between due diligence and precaution. See Tanaka, 'Reflections on the ITLOS Advisory Opinion', p. 214.
[73] Advisory Opinion on Responsibilities of States, para. 135. For a discussion of the ITLOS analysis of the precautionary approach, see Duncan French, 'From the Depths: Rich Pickings of Principles of

The precautionary approach is likely to play an important role in coastal State decision-making because of the paucity of information that is likely to exist in relation to ecosystems on the continental shelf beyond 200 nm. This principle puts the onus on the coastal State to consider protective measures if they seem likely to be necessary, even if full information is not available immediately.

(4) United Nations General Assembly resolutions and other non-binding instruments

Over a period of time, the United Nations General Assembly has repeatedly expressed a view that urgent action needs to be taken by States and international organizations to protect vulnerable marine ecosystems, including those on the seabed. The General Assembly has listed a range of activities that could have an impact on vulnerable marine ecosystems. These include over-utilization of living marine resources, the use of destructive fishing practices, the introduction of alien invasive species and pollution, the loss or release of fishing gear and the dumping of hazardous waste.[74] In 2006, Resolution 61/105 called on States individually to take action to 'sustainably manage fish populations and protect vulnerable marine ecosystems, including seamounts, hydrothermal vents and cold water corals, from destructive fishing practices, recognizing the immense importance and value of deep sea ecosystems and the biodiversity they contain'. The General Assembly also called on RFMOs to take similar steps.[75] More recently, the General Assembly has endorsed the Food and Agriculture Organization (FAO) International Guidelines for the Management of Deep-sea Fisheries in the High Seas.[76]

In some cases, vulnerable marine ecosystems will exist on the extended continental shelf of a coastal State. The General Assembly has stated in its resolutions that nothing in Resolution 61/105 prejudices sovereign rights over continental shelves or the exercise of the jurisdiction of coastal States with respect to their continental shelf under article 77 of the LOSC.[77] This reinforces that the coastal rights over the sedentary species of the extended continental shelf are exclusive, but does not excuse coastal States from the general call to protect vulnerable marine ecosystems on the continental shelf.

The FAO has issued guidelines for the management of deep-sea fisheries in the high seas, in part as a result of the focus on vulnerable marine ecosystems in the General Assembly. The FAO International Guidelines for the Management of Deep-sea Fisheries in the High Seas are designed to assist States in implementing General Assembly Resolution 61/105.[78] The Guidelines apply to fisheries beyond national jurisdiction which are likely to come into contact with the sea floor in the normal course of fishing operations.

Sustainable Development and General International Law on the Ocean Floor—the Seabed Disputes Chamber's 2011 Advisory Opinion' *International Journal of Marine and Coastal Law* 26 (2011): pp. 525–68.

[74] See, e.g. Preamble, General Assembly Resolution 59/24, A/RES/59/24 (17 November 2004); Preamble, General Assembly Resolution 61/222, A/RES/61/222 (20 December 2006).
[75] GA Res 61/105 (2006), para. 80.   [76] GA Res 63/112 (2008), para. 102.
[77] GA Res 64/72 (2009), para. 115.   [78] http://www.fao.org/fishery/topic/166308/en.

The Guidelines specifically state that they are not intended to prejudice the sovereign rights of the coastal State over the shelf. However, it is open to coastal States to apply a similar approach to areas within national jurisdiction and the Guidelines can be used as grounds for arguing in favour of the establishment of fisheries restrictions above the extended continental shelf.

Under the Guidelines, States and RFMOs should adopt measures in accordance with the precautionary approach and ecosystems-based approaches, identify areas where vulnerable marine ecosystems are likely to be present and take action using the best information available. The Guidelines help to define concepts such as 'vulnerable marine ecosystems' and 'significant adverse impacts'. They also identify the need for data collection and reporting from fishing vessels. Overall, the Guidelines are very detailed and provide significant guidance for managing deep-sea fisheries, some of which will take place above continental shelves beyond 200 nm.

(5) Summary of obligations

Broadly speaking, the coastal State has the following obligations in relation to the continental shelf:

- to protect and preserve the marine environment[79]
- to take steps to preserve marine biodiversity[80]
- to ensure that activities within its jurisdiction and control do not cause significant harm to the environment of other States, the high seas or the Area[81]
- to exercise due diligence in implementing its environmental obligations[82]
- to conduct environmental impact assessments where there is a risk of significant harm to the marine environment, a risk of transboundary harm, or risk of harm to areas beyond national jurisdiction[83]
- to apply the precautionary approach[84]
- to take steps to protect vulnerable marine ecosystems.[85]

These obligations are particularly relevant to the extended continental shelf. Ecosystems on the continental shelf beyond 200 nm will be subject to two jurisdictional regimes: the high seas in the case of non-sedentary species, and the continental shelf in the case of sedentary species. This makes the need for prior assessment more acute as the harm to the surrounding environment could involve harm to the global commons rather than another area within the coastal State's jurisdiction. States

---

[79] Article 194, LOSC.   [80] Articles 5 and 8, CBD.
[81] Article 194, LOSC; article 3, CBD; Advisory Opinion on the Legality of the Threat or Use of Nuclear Weapons.
[82] *Pulp Mills* case and Advisory Opinion on Responsibilities of States.
[83] Article 206 of the LOSC and article 14 of the CBD apply to areas within and outside the jurisdiction; the customary principles apply to harm that occurs to other States or areas beyond national jurisdiction.
[84] Advisory Opinion on Responsibilities of States.   [85] CBD, GA Res 61/105.

must exercise due diligence to avoid or mitigate harm to the marine environment, which will include taking steps to implement effective decision-making processes. Therefore, States will be obliged to have regulatory mechanisms for monitoring and evaluating the environmental impact of activities on the extended continental shelf.

Although the majority of obligations to protect the marine environment on the extended continental shelf are owed by coastal States, it should not be forgotten that other States are also under treaty and customary obligations to protect the environment. The Tribunal in the *South China Sea* arbitration found that Part XII applies to all States with respect to the marine environment in all maritime areas.[86] States whose vessels operate in the vicinity of the continental shelf will be obliged to ensure that their vessels comply with international standards such as those contained in IMO conventions and RFMO measures, as well as more general obligations contained in the LOSC and applicable coastal State laws.

(6) Creation of marine protected areas

Recent attention has turned to the value of MPAs as a tool for protecting marine habitats and biodiversity. The term 'marine protected area' covers a range of different types of protected area: from no-take zones to areas where certain activities are managed. The key commonality is that it involves 'a clearly defined geographical space, recognised, dedicated and managed … to achieve the long-term conservation of nature with associated ecosystem services and cultural values'.[87]

As mentioned above, the international community has set a number of goals for the creation of MPAs in all parts of the ocean. In 2006, the parties to the CBD agreed that ten per cent of the world's ecological regions should be effectively conserved and that areas of particular importance to biological diversity should be protected.[88] In 2010, the parties agreed to protect ten per cent of coastal and marine areas through networks of MPAs and other effective measures by 2020.[89]

Continental shelves are highly geologically diverse and support a range of types of habitat.[90] Many cold seeps and hydrothermal vent ecosystems are of very high scientific value, and calls have been made for their protection under national legislation.[91] The differences in depth and geomorphology across the entire continental

---

[86] *South China Sea* case, para. 940.
[87] IUCN World Commission on Protected Areas (IUCN-WCPA), *Establishing Marine Protected Area Networks—Making it Happen* (2008, Washington: DC). However, experts have warned that weak MPAs will be ineffective in achieving the goals of the CBD. See Alexis N Rife et al., 'When Good Intentions Are Not Enough: Insights on Networks of "Paper Park" Marine Protected Areas' *Conservation Letters* 6 (2013): pp. 200–12; Mark J Costello and Bill Ballentine, 'Biodiversity Conservation Should Focus on No-take Marine Reserves: 94% of Marine Protected Areas Allow Fishing' *Trends in Ecology and Evolution* 30 (2015): pp. 507–9.
[88] Mark D Spalding et al., 'Protecting Marine Spaces: Global Targets and Changing Approaches' *Ocean Yearbook* 27 (2013): pp. 213–48, p. 218.
[89] Spalding 'Protecting Marine Spaces'.
[90] E Ramirez-Llodra et al., 'Deep, Diverse and Definitely Different: Unique Attributes of the World's Largest Ecosystem' *Biogeosciences* 7 (2010): pp. 2851–99, p. 2857.
[91] CL Van Dover et al., 'Designating Networks of Chemosynthetic Ecosystem Reserves in the Deep Sea' *Marine Policy* 36 (2012): pp. 378–81.

shelf lead to a high level of diversity among the species found there. In a recent study, scientists compared the genetic make-up of two species found across a range of benthic habitats and found that there were significant genetic differences between one of the species found on seamounts and on continental slopes.[92] The authors concluded that the design of MPAs on the continental shelf should reflect the likelihood of differences in benthic communities in different depths and habitats.[93] It has been argued that only spatial management and the use of MPAs will be effective at protecting vulnerable sea floor organisms in the deep sea.[94] In addition, the growth of interest in the use of genetic resources may lead to a consideration that MPAs are also needed to protect potential genetic resources from exploitation by foreign researchers.[95]

It is therefore probable that many coastal States will, consistent with the CBD global targets, seek to put in place MPAs on the extended continental shelf where conservation and ecosystem values mean it is important to protect the benthic environment. It is likely that more than one MPA will be needed across the entire continental shelf to protect representative areas of the marine environment, and any network should arguably include areas on the extended continental shelf.

One option for the coastal State is to unilaterally declare the existence of an MPA on its extended continental shelf and rely on its sovereign rights under the LOSC to protect the area. This may be successful to a certain extent. The coastal State has exclusive rights over the exploitation of the resources of the shelf, and so can be certain that mining and hydrocarbon extraction will not occur in the MPA. However, not all potentially damaging activities come within the exclusive jurisdiction of the coastal State. For example, other States have the right to fish in the water column above the extended shelf and to conduct scientific research there. A coastal State may be concerned that such activities will cause environmental harm to the benthic environment. Another problem could be caused by discharges and the disposal of wastes from shipping. Again, the coastal State has no jurisdiction over these activities under Part VI of the LOSC. It may be possible for the coastal State to undertake limited interference with some high seas activities with the most impact on the shelf resources, and this is discussed in Chapter 7.

It would be more effective for a coastal State to cooperate with relevant international organizations who can then assist the coastal State to reinforce its conservation goals through the imposition of rules adopted by those organizations with which other States will be obliged to comply.[96] There are a number of possible

---

[92] EK Bors et al., 'Patterns of Deep-Sea Genetic Connectivity in the New Zealand Region: Implications for Management of Benthic Ecosystems' *PLOS One* 7(11) (2012): e49474. doi:10.1371/journal.pone.0049474.

[93] Ibid., p. 14.

[94] Malcolm R Clark et al., 'The Impacts of Deep-Sea Fisheries on Benthic Communities: a Review' *ICES Journal of Marine Science* 73 (Supp. 1) (2016) pp. i51–i69. doi: 10.1093/icesjms/fsv123. See ch 2 for a discussion about the impacts of bottom fishing on the benthic environment.

[95] JM Arrieta, S Arnaud-Haond and CM Duarte, 'What Lies Underneath: Conserving the Ocean's Genetic Resources' *Proceedings of the National Academy of Sciences* 107 (2010): pp. 18318–24.

[96] Cooperative approaches to regulating activities on the extended continental shelf are discussed in detail in ch 9.

partners for coastal States. These include regional seas arrangements, which often focus on cooperation relating to pollution and other activities.[97] In addition, the IMO can establish rules for navigation and pollution in particularly sensitive sea areas.[98] Finally, RFMOs may have authority to put in place conservation measures that would limit destructive fishing practices on or close to the extended shelf of States.

One of the first examples of a network of MPAs in areas beyond national jurisdiction involved cooperation between a coastal State wishing to protect areas of its extended continental shelf, and international organizations. Portugal is a member of the OSPAR Commission[99] and has identified a number of areas on its extended continental shelf as suitable for inclusion in a network of MPAs. These include a hydrothermal vent field and seamounts.[100] The parties to the OSPAR Commission have established, at Portugal's request, MPAs in respect of the high seas above these sensitive areas of the extended continental shelf.[101] At the same time, Portugal has implemented domestic legislation creating MPAs in respect of the seabed and subsoil of these areas.[102]

However, OSPAR's competence is limited to environmental protection and does not extend to sectoral activities such as fishing or shipping. Fishing in the OSPAR areas that are beyond national jurisdiction is managed through the North East Atlantic Fisheries Commission (NEAFC).[103] NEAFC has worked on plans to protect vulnerable marine ecosystems since 2004 and, following discussions with the

---

[97] www.unep.org/regionalseas.

[98] See, e.g. Tullio Scovazzi, 'Marine Protected Areas in Waters Beyond National Jurisdiction' in Marta Chantal Ribeiro (ed), *30 Years after the Signature of the United Nations Convention on the Law of the Sea: the Protection of the Environment and the Future of the Law of the Sea* (Coimbra: Coimbra Editoria, 2014): pp. 209–38, p. 218; Markus J Kachel, *Particularly Sensitive Sea Areas: The IMO's Role in Protecting Vulnerable Marine Areas* (Springer, 2008); Hélène Lefebvre-Chalain, 'Fifteen Years of Particularly Sensitive Sea Areas: A Concept in Development' *Ocean and Coastal Law Journal* 13 (2007): pp. 47–69.

[99] Established by the Convention for the Protection of the Marine Environment of the North-East Atlantic (opened for signature 22 September 1992, entered into force 25 March 1998) 32 ILM 1069. The OSPAR Convention is intended to provide environmental protection for an area of the North-east Atlantic Ocean. It has fifteen State parties. See www.ospar.org. Unusually for a regional seas convention, the OSPAR Convention covers areas beyond national jurisdiction, as well as areas within national jurisdiction. EJ Molenaar and AG Oude Elferink, 'Marine Protected Areas in Areas Beyond National Jurisdiction: The Pioneering Efforts Under the OSPAR Convention' *Utrecht Law Review* 5 (2009): pp. 5–20, p. 13.

[100] OSPAR Commission, *'2012 Status Report on the OSPAR Network of Marine Protected Areas'*, p. 26. See also Marta Chantal Ribeiro, 'The "Rainbow": The First National Marine Protected Area Proposed Under the High Seas' *International Journal of Marine and Coastal Law* 25 (2010): pp. 183–207.

[101] See OSPAR Decisions 2010/3, 2010/4, 2010/5 and 2010/6.

[102] Marta Chantal Ribeiro, 'Marine Protected Areas: The Case of the Extended Continental Shelf' in Marta Chantal Ribeiro (ed), *30 Years after the Signature of the United Nations Convention on the Law of the Sea: the Protection of the Environment and the Future of the Law of the Sea* (Coimbra: Coimbra Editoria, 2014): pp. 179–208, p. 195.

[103] There is significant overlap between the members and geographical coverage of the NEAFC Commission and the OSPAR Commission. See Ingrid Kvalvik, 'Managing Institutional Overlap in the Protection of Marine Ecosystems on the High Seas: The Case of the North East Atlantic' *Ocean and Coastal Management* 56 (2012): pp. 35–43.

OSPAR Commission, a number of areas were subsequently closed to bottom fishing. Kvalvik notes that the areas closed by NEAFC only partially overlap with the OSPAR MPAs, and further cooperation will be needed to streamline the protection of these areas.[104]

The work of Portugal, OSPAR and NEAFC provides a possible model for future development of MPAs on and above the extended continental shelf of coastal States. International cooperation is the best way to deal with the limitations facing coastal States seeking to protect the marine environment on the extended continental shelf. The disadvantages of relying on international cooperation include the need to achieve consensus on the need for protective measures, and the existence of competent organizations in the relevant geographical areas. These issues, and Portugal's work to establish MPAs, are discussed further in Chapter 9.

B. Bioprospecting

It is likely that marine bioprospecting on the extended continental shelf will take place in the near future.[105] Coastal States will need to consider how to deal with applications for marine bioprospecting. One of the first issues is whether to treat such applications as marine scientific research or as an extractive activity.

Secondly, coastal States must consider not just the provisions in the LOSC but also the CBD. The rights and obligations of States in relation to the exploitation of living resources of the continental shelf beyond 200 nm are the same as within 200 nm. However, when considering how to exercise these obligations, including obligations to protect marine biodiversity, the State must take into account the fact that it must not infringe or unjustifiably interfere with the rights and freedoms of other States on the high seas.[106]

(1) Distinguishing between bioprospecting and marine scientific research

One of the key concerns for coastal States who wish to control or even refuse requests for scientists to sample marine organisms on the continental shelf beyond 200 nm is to determine whether the request is for marine scientific research or an exploitative activity. The reason this is important is because, if the activity is marine scientific research, then article 246(6) of the LOSC restricts the ability of the coastal State to refuse permission on the basis that the project is of direct significance for the exploration or exploitation of natural resources. Coastal States can only prohibit such research in areas that have been designated as areas in

---

[104] Ibid., p. 37.
[105] Parts of the following discussion were previously published in Joanna Mossop, 'Marine Bioprospecting' in DR Rothwell et al., *The Oxford Handbook on the Law of the Sea* (Oxford: Oxford University Press, 2015): pp. 825–42.
[106] Article 78(2), LOSC. The interpretation of article 78(2) is discussed in ch 7.

which exploration or exploitation will be occurring within a reasonable period of time.[107] From the perspective of the coastal State, it would be convenient to characterize bioprospecting as distinct from marine scientific research because it would then fall within the coastal State's exclusive jurisdiction to determine the use and exploitation of the living resources of the extended continental shelf. The scientific community, on the other hand, would prefer bioprospecting to be defined as marine scientific research with its greater freedoms.[108]

Bioprospecting is not defined in the LOSC, and it is difficult to locate a universally agreed definition of the activity. One useful definition is 'the scientific investigation of living organisms for commercially valuable genetic and biochemical resources'.[109] A question that sometimes arises is whether the definition includes the entire process of commercializing a product. The Secretary-General of the United Nations has suggested that this is not the case:[110]

[T]he term is generally understood, among researchers, as the search for biological compounds of actual or potential value to various applications, in particular commercial applications. This involves a series of value-adding processes, usually spanning several years, from biological inventories requiring accurate taxonomic identification of specimens, to the isolation and characterization of valuable active compounds. As a mere prospecting activity, bioprospecting is only the first step towards possible future exploitation and stops once the desired compound or specific property has been isolated and characterized.

Marine scientific research has also not been defined in the LOSC and a debate has ensued about whether it includes research conducted for commercial purposes or not. Many scholars agree that applied research with no commercial intention is included in marine scientific research, but that applied research with a commercial intention is not.[111] In light of this, it has been suggested that the distinction between bioprospecting and marine scientific research turns on the commercial intent of the researchers. This is because marine scientific research is expected to be transparent and open, with an obligation to disseminate information and data and the

---

[107] See ch 6.
[108] Report of the Chairman of the Ad Hoc Working Group, United Nations General Assembly A/68/399 (2013), para. 28.
[109] LA de La Fayette, 'A New Regime for the Conservation and Sustainable Use of Marine Biodiversity and Genetic Resources Beyond the Limits of National Jurisdiction' *International Journal of Marine and Coastal Law* 24 (2009): pp. 221–80, p. 228.
[110] UN Secretary-General, '*Oceans and the Law of the Sea: Report of the Secretary-General*', UN Doc. A/60/63/Add.1 (2005), para. 202. This approach is not accepted by all scholars. See, e.g. D Leary, *International Law and the Genetic Resources of the Deep Sea* (Leiden: Martinus Nijhoff, 2007): p. 157.
[111] See Ane Jørem and Morten Walløe Tvedlt, 'Bioprospecting in the High Seas: Existing Rights and Obligations in View of a New Legal Regime for Marine Areas Beyond National Jurisdiction' *International Journal of Marine and Coastal Law* 29 (2014): pp. 321–43, pp. 327–9, concluding that commercially oriented research is included in marine scientific research; compare Paul Gragl, 'Marine Scientific Research' in Malgosia Fitzmaurice and Norman A Martínez Gutiérrez (eds), *The IMLI Manual on International Maritime Law: Volume I Law of the Sea* (Oxford: Oxford University Press, 2014): pp. 396–429, p. 402 and Alfred HA Soons, *Marine Scientific Research and the Law of the Sea* (The Hague: TMC Asser Instituut/Kluwer, 1982): p. 125, arguing that data collection concerning resources with a view to the exploitation of the resource is not included in marine scientific research.

subsequent publication outlining the results of the research.[112] Because researchers who intend to seek a patent or other intellectual property rights will not publish the results of their research until such protection is obtained, it is argued that the research cannot be considered as part of the marine scientific research regime.[113] Commercially focused research is closer to exploitation than research. Therefore, the intention of the researchers becomes the critical criteria for assessment.[114]

Some commentators have questioned the utility of distinguishing between bioprospecting and marine scientific research based on intention.[115] Scovazzi argues that a research endeavour organized with the intent to increase human knowledge may well result in the discovery of commercially valuable information. It is impossible to establish a clear-cut distinction between one activity and the other and between one purpose and the other.[116] Research trips are expensive, and may be funded by a range of organizations. Some research institutions may receive funding from commercial interests for their dives, resulting in a mix of 'commercial' and 'non-commercial' research.[117] In other cases, a research trip with non-commercial intentions may take samples that are later accessed for commercial purposes by other researchers.[118]

Despite these difficulties, it appears that many States are opting to make the distinction between bioprospecting and scientific research based on the intention of the researchers, although these regulations are primarily driven by the provisions of the CBD rather than the LOSC. In both Australia and the Philippines, regulations distinguish between the collection of specimens for commercial or non-commercial purposes.[119] In Australia's case, collection for commercial purposes creates obligations to enter into benefit-sharing agreements, whereas collection for non-commercial purposes must be accompanied by a declaration outlining the lack

---

[112] Article 244, LOSC.

[113] Study of the Relationship between the Convention on Biological Diversity and the United Nations Convention on the Law of the Sea with Regard to the Conservation and Sustainable Use of Genetic Resources on the Deep Seabed, UN Doc. UNEP/CBD/SBSTTA/8/INF/3/Rev.1 (22 February 2003), para. 39.

[114] See Report of the Secretary General on Oceans and the Law of the Sea A/60/63/Add.1 (2005), para. 202; Salvatore Arico and Charlotte Salpin, 'Bioprospecting of Genetic Resources in the Deep Seabed: Scientific, Legal and Policy Aspects' (UNU-IAS 2005) 34 http://modernncms.ecosystem-marketplace.com/repository/modernncms_documents/DeepSeabed.pdf.

[115] See Mossop, 'Protecting Marine Biodiversity on the Continental Shelf Beyond 200 Nautical Miles', p. 293.

[116] Article 246, LOSC; T Scovazzi, 'Mining, Protection of the Environment, Scientific Research and Bioprospecting: Some Considerations on the Role of the International Sea-bed Authority' *International Journal of Marine and Coastal Law* 19 (2004): pp. 383–409, p. 403. See also CH Allen, 'Protecting the Oceanic Gardens of Eden: International Law Issues in Deep-Sea Vent Resource Conservation and Management' *Georgetown International Environmental Law Review* 13 (2001): pp. 563–660, p. 644.

[117] David Leary et al., 'Marine Genetic Resources: A Review of Scientific and Commercial Interest' *Marine Policy* 33 (2009): pp 183–94, p. 184.

[118] Scovazzi, 'Mining, Protection of the Environment, Scientific Research and Bioprospecting', p. 403.

[119] Harlan Cohen, 'Some Reflections on Bioprospecting in the Polar Regions' in Davor Vidas (ed), *Law, Technology and Science for Oceans in Globalisation: IUU Fishing, Oil Pollution, Bioprospecting, Outer Continental Shelf* (Leiden: Martinus Nijhoff, 2010): pp 339–52, p. 342.

of commercial intent and controls on later commercial use.[120] In light of the practice in this area, this chapter will consider bioprospecting activities with a commercial purpose on the basis that they do not fall within the regime of marine scientific research under the LOSC.

(2) Bioprospecting and the Law of the Sea Convention

Four aspects of the LOSC are important in respect of marine bioprospecting. First, the LOSC establishes the rights that each State may exercise in maritime zones from the territorial sea to the high seas. Within national jurisdiction, coastal States exercise significant control over the harvesting of living resources, including marine genetic resources. Beyond national jurisdiction, the prevailing legal principle is that of freedom to exploit the living resources of the high seas. The intersection of these principles can be complicated in the case of bioprospecting.

The second important aspect is that the LOSC regulates marine scientific research. If research is conducted into genetic resources without a commercial intent, the activity will be marine scientific research. Where marine scientific research projects are directly relevant to the exploitation of living resources, coastal States have a significant level of control over projects in their EEZ and on their continental shelf.[121] However, as discussed in Chapter 6, on the extended continental shelf States are restricted in their ability to refuse consent for research that has implications for resources.[122] Where there is no commercial purpose to the research, the provisions of the LOSC applicable to marine scientific research will apply.

Thirdly, the LOSC contains provisions requiring States to protect and preserve the marine environment.[123] Although the majority of these provisions focus on the prevention of pollution, there are provisions relating to the protection of the environment generally, including the requirement to conduct prior assessment of activities that may cause significant and harmful changes to the marine environment.[124] Also, although it is usually accepted that bioprospecting does not generally involve great risks to the marine environment, as the samples taken are small, there may still be some disturbance of the environment which needs to be evaluated.[125] If the biotechnology process subsequently requires harvesting of marine organisms on a larger scale, the environmental effects of such activity will need to be assessed.

Finally, it is important to remember that States only have jurisdiction over the sedentary species on the extended continental shelf. The main problem is how to determine whether a particular species is 'sedentary', because species that populate hydrothermal vents and seamounts may not fit into the traditional categories

---

[120] Cohen, 'Some Reflections on Bioprospecting in the Polar Regions', p. 342; Christian Prip et al., *The Australian ABS Framework: A Model Case for Bioprospecting?* (Oslo: Fridtjof Nansens Institute, 2014).
[121] Article 246, LOSC.   [122] Article 246(6), LOSC.   [123] Article 192, LOSC.
[124] Article 206, LOSC.
[125] Robin Warner, 'Protecting the Diversity of the Depths: Environmental Regulation of Bioprospecting and Marine Scientific Regulation Beyond National Jurisdiction' *Ocean Yearbook* 22 (2008): pp. 411–43, p. 416.

established by State practice under the Continental Shelf Convention or the LOSC. Allen points out that there is a range of species often found at hydrothermal vents, including tubeworms, clams, mussels, barnacles, snails, anemones, limpets, shrimp, fish and a biomass of microbes. For some species there will be no doubt that they are sedentary—for example, tubeworms and molluscs.[126] For others there will be debate. Two main features of the sedentary species definition are likely to be problematic: what is the 'harvestable stage' of such species; and how do they move?

The 'harvestable stage' element was included in the definition to make it clear at what point the sedentary character of a species is to be ascertained.[127] This is because most marine species spend part of their time in the water column before reaching maturity. There is no interest in harvesting an oyster until it is mature. When considering that genetic material may be obtainable from organisms at any stage of their development, the definition becomes much harder to apply. Allen suggests that one approach would be to consider the capability of movement at the time it is collected, which would be easier to determine; that is, the harvestable stage is when the organism is collected, not when it is mature.[128] This approach would also make sense when the organism could be new to science and little is known about its life cycle.[129] An alternative would be to require extensive research to determine whether a particular organism is sedentary or not.

Focusing on mobility at the time of collection is not favourable to coastal States, given that most sedentary species are mobile at early stages of their lives. If this is the case, then a researcher could legitimately collect and use eggs and larvae of a species such as tubeworms or mussels whilst they are in the water column, even though they would come under the coastal State's jurisdiction if they are collected as adults. However, the alternative, as already mentioned, would be to invest considerable time and effort in determining the status of the adult organism. It is difficult to imagine how the coastal State could insist on information being shared by researchers who collect organisms in the water column of the high seas when such organisms might not be easily or immediately classifiable as sedentary species.

Allen has identified the particular difficulties in classifying marine microbes under the LOSC. Microbes might be found inside the vents; suspended above the vents; located in mats growing on rock, chimneys, sediment or even other organisms; or as symbiotic with other, larger species such as tubeworms or clams.[130] Some microbes are capable of self-locomotion; others are immobile. Allen suggests that microbes harvested from 'mats' on rocks or sediment, as well as on sedentary species, would be considered sedentary species. On the other hand, microbes collected from the water column in or near a vent would not be sedentary, even if they

---

[126] Allen, 'Protecting the Oceanic Gardens of Eden', p. 626.
[127] Richard Young, 'Sedentary Fisheries and the Convention on the Continental Shelf' *American Journal of International Law* 55 (1961): pp. 359–73, p. 368.
[128] Allen, 'Protecting the Oceanic Gardens of Eden', p. 623.
[129] Mossop, 'Protecting Marine Biodiversity on the Continental Shelf Beyond 200 Nautical Miles', p. 292.
[130] Allen, 'Protecting the Oceanic Gardens of Eden', p. 627.

are incapable of independent motion.[131] As Allen notes, the application of the sedentary species definition is 'poorly suited' to vent ecosystems. However, coastal States will need to work with this definition to determine the resources over which they can exercise jurisdiction.

It is clear that the definition of sedentary species is not particularly useful in the context of bioprospecting on seamounts and at hydrothermal vents. The coastal State may wish to ensure it has information about the species collected so that it can come to a view as to which species are sedentary or non-sedentary. Therefore, the conditions that a State imposes on the activity could include requesting information to allow the State to make its own determination. This will be difficult in cases where the bioprospecting activity is directed at both seabed and swimming organisms. It will be almost impossible legally to impose obligations on researchers who take samples of sedentary organisms while they are at juvenile stages and dispersed in the water column. However, researchers may prefer to accept such limitations in return for access to other living continental shelf resources.

(3) The Convention on Biological Diversity

The Convention on Biological Diversity 1992 (CBD) is aimed at conserving biological diversity, ensuring the sustainable use of its resources and the fair and equitable sharing of the benefits arising out of the use of genetic resources.[132] The latter objective arose out of developing States' concerns that they were losing control over genetic resources that were developed by biotechnology companies, with no benefits returned to the State in which the resources were found. Therefore, the CBD confirmed that genetic resources are under the control of the State in which they are located.[133] States are obliged to facilitate access to genetic resources by researchers in other States, but such access is to be on mutually agreed terms, based on prior informed consent, and the benefits of commercialization of genetic resources should be shared in a fair and equitable way.[134]

Article 4 of the CBD states that it applies to 'the components of biological diversity, in areas within the limits of [States'] national jurisdiction'. Therefore, the CBD will apply to marine genetic resources found in the territorial sea, in the EEZ and on the continental shelf. The CBD also applies to processes and activities carried out under the jurisdiction or control of the State, whether they are carried out within its national jurisdiction or beyond the limits of national jurisdiction. This means that the CBD has potential application to activities carried out on the high seas or in the Area.

The question of what amounts to facilitation of access, and to fair and equitable benefit sharing, has occupied the parties to the CBD for many years as States have struggled to establish appropriate legislative and regulatory frameworks.[135]

---

[131] Ibid.   [132] Article 1, CBD.   [133] Articles 3 and 15, CBD.
[134] Article 15, CBD.
[135] EC Kamau, B Fedder and G Winter, 'The Nagoya Protocol on Access to Genetic Resources and Benefit Sharing: What is New and What are the Implications for Provider and User Countries and the Scientific Community?' *Law, Environment and Development Journal* 6 (2010): pp. 246–62.

A number of attempts to clarify these requirements have been made by working groups established under the Conference of the Parties (COP). In 2002, the COP adopted the voluntary Bonn Guidelines to clarify policies on access and benefit sharing (ABS).[136] These Guidelines established expectations to be fulfilled by countries of origin, users of genetic resources, providers of genetic resources and user States in relation to prior informed consent, mutually agreed terms, provision of ABS. However, it was considered that the Guidelines did not adequately clarify the relationship between important concepts and procedures. The non-binding nature of the Guidelines was also problematic.

Soon after the conclusion of the Bonn Guidelines, the parties to the CBD agreed to work towards an international treaty on ABS. The Nagoya Protocol was adopted in 2010.[137] It establishes procedures that provide clarity and certainty to those seeking access to genetic resources. Parties must cooperate to build capacity in developing States. As with the CBD, many of the obligations are qualified by such words as 'as far as possible' or 'where applicable'.

Within areas of national jurisdiction, the CBD and Nagoya Protocol establish obligations on States; these obligations apply to both coastal States (known as provider States in CBD terms) and the States whose nationals wish to access the genetic material (often referred to as user States). First, coastal States have obligations to develop national strategies for the conservation and sustainable use of biological diversity.[138] This may involve identifying and monitoring biological diversity, and protecting it both *in situ* and *ex situ*.[139] These States must also allow for the prior assessment of the impact of projects on biodiversity, usually through environmental impact assessments.[140]

Secondly, coastal States must 'endeavour to create conditions to facilitate access to genetic resources for environmentally sound uses by other Contracting Parties and not to impose restrictions that run counter' to the objectives of the CBD.[141] Access to resources should be based on the prior informed consent of the parties, and coastal States must take steps to share 'in an equitable way the results of research and development and the benefits arising from the commercial and other utilization of genetic resources' based on mutually agreed terms.[142]

This second aspect of the CBD has proven most controversial. As already mentioned, the concepts in the CBD were not well defined, and considerable subsequent effort has been put into clarifying the role of the provider and user States. The Nagoya Protocol is intended to assist States in implementing their obligations under the CBD. Coastal States will be required to identify clearly their domestic access and benefit-sharing rules and requirements and specify how to apply for

---

[136] See, e.g. Bevis Fedder, *Marine Genetic Resources, Access and Benefit Sharing: Legal and Biological Perspectives* (London: Routledge, 2013): p. 46.
[137] Nagoya Protocol on Access to Genetic Resources and the Fair and Equitable Sharing of Benefits Arising from Their Utilization to the Convention on Biological Diversity (opened for signature 2 February 2011, entered into force 12 October 2014) www.cbd.int/abs/text (Nagoya Protocol).
[138] Article 6, CBD. See Warner, 'Protecting the Diversity of the Depths', p. 425.
[139] Articles 7–10, CBD.   [140] Article 14, CBD.   [141] Article 15(1), CBD.
[142] Article 15(7), CBD.

prior informed consent.[143] This seeks to improve the ability of foreign researchers to apply for consent to access resources. In addition, each State must designate a national focal point and national authority on ABS. These focal points again are intended to facilitate ease of information and access to decision-making procedures within the coastal State. Where genetic resources are found within the territory of more than one State, those States shall endeavour to cooperate to implement the Protocol.[144] Finally, the parties are to consider a global multilateral benefit-sharing mechanism to address the cases where genetic resources are present in transboundary situations.[145]

Of particular interest in this context is article 8, which calls on States to promote and encourage research including having simplified measures on access for non-commercial purposes. This might be seen as reflecting a similar policy goal to the LOSC, which in article 246(3) anticipates that coastal States will give their consent to research intended to increase scientific knowledge for all mankind. However, the Nagoya Protocol does not define 'non-commercial research'.

(4) The Law of the Sea Convention and the Convention on Biological Diversity

The relationship between the CBD and the LOSC is outlined in article 22 of the CBD, which states that it shall not affect the rights and obligations of any State arising under existing international agreements, 'except where the exercise of those rights and obligations would cause a serious damage or threat to biological diversity'. Contracting parties are instructed to implement the CBD 'consistently with the rights and obligations of States under the law of the sea'. Therefore, if there is an inconsistency between the provisions of the CBD and the LOSC, those of the LOSC will prevail.[146]

To the extent possible, the legal regime under the LOSC must be reconciled with the CBD. The regimes are different in that the LOSC is predominantly focused on exploitation and allocation of rights, whereas the CBD is primarily focused on species and habitat conservation.[147] However, in most respects the two conventions are compatible with each other because they encourage the facilitation of pure research and give coastal States discretion to refuse commercially focused research.[148] One

---

[143] Article 6, Nagoya Protocol. [144] Article 11, Nagoya Protocol.
[145] Article 10, Nagoya Protocol.
[146] Fedder, *Marine Genetic Resources*, p. 59; DK Anton, 'Law for the Sea's Biological Diversity' *Columbia Journal of Transnational Law* 36 (1998): pp. 341–71, p. 357. Not all authors agree. See A Boyle, 'Further Development of the Law of the Sea Convention: Mechanisms for Change' *International and Comparative Law Quarterly* 54 (2005): pp. 563–84, p. 579.
[147] Rüdiger Wolfrum and Nele Matz, 'The Interplay of the United Nations Convention on the Law of the Sea and the Convention on Biological Diversity' *Max Planck Yearbook of United Nations Law* 4 (2000): pp. 445–80, p. 464.
[148] Caroline von Kries and Gerd Winter, 'Harmonizing ABS Conditions for Research and Development under UNCLOS and CBD/NP' in Evanson Chege Kamau, Gerd Winter and Peter-Tobias Stoll (eds), *Research and Development on Genetic Resources: Public Domain Approaches in Implementing the Nagoya Protocol* (Routledge, 2015): pp. 75–90.

difference between the two regimes is that the CBD requires States to facilitate access to genetic resources, whereas the LOSC assumes, if bioprospecting is treated as exploitation of living resources, that the coastal State has discretion as to how resources are to be accessed.[149] This is not a true incompatibility. In essence, the LOSC confirms that the coastal State has sovereign rights (including sole rights of exploitation) over the resources of the seabed in a similar way as it has sovereignty over the resources found on land. That is the starting point. The CBD creates an additional layer of rights and obligations relating to the exploitation of all resources within national jurisdiction.

In the case of non-commercial projects that amount to marine scientific research, both regimes encourage access by researchers.[150] The need for coastal State consent under the LOSC is similar to the requirement for prior informed consent under the CBD. However, the LOSC does not refer to the need for access to be on mutually agreed terms.[151] Each Convention has different rules about the application for consent that will need to be reconciled by the coastal State.

Fedder has argued that the limitations on coastal State jurisdiction over marine scientific research in the LOSC mean that the CBD provides a stronger basis for refusing consent.[152] This raises the possibility that a coastal State might strictly be required to grant consent to a marine scientific research project under the LOSC when it might have been able to refuse consent under the CBD. Given that the rights under the CBD have to be exercised consistently with the rights and obligations under the LOSC, is this likely to pose a problem? Assuming the coastal State wished to refuse consent for the (non-commercial) research into genetic resources on the basis that it had direct significance for the exploration or exploitation of living resources,[153] the options for challenging the coastal State's exercise of discretion is limited. Under the dispute settlement system in the LOSC, disputes over the coastal State's application of article 246 cannot be subjected to third-party dispute settlement. The dispute can be taken to conciliation but the conciliation commission cannot question the State's discretion to withhold consent under article 246(5).[154] Therefore, in practice, any inconsistencies between the LOSC and the CBD are unlikely to pose any practical problems.

As a result of provisions in the LOSC, the CBD and the Nagoya Protocol, coastal States have obligations to establish processes for providing access to genetic resources by foreign researchers. This may require linking the regime for access to the continental shelf for marine scientific research to the regime for granting access under the CBD and the Nagoya Protocol. If coastal States believe that the marine scientific research regime is not appropriate for bioprospecting projects with a commercial purpose, there must be another, clear, route for foreign researchers to

---

[149] Fedder, *Marine Genetic Resources*, p. 56.
[150] See article 239, LOSC, which requires States to promote and facilitate the development and conduct of marine scientific research.
[151] Fedder, *Marine Genetic Resources*, p. 56.   [152] Ibid., p. 58.
[153] Article 246(5), LOSC. For further discussion about the operation of article 246, see ch 6.
[154] Article 297(2), LOSC.

seek approval for such research. This process need not differ substantially from the regime for granting consent in the EEZ and on the continental shelf, but in practice the provisions of the Nagoya Protocol are likely to impose additional procedural requirements.

There are significant complications involved in regulating bioprospecting on the continental shelf beyond 200 nm. Therefore, adopting a cooperative approach to facilitate inclusion in the research, for example by having national researchers on board the foreign research vessel, would be advisable as a way for coastal States to maximize the information available to the State about the resources of the continental shelf. Overall, the CBD and the Nagoya Protocol offer additional benefits for coastal States who wish to ensure that benefits from bioprospecting on the extended continental shelf will flow back to the coastal State through ABS arrangements.

C. Fishing

Fishing above the continental shelf may target living resources that fall into the category of sedentary species, or it may target fish swimming in the water column above the shelf. In both cases, the methods used for such fishing may be the same—using bottom trawls that impact on the benthic environment. In both situations there will be a detrimental impact on the sea floor but there is a very different legal regime that applies to each situation. There are two issues that a coastal State may have to address when considering the management of fishing on its extended continental shelf.

First, it is possible (although unlikely in most cases) that there may be some stocks of commercially valuable sedentary species that may be targeted by fishermen on the shelf beyond 200 nm.[155] One example where this has happened in the past is the fishing for scallops and crabs on the Canadian extended shelf. A coastal State will clearly have the right to regulate fishing for these species. As the fishery is governed by Part VI of the LOSC, rather than Part V, the provisions relating to the conservation and utilization of living resources contained in articles 61 and 62 do not apply to sedentary species.[156] Nor do the obligations to share any excess total allowable catch (TAC) with land-locked and geographically disadvantaged States. The reason for this is that the rights and obligations from the Continental Shelf Convention in relation to sedentary species were more or less directly transferred into the LOSC and the earlier Convention had no provisions relating to conservation.

However, the coastal State is subject to a range of other obligations for environmental protection, as already outlined in this chapter. These include the obligation

---

[155] For the development and definition of sedentary species, see ch 3.
[156] RR Churchill and AV Lowe, *The Law of the Sea*, 3rd edn (Manchester: Manchester University Press, 1999): p. 320.

to protect and preserve the marine environment in article 192 of the LOSC. In light of developments in environmental law and fisheries law, it would be difficult to argue that the coastal State was not required to take steps to conserve and manage the sedentary species of the extended continental shelf.

Sedentary species on the continental shelf may also be targeted where living resources are harvested in large amounts, not for consumption, but for the manufacture of products derived from the genetic structure of the organisms. This may be necessary in cases where large quantities of a substance are required but the relevant compounds cannot be synthesized in a laboratory. In this case, the coastal State is clearly able to regulate the exploitation of such materials in the same way as it regulates fishing.

A second, and more problematic, issue is the possibility that fishing methods targeting non-sedentary species may have a negative impact on the benthic marine environment on the extended continental shelf. Bottom trawling, for example, has been demonstrated to have significant impacts on species such as corals and sponges, which would be classified as sedentary species.[157] A bottom trawl may destroy or collect sedentary species, even if the trawl is targeting high seas species. Assuming the coastal State wishes to protect benthic marine ecosystems, what measures can it take to protect them from destructive fishing practices? Recall that in deploying this fishing gear (such as trawl nets) the foreign fishing vessel is targeting species that are not primarily within the jurisdiction of the coastal State. The vessel is exercising the freedom of the seas subject to any conservation and management measures of any regional fisheries organizations. To what extent can a coastal State prevent the fishing vessel from deploying its gear on or above the continental shelf?

Coastal States may consider a range of options to protect sedentary species, including declaring MPAs or limiting fishing within a certain distance of the sea floor. The fundamental question is whether, in seeking to prevent vessels targeting high seas species from physically impacting on the shelf ecosystems, the coastal State is infringing or unjustifiably interfering with the vessels' high-seas freedoms according to article 78(2) of the LOSC.[158] The options for coastal States to unilaterally prevent such fishing are discussed further in Chapter 7. Coastal States must carefully weigh up the relevant interests involved before taking such measures. It is only if such measures are reasonable, necessary and proportionate that the coastal State would be considered to be acting lawfully under international law.

D. Conclusion

When broad margin States argued for extending the continental shelf beyond 200 nm, it is likely that few considered the living resources of the shelf to be of

---

[157] See ch 2 for a description of the environmental impacts of bottom fishing.
[158] Daniel Owen, *The Powers of the OSPAR Commission and Coastal State Parties to the OSPAR Convention to Manage Marine Protected Areas on the Seabed beyond 200 nm from the Baseline*, Report for WWF Germany (2006): p. 41.

significance. This is because it is unlikely that commercial fisheries will target the sedentary species found on most continental shelves beyond 200 nm. However, the advancement of understanding about the ecological systems that exist on the continental shelf brings new opportunities to the fore. The most obvious of these is the potential value of genetic resources.

The legal regime that governs the high seas and the extended shelf is not particularly helpful for States that seek to manage the sampling of sedentary species for biotechnology purposes. One reason is that the definition of sedentary species in article 77(4) of the LOSC is written for the purposes of fishing, rather than for sampling marine organisms that may spend their adult lives on the seabed but which, during parts of their life cycle, will be found in the water column. In the EEZ, the coastal State has jurisdiction over the water column as well, but on the high seas the freedom of scientific research and of fishing will allow researchers to access the genetic material of a range of organisms, some of which may be sedentary species.

Where the genetic samples are obtained from organisms on or under the seabed, coastal States will be in a better position to regulate access to the material. However, practical difficulties still arise, particularly when species are discovered about which little is known in relation to their life cycle and movement. Who is to determine whether these species are sedentary or not? The coastal State may have observers on board the vessel but, if it does not, the researchers themselves will have the best—and sometimes only—information about the organisms. It may be that negotiations for an implementing agreement to the LOSC on the protection of marine biodiversity beyond national jurisdiction will develop a process for dealing with genetic resources found in the Area that could be applied to State's extended shelves if they so choose.[159] In the meantime, coastal States must ensure that any measures they adopt for the continental shelf beyond 200 nm are not an infringement on or unjustifiable interference with the freedoms of the high seas.

Coastal States must consider their environmental obligations in respect of their extended continental shelves. Although no express obligations in relation to environmental protection are found in Part VI of the LOSC, States still have obligations that derive from Part XII, other treaties and customary international law. The obligations require coastal States to consider the protection of biodiversity located on the continental shelf beyond 200 nm and to make decisions about how to ensure economic development is conducted sustainably. Amongst the range of options is the possibility of establishing MPAs to protect vulnerable marine ecosystems. There is no legal barrier to creating MPAs, although the management measures will need to be carefully designed so as not to infringe or unjustifiably interfere with the rights and freedoms of other States on the high seas. Other options include carefully planning economic activities, including fishing and oil and gas exploration, so that the effects on benthic ecosystems are minimized—something that coastal States will already be doing within 200 nm.

---

[159] See ch 10.

The peculiarity of the extended shelf is that it lies underneath the high seas. In this case, there is a particular need for coastal States to be aware of their customary international law duties to prevent significant adverse impact to the global commons (the high seas), to undertake environmental impact assessments and otherwise to exercise due diligence in evaluating activities on the continental shelf beyond 200 nm for their environmental impact.

5

Non-living Resources on the Continental Shelf Beyond 200 Nautical Miles

The desire to extend coastal-State control over the non-living resources of the seabed—hydrocarbons in particular—has driven the development of the continental shelf doctrine. The probability that revenue could be derived from the entire continental margin was a key factor in broad margin States pushing for recognition of their sovereign rights to the edge of the continental margin and the negotiation of article 76 of the Law of the Sea Convention (LOSC).[1] In most respects, the legal regime that applies to the continental shelf within 200 nm will apply equally to the continental shelf beyond 200 nm. The key issues involve the interpretation of article 82 and the protection of exploration and exploitation efforts on the continental shelf beyond 200 nm from high seas activities.

Article 82, which requires coastal States to make payments or contributions in kind in respect of exploitation on the continental shelf beyond 200 nm, will be one of the key factors that coastal States will need to consider when authorizing oil, gas and mineral exploitation on the extended shelf. This chapter examines the obligations in article 82 in light of discussions taking place within the International Seabed Authority (ISA) in preparation for when exploitation on the extended shelf begins. Without State practice in this area, it is only possible to refer to the various issues that have been highlighted by such discussions and academic writings.

The chapter concludes by evaluating some other possible issues that may arise on the continental shelf beyond 200 nm related to mining and hydrocarbon activities.

A. Mining and the Article 82 Obligation to Make Payments

Article 82 of the LOSC requires States to make payments or contributions in respect of exploitation of any area of a State's continental shelf that is beyond the 200 nm zone. The article states that:

1. The coastal State shall make payments or contributions in kind in respect of the exploitation of the non-living resources of the continental shelf beyond 200 nautical miles from the baselines from which the breadth of the territorial sea is measured.

---

[1] United Nations Convention on the Law of the Sea (opened for signature 10 December 1982, entered into force 16 November 1994) 1834 UNTS 397.

2. The payments and contributions shall be made annually with respect to all production at a site after the first five years of production at that site. For the sixth year, the rate of payment or contribution shall be 1 per cent of the value or volume of production at the site. The rate shall increase by 1 per cent for each subsequent year until the twelfth year and shall remain at 7 per cent thereafter. Production does not include resources used in connection with exploitation.
3. A developing State which is a net importer of a mineral resource produced from its continental shelf is exempt from making such payments or contributions in respect of that mineral resource.
4. The payments or contributions shall be made through the Authority, which shall distribute them to States Parties to this Convention, on the basis of equitable sharing criteria, taking into account the interests and needs of developing States, particularly the least developed and the land-locked among them.

This provision has not yet been triggered; however, it is expected that there may be some exploitation undertaken on the continental shelf beyond 200 nm in the near future.[2] Canada has granted a number of petroleum exploration licences on its extended continental shelf. In the Jeanne d'Arc and Eastern Newfoundland area, the Canada–Newfoundland Offshore Petroleum Board has issued eight exploration licences and two significant discovery licences. In four of these cases, a validating well has been drilled, which allows progress to be made towards a production licence.[3] Reports indicate that a joint venture operation in the Flemish Pass, 300 nm from the Newfoundland and Labrador coast, has identified 300 to 600 million barrels of recoverable oil, although exploitation has not yet begun.[4] Norway has issued licences for exploration and production in three areas that are on the extended continental shelf, or which straddle the extended shelf and the exclusive economic zone (EEZ).[5] The United States has issued licences to explore the resources in the Western Gap area established by a maritime boundary agreement with Mexico.[6] These licences are estimated to apply to 20 per cent of the available area on the United States' continental shelf beyond 200 nm. New Zealand has issued a petroleum prospecting permit allowing seismic surveys on the extended continental shelf.[7] In 2013, it was estimated that up to sixteen countries have 'issued and/or are offering' offshore oil and gas exploration concession licences on their extended continental shelves.[8]

---

[2] International Seabed Authority, 'Issues Associated with the Implementation of Article 82 of the United Nations Convention on the Law of the Sea', Technical Study No. 4 (2009): p. 12.
[3] Details of the licences can be found at www.cnlopb.ca.
[4] Paul Radevski, 'Canada's Potential International Oil and Gas Royalty Obligations' *BOE Report*, 25 November 2015 http://boereport.com/2015/11/25/law-of-the-sea-and-canadas-potential-international-oil-and-gas-royalty-obligations/.
[5] 'Licensing System in Norway', presentation to ISA Workshop in 2012 http://www.isa.org.jm/files/documents/EN/Workshops/2012/Norway.pdf.
[6] House Report. 113-101, Part 1—Outer Continental Shelf Transboundary Hydrocarbon Agreements Authorization Act, 6 June 2013, p. 4.
[7] http://www.shell.co.nz/future-energy/exploration/new-caledonia-basin.html.
[8] Clive Schofield, 'New Marine Resource Opportunities, Fresh Challenges' *University of Hawaii Law Review* 35 (2013): pp 715–33, pp. 723–4.

This activity makes the need to clarify some of the ambiguities in article 82 a concrete issue. To date, the ISA has been the forum for discussions about how to implement article 82. However, there is general recognition that the ISA cannot impose an interpretation of article 82 onto States, and it will be up to coastal States to determine how they calculate their payments or contributions.[9] At the time of writing, coastal States with offshore exploration had not announced how they would deal with their article 82 obligations.[10]

From a practical perspective, the hydrocarbon industry will generally prefer to exploit resources closer to shore where it is cheaper and easier to do so. However, some clarity about how article 82 will be applied is important to encourage development on the extended continental shelf. Companies prefer certainty before investing in exploration that will cost millions of dollars. Until a coastal State determines its approach to article 82, companies will prefer to focus on areas within the 200 nm limit. It may also be that the additional requirements in article 82 will discourage companies from exploring and exploiting the extended continental shelf. For example, one practical issue of interest to the industry will be whether the article 82 payments are made by the coastal State directly out of its taxation and royalty revenue, or whether the production companies will be required to make the additional payment out of their revenue.[11] This will have an impact on the cost–benefit analysis made by private investors when considering the development of areas beyond 200 nm.

(1) Background to article 82

Because it was clear that the LOSC would apply the principle of common heritage of mankind to the Area, a crucial issue during negotiations became whether the continental shelf could extend beyond 200 nm, and in what circumstances. Some countries insisted that the continental shelf should not extend beyond 200 nm, in order to maximize the resources that would be subject to management by the ISA.[12]

---

[9] International Law Association, '*Report on Article 82 of the 1982 UN Convention on the Law of the Sea*' (2008), para. 2.7 (ILA Report on Article 82). Hereafter 'payments' will refer to both payments and contributions.

[10] Canadian offshore bid documents contain a proviso that additional terms may be applied to any licences which will allow Canada to respond to its article 82 obligations. Erik Arnold, 'Canada East Coast (Newfoundland and Labrador) Oil and Gas Update', 21 April 2016 http://www.globalenergyblog.com/canada-east-coast-newfoundland-and-labrador-oil-and-gas-update.

[11] George Mingay, 'Article 82 of the LOS Convention—Revenue Sharing—The Mining Industry's Perspective' *International Journal of Marine and Coastal Law* 21 (2006): pp. 335–46, p. 341.

[12] Aldo Chircop and Bruce A Marchand, 'International Royalty and Continental Shelf Limits: Emerging Issues for the Canadian Offshore', *Dalhousie Law Journal* 26 (2003): pp. 273–302, p. 283; Michael D Morin, 'Jurisdiction Beyond 200 Miles: A Persistent Problem' *California Western International Law Journal* 10 (1980): pp. 514–35, p. 522. See, e.g. LOSC Records, Report of 25th Meeting, 5 August 1974, p. 201, A/CONF.62/C.2/SR.25 Summary records of meetings of the Second Committee, 25th Meeting. Extract from the Official Records of the Third United Nations Conference on the Law of the Sea, Volume II (Summary Records of Meetings of the First, Second and Third Committees, Second Session), p. 201.

The larger the area of the seabed that came under national jurisdiction, the smaller the area that would be available to the ISA. Once it was apparent that the extended continental shelf was likely to become a reality, focus shifted towards ensuring that some benefit from exploiting the resources of the shelf would be returned to the international community.[13] Broad margin States found that they were required to accept this idea in return for the expansion of their jurisdiction. They would receive recognition of their extended shelf, but in return a portion of the revenue would be shared with the international community through the ISA.[14] This compromise allowed for progress to be made in the negotiations relating to article 76 and the outer limits of the shelf.[15]

One issue during negotiations was the area in respect of which the obligation to make payments would be required. The expansion of coastal State jurisdiction to 200 nm inspired some States to argue that coastal States should share a portion of the financial benefits of the exploitation of non-living resources in the EEZ and on the continental shelf outside the territorial sea.[16] For example, in 1974 the United States proposed that payments be made in respect of seabed exploitation seaward of the territorial sea or the 200 metre isobaths, whichever is further from shore.[17] Land-locked States argued for access to the resources of the shelf thus limiting coastal State jurisdiction.[18] However, the main discussion focused on establishing some form of revenue sharing, with payments made to an international organization that would distribute them equitably. Coastal States were mostly willing to discuss the possibility of sharing benefits of exploitation from the outer part of the shelf. By 1975 most States were focusing on revenue sharing on the continental shelf beyond 200 nm.[19]

The terms of the payment were a significant focus for negotiations. First, early suggestions that the payment be based on revenues were rejected in favour of focusing on the production at a particular site. It was considered that payments based on net production, or profits, was too problematic to be practical.[20] Different States would have different criteria for calculating profit, and in addition,

---

[13] Shigeru Oda, 'The Ocean: Law and Politics' *Netherlands International Law Review* 25 (1978): pp. 149–58.

[14] Ted McDorman, 'The Entry into Force of the 1982 LOS Convention and the Article 76 Outer Continental Shelf Regime' *International Journal of Marine and Coastal Law* 10 (1995): pp. 165–87, p. 174; DP O'Connell, *The International Law of the Sea: Volume I* (Oxford: Clarendon Press, 1982): p. 507.

[15] ISA, Technical Study No. 4, p. 16.

[16] SC Vasciannie, *Land-locked and Geographically Disadvantaged States in the International Law of the Sea* (Oxford: Clarendon Press, 1990): p. 88.

[17] Myron H Nordquist (ed), *United Nations Convention on the Law of the Sea 1982: A Commentary, Volume II*, (The Hague: Martinus Nijhoff, 1993): p. 934. See also Shigeru Oda, *Fifty Years of the Law of the Sea, With a Special Section on the International Court of Justice* (Dordrecht: Martinus Nijhoff, 2003): p. 523.

[18] Chircop and Marchand, 'International Royalty and Continental Shelf Limits', p. 287; Vasciannie, *Land-locked and Geographically Disadvantaged States*, pp. 89–93.

[19] Chircop and Marchand, 'International Royalty and Continental Shelf Limits', p. 288.

[20] Ibid., p. 290.

deep-sea oil wells would be expensive to establish, thus minimizing the overall profits. Therefore, the preferable approach was to focus on the gross volume of production.

Secondly, proposals for the level of payment settled on an escalating scale over time, reflecting the high costs of exploration and exploitation at an early stage. Most States agreed that there should be a grace period to allow operators to recover the costs of developing the site.

Thirdly, the role of an international organization in distributing the payments from coastal States was generally agreed early in the negotiations. The draft texts referred to an 'International Authority' until 'Authority' appeared in a draft negotiating text in 1977, presumably referring to the ISA.[21] Early drafts also referred to the need for agreement between the 'International Authority' and the coastal State on the terms of payment.[22] This requirement disappeared from draft texts after 1977.

Fourthly, the position of developing coastal States was a concern from the beginning of the negotiations. It was felt that it would be unfair to levy payments from underdeveloped coastal States. Initially, the draft texts left the question to an 'International Authority', which would determine whether a developing coastal State should be required to make the article 82 payments.[23] Following considerable disagreement about the nature of any exemption given to developing States, the negotiators settled on a formula for exemption that was confined to the circumstances where a developing country is a net importer of the resource.[24] This had the benefit of clarity as opposed to leaving the discretion to the International Authority.

(2) Approach to interpretation of article 82

Because so few terms in article 82 are defined in the LOSC, it is necessary to rely on principles contained in articles 31 and 32 of the Vienna Convention on the Law of Treaties (VCLT).[25] Article 31 provides that: 'a treaty shall be interpreted in good faith in accordance with the ordinary meaning to be given to the terms of the treaty in their context and in the light of their object and purpose'. Article 32 allows for recourse to supplementary means of interpretation, such as the preparatory work of the treaty. The requirement of good faith was heavily emphasized by the International Law Association (ILA) in its report on article 82. An important, but often overlooked, feature of the LOSC is article 300, which also imposes an

---

[21] Nordquist (ed), *United Nations Convention on the Law of the Sea 1982: A Commentary, Volume II*, p. 942.
[22] Ibid. pp. 936 ff.
[23] See, e.g. the Informal Single Negotiating Text/Part II text (1975); Nordquist (ed), *United Nations Convention on the Law of the Sea 1982: A Commentary, Volume II*, p. 937.
[24] The net importer criteria appeared in the Informal Composite Negotiating Text in 1977. Nordquist (ed), *United Nations Convention on the Law of the Sea 1982: A Commentary, Volume II*, p. 941.
[25] Vienna Convention on the Law of Treaties (opened for signature 23 May 1969, entered into force 27 January 1980) 1155 UNTS 331. See ILA Report on Article 82, para. 2.4.

obligation on States to fulfil their obligations under the LOSC in good faith, and not in a manner that would amount to an abuse of right.[26] The reason that this was considered significant by the ILA is that coastal States will be the ultimate decision-makers as to how to implement article 82. A requirement of good faith could go a long way to countering extreme interpretations by coastal States.

A further question, that is largely academic but may affect interpretation of article 82, is the role of common heritage of mankind on the extended shelf. The ILA, in its report on article 82, pointed to the fact that the article was the result of a compromise in the negotiations that allowed coastal States to extend their jurisdiction over resources beyond 200 nm in return for a share of the proceeds of exploiting those resources. The report notes that:[27]

Article 82 thus provides for the application of the Common Heritage of Mankind (CHM) principle within the OCS [outer continental shelf], even though the OCS is within the coastal State's maritime jurisdiction. As Oda points out, this provision was 'instituted in such a manner that the concept of the common heritage of mankind plays a role in controlling over-expansion of the exclusive interests of coastal States in their continental shelves'.

Barbara Kwiatkowska wrote that the limitations in article 82 and article 246(6) 'give recognition to the legitimate concerns of the international community in protecting the fundamental principles of the common heritage of mankind and freedom of scientific research against maritime expansionism of the broad margin States'.[28]

However, the fact that article 82 represents a compromise between broad margin States and the international community does not mean that the principle of common heritage of mankind applies to the extended continental shelf. It is true that a portion of revenue is shared with other States through the ISA, which clearly resembles the approach in Part XI of the LOSC. The negotiating position of some States had been that the Area should begin at the 200 nm limit; thus the concept of the extended shelf has narrowed the geographical extent of the Area. However, two points refute the idea that common heritage of mankind is relevant to article 82.

First, article 136 states that '[t]he Area and its resources are the common heritage of mankind'. This in no way applies to the extended continental shelves of coastal States.[29] Instead, the rights of the coastal State in relation to the continental shelf

---

[26] Ibid., paras. 2.4–2.5.
[27] Ibid., para. 1.3, citing Shigeru Oda, *International Control of Sea Resources*, reprint with new Introduction (Dordrecht: Martinus Nijhoff, 1989): p. xxxii. However, Oda made reference in the same paragraph to the fact that earlier proposals of the United States for international management of the extended continental shelf were not reflected in article 82. Thus, Oda was possibly simply referring to the final compromise between proponents of exclusive coastal State interests and the common heritage of mankind on the extended shelf. See also Barbara Kwiatkowska, 'Creeping Jurisdiction beyond 200 miles in Light of the 1982 Law of the Sea Convention and State Practice' *Ocean Development and International Law* 22 (1991): pp. 153–87, p. 158; David M Ong, 'A Legal Regime for the Outer Continental Shelf? An Inquiry as to the Rights and Duties of Coastal States within the Outer Continental Shelf', Paper presented to the Third ABLOS Conference (2003), p. 3 http://www.iho.int/mtg_docs/com_wg/ABLOS/ABLOS_Conf3/PAPER7-4.PDF.
[28] Barbara Kwiatkowska, *The 200 Mile Exclusive Economic Zone in the New Law of the Sea* (Dordrecht: Martinus Nijhoff, 1989): p. 18.
[29] ISA, Technical Study No. 4, p. 23.

are exclusive.³⁰ On the face of the LOSC, there is no indication that the common heritage of mankind has any role on the continental shelf regime in Part VI.

Secondly, almost none of the elements traditionally associated with the principle of the common heritage of mankind is found in article 82. Pardo and Christol defined the principle of the common heritage of mankind as:³¹

[F]irst, the common heritage cannot be appropriated—it could be used but not owned; second, the use of the common heritage required a system of management in which all users must share; third, it implied an active sharing of benefits, including not only financial benefits but also benefits derived from shared management and exchange and transfer of technologies; fourth, the principle of common heritage implied eventual reservation for peaceful purposes; and finally, it implied transmission of the heritage substantially unimpaired to future generations.

Of these elements, only the sharing of benefits is found in article 82. The coastal State retains exclusive jurisdiction over the resources of the continental shelf. There is no system of shared management on the extended shelf, there is no express statement relating to peaceful purposes³² and the only requirements for conservation of the resources or the environment are derived from general principles that apply to many other maritime zones. The only role for the ISA is as a conduit for the sharing of payments.³³ Indeed, as mentioned above, during negotiations States rejected the idea that there would need to be agreement between the coastal State and the ISA, further limiting its potential role.³⁴

A more accurate view is that article 82 reflects a limitation on the rights of the coastal State as a result of the negotiating position—a situation that echoes many other compromise positions reached during negotiations.³⁵ Vasciannie has argued that article 82 was an essential element of the 'package' and its primary purpose was to compensate developing and land-locked States for the fact that article 76 extended coastal State jurisdiction to an area that some States had considered should

---

³⁰ Article 77(2), LOSC.
³¹ A Pardo and CQ Christol, 'The Common Interest: Tension Between the Whole of the Parts' in R St J MacDonald and DM Johnston (eds), *The Structure and Process of International Law: Essays in Legal Philosophy Doctrine and Theory* (The Hague: Martinus Nijhoff, 1983): p. 654. See also John E Noyes, 'The Common Heritage of Mankind: Past, Present and Future' *Denver Journal of International Law and Policy* 40 (2011–2012): pp. 447–71, p. 450; Rüdiger Wolfrum, 'The Principle of the Common Heritage of Mankind' *Zeitschrift für Ausländisches Öffentliches Recht und Völkerrecht* 43 (1983): pp. 312–37, pp. 333–4.
³² Compare articles 88 and 141, LOSC.
³³ ISA, Technical Study No. 4, p. 23. The author of this study also points out that the ISA has other tasks under the LOSC that are not connected with the primary mandate of controlling activities in the Area.
³⁴ Atsuko Kanehara, 'The Revenue Sharing Scheme with Respect to the Exploitation of the Outer Continental Shelf under Article 82 of the United Nations Convention on the Law of the Sea—A Plethora of Entangling Issues', Seminar on the Establishment of the Outer Limits of the Continental Shelf Beyond 200 Nautical Miles under UNCLOS, Ocean Policy Research Foundation, 2008 https://www.sof.or.jp/en/topics/pdf/aca.pdf.
³⁵ ISA, Technical Study No. 4; Frida Armas-Pfirter, 'Potential Options on Equitable Distribution of Payments and Contributions', Annex 6; ISA, '*Implementation of Article 82 of the United Nations Convention on the Law of the Sea: Report of an International Workshop*', Technical Study No. 12 (2013): p. 87.

be governed by common heritage.[36] Arguably, the common heritage of mankind principle is irrelevant in interpreting article 82. However, the fact that the article was motivated by a desire to share some of the shelf resources with the international community should give further weight to the need for the application of good faith when coastal States interpret and implement article 82.

(3) Ambiguities in article 82

The provisions of article 82 are relatively brief in comparison to Part XI of the LOSC relating to activities in the Area, which was also supplemented by Annex III. It is inevitable, then, that questions will arise as to how the article should be interpreted and implemented. The next sections address some of the more significant issues identified to date.

(a) Payments or contributions in kind

Article 82(1) requires coastal States to make 'payments or contributions in kind' in order to meet their obligations. One report has described this obligation as 'a type of international royalty' payment.[37] The principle, while apparently simple, raises a number of questions.

First, the article is silent on who has the responsibility for determining the amount of the payment or contribution.[38] Given that the coastal State will be in a better position than the ISA to make any assessments, it would be plausible to infer that the State determines the amount of the payment. Lodge has suggested that because the ISA has a fiduciary duty to mankind as a whole, it should be consulted and should agree with the amount of the payments to enable it to discharge its duty.[39] In light of the negotiating history, which saw an abandonment of a requirement for agreement on the part of the ISA,[40] it seems that there can be no feasible argument that there is an obligation on coastal States to reach agreement with the ISA on the value of payments. Despite this, as a practical matter the ISA should be provided with information as to how the amount was determined, and preferably agreement can be reached on how States will apply the article 82 obligation.[41]

---

[36] Vasciannie *Land-locked and Geographically Disadvantaged States in the International Law of the Sea*, p. 127. Vasciannie describes the revenue-sharing obligations as the quid pro quo for the loss of potential common heritage resources, p. 128.

[37] ISA, Technical Study No. 4, p. 25.   [38] Ibid., p. 29.

[39] Michael W Lodge, 'The International Seabed Authority and Article 82 of the UN Convention on the Law of the Sea' *International Journal of Marine and Coastal Law* 21 (2006): pp. 323–33, p. 328. The ILA considers that there is a practical (rather than legal) requirement to negotiate with the ISA. ILA Report on Article 82, para. 2.13.

[40] See text above at Section A(1).

[41] ISA, Technical Study No. 4, p. 29; Lodge, 'The International Seabed Authority and Article 82', p. 328.

Secondly, the principle that States could choose between payments or contributions in kind appeared in the negotiating texts from 1975.[42] The ILA has argued that the use of 'payments *or* contributions' rather than 'payments *and* contributions' implies that once States have made their choice between payments or contributions in kind, they are not permitted to make a combined payment and contribution.[43] Another report has suggested that States could choose different forms of payment in different years, but does not consider that a State could make a combined payment and contribution.[44] From a practical perspective, this would make it easier for the ISA, which would only have to deal with one form of settlement rather than two. However, from the perspective of the purpose and intent of the LOSC which was to make it easier for States to fulfil their obligations in article 82, it seems more logical that the ISA should also accept a combination of payments and contributions in kind. In addition, while the ILA's argument relied on the use of 'or' in paragraphs 1, 2 and 4 of article 82, the report did not reconcile this argument with the use of 'and' in paragraph 2. In fact, article 82 is not entirely consistent in its use of terminology.

It is consistent with the object and purpose of the LOSC to allow States to make a combined payment and contribution.[45] Nothing in article 82 expressly prevents States from making a combined payment and contribution, so long as the total value accords with paragraph 2. In practice, it seems unlikely that the ISA would insist on a single payment or contribution rather than a combination if that is how a State chooses to fulfil its obligation. The important goal of article 82 is that the benefits of exploiting the extended continental shelf are shared with the international community, and the article reflects some flexibility in how States will achieve that.

Linked to this question is whether States can alter whether they make payments or contributions from year to year. On its face the article does not prohibit this, but it appears likely that the easiest option would be for States to discharge their obligations in the same way each year.[46]

Thirdly, although contributions in kind can generally be accepted as the provision of resources to the equivalent monetary value, article 82 does not stipulate what sorts of resources must be used. Must it be in the form of resources as extracted from the extended shelf? Or can another type of resource be used instead?[47] The

---

[42] Nordquist, *United Nations Convention on the Law of the Sea 1982: A Commentary*, Volume II, p. 937. A report has suggested that 'the intention behind insertion of contributions in kind was to secure resource access to State Party beneficiaries'. Annex 1, ISA, Technical Study No. 12, p. 20.

[43] ILA Report on Article 82, paras. 2.13–2.14. [44] ISA, Technical Study No. 4, p. 30.

[45] The ILA accepted, in relation to another part of the article, that the object and purpose of article 82 could be used to defeat an argument based on a strict reading of the text. This was in relation to the discussion as to whether 'mineral resources' mentioned in article 82(3) are more restricted than 'non-living resources' and therefore could be used to limit the exemption that applies to developing countries. See para. 2.20.

[46] ISA, Technical Study No. 4, p. 30.

[47] ILA Report on Article 82, para. 211. See ISA, Technical Study No. 12, 'Workshop Report', para. 16, where the participants assumed that the contributions would be a share of the resource. Paskal and Lodge raised the possibility that contributions in kind might also include technology transfers or provision of Official Development Assistance, although this seems unlikely to be consistent with article 82.

intent of the provision must have been that the contributions in kind would be sourced from the resource that is being extracted, rather than allowing States a broad choice of commodities with which to meet their obligations. Allowing States to make their contribution with, for example, grain of an equivalent value, would pose serious challenges for the ISA.

Fourthly, the article does not specify how the payment or contribution will be made in practice. There is a strong argument that the payment must be made in a freely convertible currency or payment unit.[48] If local currencies that are not widely traded are used, this will cause practical problems for the ISA in receiving and distributing the benefits of the payments.

Although article 82 provides choices for the coastal State in terms of making payments or contributions, it will clearly be preferable from the ISA's perspective that States choose to make payments rather than contributions. If the ISA took receipt of a large amount of minerals it would have to arrange for those minerals to be converted into a monetary form before distributing the income to the beneficiary States. This will impose significant additional costs on the ISA and is certainly beyond its current capacity. Indeed, one of the recommendations from a 2012 workshop on the implementation of article 82 was that States should be recommended to opt to make payments rather than make contributions.[49] Accepting contributions in kind would require the ISA to acquire considerable additional expertise that it currently lacks.

(b) Determining the value of payments and contributions

The potential for conflict as a result of ambiguities is also well demonstrated in article 82(2). The important elements in this paragraph are:

1. Payments must be made annually.

2. Payments are made in respect to all (gross) production at a site, although this does not include resources used in connection with exploitation.

3. Payments begin after the first five years of production are completed.

4. The rate of payments begins at 1 per cent of the 'value or volume of production' at the site in year six, and increase by 1 per cent each subsequent year of production until the twelfth year and shall remain at 7 per cent thereafter.

It has been suggested that the timing of annual payments could potentially alter the value of the payment.[50] This is because currencies and commodity prices fluctuate

---

Cleo Paskal and Michael Lodge, 'A Fair Deal on Seabed Wealth: The Promise and Pitfalls of Article 82 on the Outer Continental Shelf', Chatham House Briefing Paper, February 2009, p. 5.

[48] Lodge, 'The International Seabed Authority and Article 82', p. 326; ILA Report on Article 82, para. 2.10.

[49] 'Outcomes of the International Workshop on Further Consideration of the Implementation of Article 82 of the United Nations Convention on the Law of the Sea', Report of the Secretary General, 6 May 2013, ISBA/19/A/4, para. 4(a).

[50] ISA, Technical Study No. 4; Chircop and Marchand, 'International Royalty and Continental Shelf Limits'.

over the year. Therefore, States should be encouraged to set a regular payment schedule.

As mentioned above, during negotiations it was decided that calculation of the value of payments should be based on gross, rather than net, production. There appears to be little dispute about this approach.[51] Other issues do remain, however. In particular, the meaning of 'value or volume of production' poses a number of questions about how a coastal State will determine the value of production. It has been suggested that this could be considered to be the well-head value or market price per barrel in the case of oil, but will not be easily identified in the case of other non-living resources.[52] The ILA concluded that it was up to the coastal State to decide on the form, method and timing of the payments, subject to its duty of good faith, but that the ISA 'could express its view' in relation to the method and timing aspects.[53] Again, as a practical matter, it is preferable that general agreement is reached between States and the ISA on methods of valuing production that are acceptable to all concerned in order to avoid disputes.

The ILA also concluded that the coastal State should report to the ISA how it reached its determination of the value. The basic information required would include the date on which exploitation or production began, the total annual production of the non-living resources and the method which the coastal State used to determine the value or volume of the payments.[54] Although this obligation is not expressly set out in article 82, the argument is that the ILA should be provided with the information required to come to its own assessment as to the appropriateness of the payment. While the final determination is within the coastal State's control, providing this detail would be consistent with a coastal State's obligation to act in good faith.

The lack of detail in article 82 is in stark contrast to the provisions of Part XI and the annexes to the LOSC in relation to deep seabed mining. For example, Annex III elaborates the obligations of States, contractors and the ISA, and defines important terms in connection with mining in the Area. It establishes how prices including market value will be determined. Annex III does not directly apply to article 82 payments, but it is possible that this Annex might be able to be used as a guide for how some aspects of article 82 should be implemented. For example, in relation to the currency of payments made by contractors, article 13(12) requires that payments shall 'be made in freely usable currencies or currencies which are freely available and effectively usable on the major foreign exchange markets'.

(c) Exemption from payments for developing States

Paragraph 3 of article 82 is intended to relieve developing States of the burden of providing payments in situations where they are in a vulnerable economic

---

[51] For example, see ILA Report on Article 82, para. 2.8; Lodge, 'The International Seabed Authority and Article 82', p. 327; ISA, Technical Study No. 4, p. 33. See also Section A(1).
[52] ISA, Technical Study No. 4, p. 33. [53] ILA Report on Article 82, Conclusion 6.
[54] ILA Report on Article 82, Conclusion 7.

situation. As discussed above, the initial proposal to exempt all developing States from the obligation to make payments was rejected in favour of a narrower exemption. Under article 82(3), two criteria must be met. First, the coastal State must be a developing State. However, it is unclear how a 'developing State' should be defined, and by whom.[55] Practice on the definition of developing States varies depending on the context. Under the WTO, States can choose whether to identify themselves as developing States. Other States are able to challenge that State's use of provisions that provide advantages for developing States. The World Bank applies the term 'developing' to countries with low or middle incomes based on gross national income per capita.[56] The United Nations Statistics Division states that there is no established system for the designation of developed and developing countries in the UN system.[57]

More guidance is available in relation to least developed countries (LDCs), which are mentioned in article 82(4). The United Nations Economic and Social Council publishes a list of LDCs every three years. LDCs are determined using three criteria: income based on gross national income per capita; weak human resources based on information including nutritional, health, education and literacy information; and high economic vulnerability based on a range of demographic and economic information.[58]

The difficulty posed by this question is exemplified by Brazil, which has an extended continental shelf capable of hydrocarbon exploration and production.[59] Although it is a net exporter of oil, Brazil is a net importer of gas.[60] Da Silva has observed that, although Brazil has the sixth largest economy in the world, it is eighty-fifth out of 187 on the Human Development Index and is classified as a developing State in the World Trade Organization.[61] Whether it would be a developing State for the purposes of article 82(3) is unclear.

It is also not clear from the terms of article 82 whether the coastal State or the ISA determines whether the State is a developing country for the purposes of paragraph 3. However, it is logical that, just as the coastal State is the final arbiter of the amount of the payment under paragraphs 1 and 2, it should also determine whether it is a developing State under paragraph 3. As with all of its actions under article 82, the coastal State is required to act in good faith. If the ISA published a list of States it considers to be developing, it may be arguable that a State not on that list that claimed developing State status may be acting in bad faith unless there is a

---

[55] Lodge, 'The International Seabed Authority and Article 82', p. 329.
[56] http://data.worldbank.org/about/country-and-lending-groups.
[57] http://unstats.un.org/unsd/methods/m49/m49regin.htm#ftnc.
[58] UN Office of the High Representative for the Least Developed Countries, Landlocked Developing Countries and Small Island Developing States, http://unohrlls.org/about-ldcs/criteria-for-ldcs/.
[59] ISA, '*Non-living Resources of the Continental Shelf Beyond 200 Nautical Miles: Speculations on the Implementation of Article 82 of the United Nations Convention on the Law of the Sea*', Technical Study No. 5 (2010): p. 43.
[60] Alexandre Pereira da Silva, 'Dealing with Articles 76 and 82 of the United Nations Convention on the Law of the Sea: Legal and Political Challenges for Brazil' *Ocean Yearbook* 28 (2014): pp. 145–72, p. 170.
[61] Ibid., pp. 170–1.

reasonable argument in its favour. Therefore, there is a role for the ISA to address this issue.

Secondly, the State must be a 'net importer of a *mineral* resource produced from its continental shelf'. Paragraph 3 refers to 'mineral' resources, in contrast to paragraph 1, which establishes an obligation to make payments in respect of 'non-living' resources. In article 77(4), the natural resources of the shelf are stated to include mineral resources, other non-living resources and sedentary species. This distinction implies that mineral resources are a sub-category of non-living resources and raises the question whether the exemption for developing States in article 82(3) only applies in relation to minerals and not to other non-living resources such as hydrocarbons. If this interpretation is taken, then developing States would always have to make payments in relation to non-mineral, non-living resources (i.e. hydrocarbons) but would be exempt in relation to mineral resources if they are net importers of the mineral. However, this outcome appears unusual in light of the policy behind article 82(3).

There is no explanation for this potential distinction in the records of the negotiations. One commentary has referred to the inconsistency as a 'drafting problem'.[62] Both the ILA Report and the ISA Technical Study No. 4 agree that there appears to be a gap between a strict interpretation of article 82(3) and the purpose of the provision.[63] They therefore conclude that it would be appropriate to interpret article 82(3) as creating an exemption when the State is a net importer of mineral *and* other non-living resources.

This is a reasonable response to the apparent discrepancy created by the strict wording of article 82(3). Article 31(1) of the VCLT requires a treaty to be interpreted in accordance with the ordinary meaning of the terms of the treaty in their context and in the light of its object and purpose. This has been expressed through the principle of effectiveness. According to this principle, when a treaty is open to two interpretations, one of which does and the other does not enable the treaty to have appropriate effects, good faith and the objects and purpose of the treaty demand that the former interpretation should be adopted.[64]

It is worth noting that this principle of effectiveness could also apply to the question of whether coastal States may make payments to the ISA in a combined form, as discussed above. In that case, so long as the ISA receives the total value of the contribution, it would seem contrary to the purpose of the article to insist on either payments or contributions and not both.

(d) Role of the ISA

Article 82(4) provides that the 'payments or contributions shall be made through the Authority, which shall distribute them to States Parties to this Convention,

---

[62] ISA, Technical Study No. 4, p. 36.
[63] ILA Report on Article 82, para. 2.20; ISA, Technical Study No. 4, p. 36.
[64] Richard Gardiner, 'The Vienna Convention Rules on Treaty Interpretation' in Duncan B Hollis (ed), *The Oxford Guide to Treaties* (Oxford: Oxford University Press, 2014): pp. 475–506, p. 496. See also Territorial Dispute *(Libyan Arab Jamahiriya v Chad)* (Judgment) [1994] ICJ Reports 6, pp. 25–6.

on the basis of equitable sharing criteria, taking into account the interests and needs of developing States, particularly the least developed and the land-locked among them'. The Council of the ISA has the responsibility to develop rules, regulations and procedures for the equitable sharing of the payments and contributions received pursuant to article 82.[65] These rules, regulations and procedures will then be considered by the Assembly. If the Assembly does not approve them, they must be sent back to the Council for reconsideration in light of the Assembly's views.[66]

From the coastal State's perspective, the way in which the ISA distributes the payments once it has received them is of little relevance. However, other States will be keenly interested in the approach that the Authority takes to distribution. The issues and options facing the ISA have been canvassed elsewhere[67] and it is not within the scope of this study to explore these in depth. However, it is clear that the ISA will need, in the near future, to resolve the issue of how article 82 will be dealt with. This will require agreement on equitable sharing criteria, and how the Authority will cooperate with coastal States. One proposal has been for a model agreement to be developed by the ISA that would form the basis for a bilateral treaty between a coastal State with obligations under article 82 and the Authority.[68] More recent discussions have focused on developing memoranda of understanding between the parties.[69]

An additional issue for the ISA to determine is how it deals with the costs of receiving payments and contributions and then distributing them to State parties. Nothing in article 82 indicates that the Authority may absorb the costs of the transaction before distributing the proceeds. Therefore, the ISA will have to either fund the process from the regular budget of the Authority or retain an agreed portion of the amounts collected.[70] This would need to be addressed by the members of the ISA.

(4) Options for settling article 82 disputes

It is likely that, unless coastal States and the ISA come to an understanding about the interpretation and implementation of article 82, disputes may arise over the way coastal States apply their obligations under article 82. This could, for example, be in the form of an allegation by the ISA that the coastal State has undervalued the payment or contribution, or incorrectly valued an 'in-kind' calculation. The ISA

---

[65] Article 162(o)(i), LOSC.   [66] Article 160(f)(i), LOSC.
[67] In particular, ISA, Technical Study No. 4, pp. 37–44.
[68] Aldo Chircop, 'Development of Guidelines for the Implementation of Article 82', Working Paper prepared for the International Workshop on Further Consideration of the Implementation of Article 82 of the United Nations Convention on the Law of the Sea, 1982, Beijing, 26–30 November 2012 http://www.isa.org.jm/files/documents/EN/Workshops/2012/AChircop.pdf.
[69] 'Outcomes of the International Workshop on Further Consideration of the Implementation of Article 82 of the United Nations Convention on the Law of the Sea', Report of the Secretary General, 6 May 2013, ISBA/19/A/4, para. 4(d).
[70] Ibid., para. 4(f).

may disagree with the coastal State's assessment of the portion of an oil field that lies under the extended continental shelf as opposed to within 200 nm. An extreme situation would be where a coastal State refuses to make any payments at all in relation to mineral exploitation on its extended shelf. These possibilities raise the question whether the ISA has any ability to require the coastal State to make payment, or whether there is any dispute settlement mechanism available.

First, nothing in article 82 indicates that the ISA has any special power to coerce payment from a coastal State.[71] The article simply refers to the payment being made 'through' the Authority and does not confer a compliance power.

The LOSC established a Seabed Disputes Chamber (SDC) as part of the International Tribunal for the Law of the Sea (ITLOS) with jurisdiction 'with respect to activities in the Area'.[72] The purpose of the SDC was to resolve certain disputes between States, the ISA and contractors.[73] It is unusual in that non-State parties and an intergovernmental organization may be parties to a dispute.[74] Parties to a category of dispute set out in article 187 may submit their dispute to a special chamber of ITLOS or to an ad hoc chamber of the SDC.[75] The SDC may not challenge the discretionary powers of the ISA, but is restricted to considering the contractual obligations of the parties, claims concerning the excess of jurisdiction or misuse of power and claims for damages.[76] It seems unlikely that either the ISA or a coastal State could use article 187 as the basis for a contentious case before the SDC in relation to an article 82 dispute. The primary obstacle is that such a dispute is not 'with respect to activities in the Area'.[77] Nor does article 82 appear in the list of categories of dispute in article 187 over which the SDC has jurisdiction.

The ISA may be able to make use of the SDC's jurisdiction to give advisory opinions. The Assembly or the Council of the ISA may request advisory opinions 'on legal matters arising within the scope of their activities'.[78] There can be no doubt that receipt of the payment or contribution from a coastal State is within the scope of the activities of the ISA. In addition to article 82, the Assembly of the ISA is given power to establish rules on the equitable sharing of the income derived from article 82 on the basis of recommendations from the Council.[79]

Significantly, the jurisdiction to give advisory opinions does not appear to be subject to the same subject-matter limitations as contentious cases. This is expressly set out in article 189, where the limitation in relation to considering rules, regulations and procedures of the ISA are 'without prejudice' to the right to request an advisory opinion. The SDC has only issued one previous advisory opinion, in relation to the obligations resulting from sponsorship of operators conducting mining

---

[71] ILA Report on Article 82, para. 4.1.   [72] Article 187, LOSC.
[73] AO Adede, *The Systems for Settlement of Disputes under the United Nations Convention on the Law of the Sea* (Leiden: Martinus Nijhoff, 1987): p. 246.
[74] Alan Boyle, 'Dispute Settlement and the Law of the Sea Convention: Problems of Fragmentation and Jurisdiction' *International and Comparative Law Quarterly* 46 (1997): pp. 37–54, p. 39.
[75] Article 188, LOSC.   [76] Article 189, LOSC.
[77] See ILA Report on Article 82, para. 4.4; ISA, Technical Study No. 4, p. 64.
[78] Article 190, LOSC.   [79] Article 160(2)(f)(i) and 162(2)(o)(i), LOSC.

in the Area.⁸⁰ The disadvantage of advisory opinions, of course, is that they are not binding on State parties. One possible objection to the ISA requesting such an advisory opinion would be that it is an attempt to bring a contentious case under the guise of an advisory opinion. However, a State party that disregarded the legal view of the SDC expressed in an advisory opinion would face international condemnation.

Another option for dispute settlement would be for a potential beneficiary State to bring a contentious case against a coastal State under the general dispute settlement provisions under Part XV of the LOSC. Part XV establishes a process that allows for disputes to be settled peacefully by States followed by the possibility of compulsory procedures.⁸¹ Such disputes must be 'concerning the interpretation or application' of the LOSC.⁸² Article 297(1) stipulates that disputes about the exercise by a coastal State of its sovereign rights may be subject to compulsory jurisdiction when it is alleged that a coastal State has acted illegally 'in regard to the freedoms and rights of navigation, overflight or the laying of submarine cables and pipelines, or in regard to other internationally lawful uses of the sea specified in article 58'. The exercise of the coastal State's obligations under article 82 do not appear to fall into that category. However, the fact that a dispute does not fall into article 297(1) does not prevent compulsory procedures from being available.⁸³ The article was not intended to create a broad exclusion of the application of the dispute settlement provisions.⁸⁴ Therefore, it is arguable that compulsory procedures would be available for a dispute about article 82. In addition, the State would need to establish *locus standi* for the judicial body to hear the case.⁸⁵ However, the fact that a State was a potential beneficiary under article 82(4) could be sufficient to prove standing.

Any dispute arising among parties to the ISA as to the manner in which the ISA has chosen to distribute the proceeds of article 82 payments would have to be resolved internally within the ISA. Although article 187 allows for disputes between the ISA and State parties to be resolved by the SDC of the ITLOS, its jurisdiction is limited. In particular, the SDC cannot substitute its decision for that of the ISA and may not rule on the conformity of rules, regulations and procedures with the LOSC.⁸⁶

(5) Cross-boundary issues

Complications arise in a number of possible situations when article 82 needs to be applied where oil or gas fields lie under two or more juridical boundaries. First, a single hydrocarbon field may extend across the 200 nm 'boundary' between the

---

⁸⁰ Responsibilities and Obligations of States Sponsoring Persons and Entities with Respect to Activities in the Area (Advisory Opinion) (2011) 50 ILM 458.

⁸¹ For the history of the development of Part XV, see AO Adede, *The Systems for Settlement of Disputes under the United Nations Convention on the Law of the Sea*. See also Natalie Klein, *Dispute Settlement in the UN Convention on the Law of the Sea* (Cambridge: Cambridge University Press, 2005).

⁸² Articles 279 and 286, LOSC.   ⁸³ Klein, *Dispute Settlement*, p. 142.   ⁸⁴ Ibid.

⁸⁵ RR Churchill and AV Lowe, *The Law of the Sea*, 3rd edn (Manchester: Manchester University Press, 1999): p. 460.

⁸⁶ Article 189, LOSC.

inner and extended continental shelf. Secondly, the field may be located in the continental shelf of two or more States. Thirdly, and less likely owing to geomorphological characteristics, the field may lie beneath a coastal State's continental shelf and the Area, which is governed by the ISA under Part XI of the LOSC. The following discussion explores the legal issues raised by each situation. The discussion is based on an assumption that a boundary or clear demarcation is in place—a short discussion on disputed areas is included later in the analysis.

(a) Fields located across the 200 nm line on the continental shelf

It is likely that there will be some hydrocarbon fields that straddle the demarcation line between the 200 nm EEZ and the extended continental shelf within the jurisdiction of a single State. The practical difficulty with this situation is that the resource located on the continental shelf beyond 200 nm engages the State's responsibility to make payments under article 82 but the resource located within 200 nm does not. Most hydrocarbons are exploited by a well in one location, which is capable of extracting the resource located in an entire field. If the State situates a well within 200 nm, which extracts a resource that was originally located beyond 200 nm, what are its obligations under article 82?

The ISA Technical Study No. 4 considered this problem, and concluded that the obligation to make payments exists in relation to the resource that was situated under the extended continental shelf, even though it is extracted from within 200 nm.[87] This must be correct, even though the practicalities of calculating such payment may be difficult. It might also be difficult for the ISA to verify the existence of a transboundary field.

(b) Transboundary resources that cross the maritime boundaries of two or more States

Cross-boundary hydrocarbon resources can cause a range of legal and political difficulties, even assuming that a maritime boundary is in place. Oil and gas fields are highly pressurized and, as extraction takes place, the hydrocarbons 'migrate' towards the exploitation point as a result of changes in the pressure. Oil and gas are therefore sometimes referred to as migratory resources. This means that an operator can extract resources from a field that is located under one or more boundaries without crossing those boundaries. The objection is that, in that case, the State has appropriated resources that more properly belong to the neighbour because they were originally under the neighbour's land.[88] It also may be the case that one State's

---

[87] ISA, Technical Study No. 4, p. 59.
[88] In the *North Sea Continental Shelf* cases, the ICJ observed that common deposits often lie across boundaries 'and since it is possible to exploit such a deposit from either side, a problem immediately arises on account of the risk of prejudicial or wasteful exploitation by one or other of the states concerned'. North Sea Continental Shelf *(Federal Republic of Germany v Denmark and the Netherlands)* [1969] ICJ Reports 3, para. 97.

extraction of the portion originally under its shelf may change the conditions in the whole field to make it more difficult, or impossible, for the neighbour to extract its portion of the deposit.[89]

It is contrary to good practice for the neighbour to begin drilling into the common field from the other side of the boundary. The existence of a second or subsequent exploitation point affects the pressure in the field and means that, overall, the level of recoverable resource drops. In addition, hydrocarbon exploitation often requires the reinjection of some of the resource to keep the field adequately pressurized.[90] An operator may be reluctant to reinject the resource into the field if it might subsequently be recovered by the neighbour.[91] Therefore, it is to no actor's benefit for multiple points of exploitation to be established in the same area.

In many countries, the response has been to 'unitize' the common field. This involves appointing one operator to manage the common deposit on behalf of all parties.[92] The purpose is to exploit the resource rationally, so as to protect the rights of owners of the resource and avoid disputes. It is a concept that emerged in the context of oil development in the United States, but is now used in other States.[93] It is also becoming increasingly common for unitization to be a feature of bilateral treaties, where the parties agree to develop a transboundary deposit as a single unit.[94] The treaties vary in the level of detail but may deal with matters such as the appropriate share of the proceeds of exploitation of the resources, the method by which exploitation will be undertaken, how taxation will be dealt with by each party, the establishment of institutional mechanisms, and dispute settlement.[95] The treaty may deal with an identified resource or set out a process by which future discoveries will be managed.

---

[89] Rainer Lagoni, 'Oil and Gas Deposits Across National Frontiers' *American Journal of International Law* 73 (1979): pp. 215–43, p. 217.

[90] Other substances such as water or carbon dioxide may also be used. Bernard Taverne, *Petroleum, Industry and Governments: A Study of the Involvement of Industry and Governments in the Production and Use of Petroleum*, 2nd edn (The Hague: Wolters Kluwer, 2008): p. 10.

[91] Vasco Becker-Weinberg, *Joint Development of Hydrocarbon Deposits in the Law of the Sea* (Heidelberg: Springer, 2014): p. 9.

[92] Lagoni, 'Oil and Gas Deposits Across National Frontiers', p. 223. Unitization can occur within one country where the field crosses boundaries, as well as at an international level.

[93] See, generally, Becker-Weinberg, *Joint Development of Hydrocarbon Deposits*, ch 2; David Asmus and Jacqueline Lang Weaver, 'Unitizing Oil and Gas Fields Around the World: A Comparative Analysis of National Laws and Private Contracts' *Houston Journal of International Law* 28(3) (2006): pp. 1–197.

[94] Ian Townsend-Gault, 'Rationales for Zones of Cooperation' in R Beckman, CH Schofield, I Townsend-Gault, T Davenport and L Bernard (eds), *Beyond Territorial Disputes in the South China Sea: Legal Frameworks for the Joint Development of Hydrocarbon Resources* (Cheltenham: Edward Elgar Publishers, 2013): pp. 114–38, p. 125; Ana E Bastida et al., 'Cross-Border Unitization and Joint Development Agreements: An International Law Perspective' *Houston Journal of International Law* 29 (2006–2007): pp. 355–422.

[95] For a discussion of such agreements, see David M Ong, 'Joint Development of Common Offshore Oil and Gas Deposits: "Mere" State Practice or Customary International Law?' *American Journal of International Law* 93 (1999): pp. 771–804; Becker-Weinberg, *Joint Development of Hydrocarbon Deposits*; Townsend-Gault, 'Rationales for Zones of Cooperation'; Asmus and Weaver, 'Unitizing Oil and Gas Fields Around the World'; and Lagoni, 'Oil and Gas Deposits Across National Frontiers'. Examples of unitization treaties include: Agreement between United Kingdom of Great Britain and Northern Ireland and Norway Relating to the Exploitation of the Frigg Field Reservoir and the Transmission of Gas therefrom to the United Kingdom (10 May 1976) 1978 UNTS 4; Agreement between the Government of Australia and the Government of the Democratic Republic of Timor-Leste

Mining and the Article 82 Obligation to Make Payments   141

Another common response is 'joint development' of transboundary resources. There is no uniform understanding or definition of joint development.[96] Joint development treaties include unitization agreements but may cover a wider range of arrangements and subject matters.[97] Becker-Weinberg suggests that joint development agreements have a broader scope and regulate matters regarding hydrocarbon resources, but also matters such as the protection and preservation of the marine environment or the safety of navigation.[98] Such agreements are increasingly being used where States wish to exploit resources found in an area of overlapping claims to jurisdiction.[99]

A feature of delimitation agreements is to include a provision for dealing with transboundary resources that are discovered in the future. The oft-cited example is article 4 of the 1964 Agreement between the United Kingdom and Norway.[100] The article provides that, if a common field is discovered, the two States must seek agreement as to the manner in which the reservoir should be exploited and the manner in which the proceeds will be apportioned. This article has been repeated in many other delimitation agreements.[101]

A State that identifies a common hydrocarbon field that crosses a boundary is under some important obligations under international law. These apply to the continental shelf as a whole, and not just to the extended continental shelf.

First, because a State cannot exploit a common field without affecting the neighbour's share of the resource, there is an obligation to exercise mutual restraint. If a State drilled into a common hydrocarbon field and extracted the resource, this would negatively impact on the ability of the other State to retrieve its share of the resource, and also possibly its share of revenue from the operation. Therefore, a number of commentators have agreed that such States sharing transboundary hydrocarbons are required to refrain from unilateral exploitation of the resource.[102] However, some activities, such as seismic surveying, which do not cause irreparable harm, may be permitted.[103]

---

relating to the Unitization of the Sunrise and Troubadour Fields (6 March 2003) 2483 UNTS 317; Agreement between the United States of America and the United Mexican States Concerning Transboundary Hydrocarbon Reservoirs in the Gulf of Mexico (20 February 2012) TIAS 14-0718. The latter agreement relates to areas including the continental shelf of each party beyond 200 nm.

[96] Ong, 'Joint Development of Common Offshore Oil and Gas Deposits', p. 772.
[97] Masahiro Miyoshi, 'The Basic Concept of Joint Development of Hydrocarbon Resources on the Continental Shelf' *International Journal of Estuarine and Coastal Law* 3 (1988): pp. 1–18, p. 5.
[98] Becker-Weinberg, *Joint Development of Hydrocarbon Deposits*, p. 20.
[99] Ong, 'Joint Development of Common Offshore Oil and Gas Deposits', p. 773; Clive Schofield, 'Blurring the Lines? Maritime Joint Development and the Cooperative Management of Ocean Resources' Issues in Legal Scholarship, *Berkeley Electronic Press* 8(1) (Frontier Issues in Ocean Law: Marine Resources, Maritime Boundaries, and the Law of the Sea), Article No. 3.
[100] Agreement between the Government of the United Kingdom of Great Britain and Northern Ireland and the Government of the Kingdom of Norway relating to the Delimitation of the Continental Shelf between the Two Countries (10 March 1965) 551 UNTS 214. See Lagoni, 'Oil and Gas Deposits Across National Frontiers', pp. 229–30.
[101] Lagoni, 'Oil and Gas Deposits Across National Frontiers', p. 229.
[102] Miyoshi, 'Joint Development of Hydrocarbon Resources on the Continental Shelf', p. 10; Becker-Weinberg, *Joint Development of Hydrocarbon Deposits*, p. 73.
[103] Tara Davenport, 'The Exploration and Exploitation of Hydrocarbon Resources in Areas of Overlapping Claims' in R Beckman, CH Schofield, I Townsend-Gault, T Davenport and L Bernard

Secondly, such States are under a duty to cooperate with the neighbouring State, and to negotiate in good faith in order to seek agreement on how to deal with transboundary resources.[104] The obligation to negotiate in good faith is an accepted principle of customary international law.[105] However, in the absence of agreement, it is arguable that the principle of mutual restraint will prevail because of the potential harm to another State's exclusive sovereign rights if unilateral exploitation takes place.[106]

Although States are under an obligation to cooperate, they are not under a customary international law obligation to create a joint development project.[107] There are many models of cooperation and despite the increasing prevalence of joint development agreements, there is no indication that States have entered into these agreements out of a sense of legal obligation. There are also a number of different types of agreement that have been concluded meaning that State practice is not consistent as to the form of agreement.[108]

In respect of the continental shelf beyond 200 nm, any unitization or joint development agreements in relation to the development of reserves beneath the extended continental shelf should address the issue of how article 82 will apply to those agreements.

An example of an arrangement that applies to the continental shelf beyond 200 nm is the treaty between the United States and Mexico. The United States and Mexico concluded an agreement for management of transboundary hydrocarbon reservoirs in the Gulf of Mexico in 2012 (2012 Agreement).[109] The basis for this agreement was a clause in the 2000 treaty that delimited the continental shelf beyond 200 nm in the Western Gulf of Mexico, requiring the parties to reach agreement on the efficient and equitable exploitation of transboundary resources.[110] Therefore, this agreement is the first relating to transboundary hydrocarbon resources on the extended continental shelf, although the agreement also applies to

(eds), *Beyond Territorial Disputes in the South China Sea: Legal Frameworks for the Joint Development of Hydrocarbon Resources* (Cheltenham: Edward Elgar Publishers, 2013): pp. 93–113.

[104] Miyoshi, 'Joint Development of Hydrocarbon Resources on the Continental Shelf', p. 12.

[105] For example, see Lac Lanoux Arbitration *(France v Spain)* 12 RIAA 306; 24 ILR 101 (1958); *North Sea Continental Shelf* cases; Fisheries Jurisdiction *(United Kingdom v Iceland)* (Jurisdiction) [1974] ICJ Reports 3.

[106] Miyoshi, 'Joint Development of Hydrocarbon Resources on the Continental Shelf', p. 14.

[107] Ibid., pp. 9–10. For an opposing view, see David Ong, 'The New Timor Sea Arrangement 2001: Is Joint Development of Common Offshore Oil and Gas Deposits Mandated Under International Law?' *International Journal of Marine and Coastal Law* 17 (2002): pp. 79–122.

[108] See Lagoni, 'Oil and Gas Deposits Across National Frontiers'; Becker-Weinberg, *Joint Development of Hydrocarbon Deposits*; Karla Urdaneta, 'Transboundary Petroleum Reservoirs: A Recommended Approach for the United States and Mexico in the Deepwaters of the Gulf of Mexico' *Houston Journal of International Law* 32 (2010): pp. 333–91, p. 378.

[109] Agreement between the United States of America and the United Mexican States Concerning Transboundary Hydrocarbon Reservoirs in the Gulf of Mexico (20 February 2012) TIAS 14-0718 (2012 US–Mexico Agreement).

[110] Treaty between the Government of the United States of America and the Government of the United Mexican States on the Delimitation of the Continental Shelf in the Western Gulf of Mexico beyond 200 Nautical Miles (9 June 2000) 2143 UNTS 417, article 5(1)(b) (2000 US–Mexico Treaty).

the continental shelf boundaries within 200 nm. It appears that one of the motivating factors for Mexico to seek an agreement with the United States was a fear that, without an agreement, oil companies would drill into transboundary reserves and appropriate Mexico's share of the common field.[111] However, had the United States permitted such activities, it would have been in breach of the principles of mutual restraint and cooperation, as set out above.[112]

One feature of the 2000 US–Mexico Treaty was that the parties were prohibited from authorizing drilling or exploitation of the continental shelf within 1.4 nm of the boundary for ten years from the date of the treaty.[113] This was to ensure that no transboundary resources were exploited by either party until an agreement could be reached on how such resources should be dealt with. This 'buffer zone' is a feature that is often found in such agreements.[114] The moratorium was overridden by the 2012 Agreement.[115]

The 2012 Agreement envisages that transboundary resources will be exploited pursuant to a unitization agreement.[116] The Agreement also makes provision for management in cases where a unitization agreement cannot be agreed upon.[117] A Joint Commission is established to attempt to resolve any disputes or differences arising from the Agreement.[118] However, in the case that the Joint Commission cannot resolve such disputes, the Agreement includes a system for settlement of disputes.[119] Finally, the Agreement provides for inspections and environmental and safety standards.[120]

It is noteworthy that the 2012 Agreement makes no reference to the respective States' obligations under article 82. Arguably, because the Agreement covered an area of the extended continental shelf, it should establish the responsibilities of each State for payments under article 82.[121] There may be several reasons for this omission. First, the United States is not yet a party to the LOSC and may not have wished article 82 to be included. It would not want any suggestion that the obligation under article 82 has entered customary international law, and inclusion of the obligation might have been used by others as evidence of State practice. Secondly, it is possible that the parties did not consider the obligation under article 82 as relevant to the operation of the Agreement. This second argument is weaker when one considers that the Agreement stipulates how taxation arising from the income from

---

[111] Miriam Grunstein, 'Unitized we Stand, Divided we Fall: A Mexican Response to Karla Urdaneta's Analysis of Transboundary Petroleum Reservoirs in the Deep Waters of the Gulf of Mexico' *Houston Journal of International Law* 33 (2011): pp. 345–67, p. 354.
[112] See Jorge A Vargas, 'The 2012 US-Mexico Agreement on Transboundary Hydrocarbon Reservoirs in the Gulf of Mexico: A Blueprint for Progress or a Recipe for Conflict?' *San Diego International Law Journal* 14 (2012): pp. 3–70, p. 61.
[113] Article 4, 2000 US–Mexico Treaty.
[114] See Lagoni, 'Oil and Gas Deposits Across National Frontiers', p. 217.
[115] Article 24, 2012 US–Mexico Agreement.   [116] Ibid., article 6.
[117] Ibid., articles 7, 8 and 9.   [118] Ibid., article 14.   [119] Ibid., ch 5.
[120] Ibid., articles 18–19.
[121] Becker-Weinberg, *Joint Development of Hydrocarbon Deposits*, p. 131. Becker-Weinberg notes that an agreement between Mauritius and the Seychelles applicable to the shelf beyond 200 nm also does not contain reference to article 82.

exploitation will be dealt with.[122] It would therefore have been possible to deal with article 82 in the Agreement.

In the future, agreements that include areas of the extended continental shelf would benefit from consideration of article 82. The obligation in respect of article 82 is not excused in cases of joint development.[123] Becker-Weinberg suggests that a number of issues may arise if article 82 obligations are not dealt with in joint development agreements. These include questions of liability if payments are not made, and responsibility if one or both States are developing States and exempt from making payments.[124] It may be that the parties simply state that each State is responsible for payments based on its share of the income/production. However, given that unitization is often used to develop these common fields, it might be efficient to calculate the payments as a whole. This would be complicated if one of the parties is a developing State which is exempt from making payments under article 82(3). It may be necessary for the States to reach agreement on how production is calculated and regarding other ambiguities already discussed above.

(c) Transboundary fields where there are disputed claims

The potential for a dispute to arise among States over maritime boundaries is exacerbated when there are proven or suspected hydrocarbon resources in the vicinity.[125] Indeed, States are increasingly focused on mineral and hydrocarbon resources during the negotiation of delimitation agreements, sometimes to the exclusion of other interests.[126] States may therefore be inclined to seek the maximum possible maritime area, leading to overlapping claims.[127]

The LOSC reinforces the general obligation for States to cooperate in respect of transboundary resources, particularly when there are overlapping claims. Articles 74(3) and 83(3) state that, pending delimitation of EEZs or continental shelves, 'the States concerned, in a spirit of understanding and cooperation, shall make every effort to enter into provisional arrangements of a practical nature and, during this transitional period, not to jeopardize or hamper the reaching of the final agreement'. This would apply equally to the delimitation of the extended continental shelf. The articles do not stipulate the form of any such agreement. A proposal that unilateral measures in respect of the disputed area be expressly prohibited was rejected during the negotiations. Rather, the obligation is to 'exercise mutual

---

[122] Article 13, 2012 US–Mexico Agreement.
[123] Becker-Weinberg, *Joint Development of Hydrocarbon Deposits*, p. 132.
[124] Ibid., pp. 131–2.
[125] Ong, 'Joint Development of Common Offshore Oil and Gas Deposits'; Clive Schofield, 'The El Dorado Effect: Reappraising the "Oil Factor" in Maritime Boundary Disputes' in Clive Schofield, Seokwoo Lee and Moon-Dang Kwon (eds), *The Limits of Maritime Jurisdiction* (Leiden: Martinus Nijhoff, 2014): pp. 111–26.
[126] John W Donaldson, 'Oil and Water: Assessing the Link between Maritime Boundary Delimitation and Hydrocarbon Resources' in Clive Schofield, Seokwoo Lee and Moon-Dang Kwon (eds), *The Limits of Maritime Jurisdiction* (Leiden: Martinus Nijhoff, 2014): pp. 127–43, p. 143.
[127] Ong, 'Joint Development of Common Offshore Oil and Gas Deposits', p. 774; Davenport, 'The Exploration and Exploitation of Hydrocarbon Resources in Areas of Overlapping Claims', p. 99.

restraint in a difficult situation'.[128] In 1984, Lagoni suggested that articles 74(3) and 83(3) represented a negotiated deal and did not reflect customary international law, although he also suggested that the articles could be the basis for an emerging customary rule.[129] His conclusion was based on the fact that the articles represented progressive development within the context of rules on delimitation in the LOSC that did not in themselves go beyond existing customary international law.[130]

These articles were considered by an arbitral tribunal established to hear a dispute between Guyana and Suriname.[131] In that case, the parties had not entered into a provisional arrangement. Guyana authorized exploratory drilling, and Suriname responded by using threats of force against the operators to force them to leave the area of overlapping claims. The Tribunal found that the obligations in articles 74(3) and 83(3) entailed a duty to negotiate in good faith, using a conciliatory approach.[132] Suriname had failed to enter into negotiations in good faith while Guyana had failed to notify Suriname of the planned exploratory drilling.[133]

The obligation not to jeopardize or hamper the final agreement meant that States could not undertake unilateral activities 'having a permanent physical impact on the marine environment'.[134] The Tribunal based its conclusion in part on the *Aegean Sea Continental Shelf* case in which the ICJ found that seismic exploration did not risk permanent physical damage to the seabed and therefore it refused to issue interim orders preventing the activity.[135] The Tribunal found that there is a substantive legal difference between the oil exploration activities of seismic testing and exploratory drilling.[136] It held that unilateral exploratory drilling in a disputed area may jeopardize or hamper the reaching of a final agreement because it could result in a perceived change to the status quo.[137]

It is noteworthy that this conclusion differs slightly from the duty to exercise mutual restraint in the case of a boundary. Articles 74(3) and 83(3) do not apply where a boundary has been resolved between the parties. If there is no applicable treaty setting out methods for exploiting transboundary resources, the principle of mutual restraint would prevent one State from taking actions that would affect the resources of the other State. This is the primary obligation. Some activities that have a permanent physical impact on the marine environment could be permitted so long as they do not negatively affect the joint resource. The difference is explicable by the recognition in articles 74(3) and 83(3) that until a claim is resolved, both States have equally valid claims to the hydrocarbon resource as well as obligations

---

[128] Rainer Lagoni, 'Interim Measures Pending Maritime Delimitation Agreements' *American Journal of International Law* 78 (1984): pp. 345–68, p. 362.
[129] Ibid., pp. 354, 367.   [130] Ibid., p. 368.
[131] *Guyana v Suriname* (Arbitration) 139 ILR 566.   [132] Ibid., para. 461.
[133] Ibid., paras. 476 and 478.   [134] Ibid., para. 470.
[135] Aegean Sea Continental Shelf *(Greece v Turkey)* (Interim Protection) [1976] ICJ Reports 3. Lagoni adopted a similar, albeit not identical, interpretation of this case, distinguishing between activities of a 'transitory character, such as seismic investigation, and those which are permanent and employ stationary means'. Lagoni, 'Interim Measures', p. 366. However, both interpretations are based on the principle that one State may not jeopardize the final agreement.
[136] *Guyana v Suriname*, para. 479.   [137] Ibid., para. 480.

in relation to the marine environment. Therefore, where boundaries are not delimited, the LOSC imposes additional obligations on States to refrain from certain unilateral activities.

Provisional joint development arrangements have been reached in a number of situations where overlapping claims exist. Examples include agreements between Malaysia and Thailand,[138] and Australia and Timor-Leste.[139] Such agreements allow for utilization of a hydrocarbon resource pending delimitation, but will not always be appropriate in every situation, because it will depend on the existence of a conducive political and legal situation.[140]

(d) Fields lying under the continental shelf and the Area

Parties to the LOSC anticipated that there might be hydrocarbon or other mineral deposits that cross the boundary between the Area and the outer limit of a coastal State's continental shelf. Article 142 of the LOSC requires activities in the Area to be conducted with due regard to the rights and legitimate interests of any coastal State across whose jurisdiction the deposit lies. The ISA is required to establish consultations with the coastal State, but the obligation goes beyond the customary international law requirement to cooperate. The ISA is not permitted to undertake an activity that may result in the exploitation of resources lying within national jurisdiction without the consent of the coastal State.[141]

The obligation to obtain consent from the coastal State is clearly aimed at exploitation of hydrocarbon resources, and reflects the general approach to transboundary hydrocarbon resources where States must exercise mutual restraint and not harm the interests of the other State. Article 142 does not impose an obligation on a coastal State to consult with the ISA and seek agreement in this situation. However, Chircop has argued that customary international law obligations in the case of transboundary resources, such as cooperation, would apply between the

---

[138] Memorandum of Understanding between the Kingdom of Thailand and Malaysia on the Establishment of the Joint Authority for the Exploitation of the Resources of the Seabed in a Defined Area of the Continental Shelf of the Two Countries in the Gulf of Thailand, 21 February 1979 http://cil.nus.edu.sg/rp/il/pdf/1979%20MOU%20between%20Malaysia%20and%20Thailand-pdf.pdf. See Becker-Weinberg, *Joint Development of Hydrocarbon Deposits*.

[139] Timor Sea Treaty between the Government of East Timor and the Government of Australia (20 May 2002) 2258 UNTS 3; Treaty between Australia and the Government of Democratic Republic of Timor-Leste on Certain Maritime Arrangements in the Timor Sea (12 January 2006) 2483 UNTS 359. See Clive Schofield, 'Minding the Gap: The Australia-East Timor Treaty on Certain Maritime Arrangements in the Timor Sea' *International Journal of Marine and Coastal Law* 22 (2007): pp. 189–234; Stuart Kaye, 'Joint Development in the Timor Sea' in R Beckman, CH Schofield, I Townsend-Gault, T Davenport and L Bernard (eds), *Beyond Territorial Disputes in the South China Sea: Legal Frameworks for the Joint Development of Hydrocarbon Resources* (Cheltenham: Edward Elgar Publishers, 2013): pp. 249–67; Becker-Weinberg, *Joint Development of Hydrocarbon Deposits*.

[140] Clive Schofield, 'No Panacea? Challenges in the Application of Provisional Arrangements of a Practical Nature' in Myron H Nordquist and John Norton Moore (eds), *Maritime Border Diplomacy* (Leiden: Martinus Nijhoff, 2012): pp. 151–69.

[141] Article 142(2), LOSC.

State and the ISA.[142] Arguably, the principle of mutual restraint would also apply even though it is not a State that would be harmed by exploitation, but the ISA. It should be borne in mind that the ISA has international legal personality and is acting for the benefit of all mankind.[143]

It is conceivable that the ISA and the coastal State could enter into a unitization agreement for the exploitation of the transboundary resource in a similar manner to those concluded between States.[144] Such an agreement would need to set out how the article 82 obligations would be applied, if the resource straddles the Area and an extended continental shelf.

The situation is much simpler in relation to mineral resources, such as manganese nodules, that are not mobile when extracted. The obligation on the ISA to consult and give prior notification still exists, but the ISA will be able to exploit resources in the Area without the coastal State's consent, so long as it does not infringe on the sovereign rights of the coastal State.

In a situation where the outer limits of the coastal State have not been concluded according to article 76 of the LOSC, both coastal States and the ISA must be particularly careful not to potentially exploit resources that may ultimately be found to belong to the other side. For example, a coastal State that exploited mineral resources later found to be in the Area would be responsible for a breach of its international obligations under the LOSC.

B. Installations and Structures on the Extended Continental Shelf

Article 60, which sets out coastal States' rights and responsibilities in relation to artificial islands, installations and structures (together, structures), is based on article 5 of the 1958 Geneva Convention on the Continental Shelf. That Convention established many of the important features of article 60, including the right to construct installations, establish safety zones of up to 500 metres and respect for navigation in high-use areas.

The jurisdictional principles surrounding structures are clearly drawn in article 60 of the LOSC. This has not prevented some debate about the precise content of the right in article 60.[145]

---

[142] Aldo Chircop, 'Managing Adjacency: Some Legal Aspects of the Relationship Between the Extended Continental Shelf and the International Seabed Area' *Ocean Development and International Law* 42 (2011): pp. 307–16, p. 313. See also ISA, Technical Study No. 4, p. 62.
[143] Articles 140 and 176, LOSC. See also Reparation for Injuries Suffered in the Service of the United Nations, Advisory Opinion [1949] ICJ Reports 174.
[144] The ISA would have power to do this because it has such international personality as is necessary for the exercise of its functions and the fulfilment of its purposes: see article 176, LOSC.
[145] For example, the coastal State's right to respond to protests: see Alex G Oude Elferink, 'The *Arctic Sunrise* Incident: A Multi-faceted Law of the Sea Case with a Human Rights Dimension' *International Journal of Marine and Coastal Law* 29 (2014): pp. 244–89. Sebastian tho Pesch has argued that the coastal State's rights in the safety zone must be read restrictively: 'Coastal State Jurisdiction Around

Article 80 stipulates that article 60 applies to the continental shelf 'mutatis mutandis'.[146] There are two questions that arise in relation to the extended shelf. First, are any necessary changes required as a result of placing structures on the extended shelf? This question can generally be answered in the negative. Structures within 200 nm of the coast primarily interfere with fishing (under the State's jurisdiction) and navigation (a freedom of the high seas) and there are rules that are aimed at minimizing the interference to navigation. Beyond 200 nm there is the potential to interfere with the freedom of fishing on the high seas, but there is no indication that the negotiators of the LOSC intended this fact to limit the coastal State's right to authorize the construction of a structure for the purposes of exploiting the continental shelf resources.

A far more difficult question arises in respect of activities that explore or exploit the resources of the continental shelf in a way other than using a traditional structure, or where the sea floor infrastructure extends beyond 500 metres from the platform. The hydrocarbon industry is now making use of 'subsea developments'. In these structures, individual wells drilled into small reservoirs are established on the sea floor and pipelines run from the wells to a centralized platform.[147] This type of structure is useful in deep-water locations as it minimizes the number of platforms needed. It also makes the exploitation of smaller reservoirs more economic.[148] The problem for coastal States is that the footprint of the sea floor infrastructure (such as well heads and pipelines) will usually extend well beyond the 500 metre safety zone that is established around the platform on the surface. Pipelines can be damaged by anchors dropping or hooking them. Bottom trawling can damage the pipelines when the trawl gear hits the pipes.[149] Some protection can be afforded by ensuring that the pipelines are trenched or buried, but this may not always be possible.[150] Coating the pipelines with concrete can also assist in protecting the pipelines from impacts. However, the equipment above each well cannot be buried. In some cases, the subsea installations will be sufficiently deep that fishing and anchoring may not be practical at all.

In many States, pipelines in the EEZ are protected by bans on anchoring and fishing in the area around the pipelines and platform.[151] There is an obvious question whether coastal States can impose similar restrictions for pipelines leading to

---

Installations: Safety Zones in the Law of the Sea' *International Journal of Marine and Coastal Law* 30 (2015): pp. 512–32.

[146] Meaning: with the necessary changes.

[147] See, e.g. the Independence Hub, Gulf of Mexico http://www.offshore-technology.com/projects/independence/.

[148] Amin Amin et al., 'Subsea Development from Pore to Process' *Oilfield Review* 17 (1999): pp. 4–17, p. 6.

[149] SJ de Groot, 'The Impact of Laying and Maintenance of Offshore Pipelines on the Marine Environmental and the North Sea Fisheries' *Ocean Management* 8 (1982): pp. 1–27, p. 10.

[150] AA Ryder and SC Rapson, 'Pipeline Technology' in Stefan T Orszulik (ed), *Environmental Technology in the Oil Industry* (Berlin: Springer, 2008): pp. 229–80, p. 259.

[151] Geir Ulfstein 'The Conflict between Petroleum Production, Navigation and Fisheries in International Law' *Ocean Development and International Law* 19 (1988): pp. 229–62, pp. 233 and 246.

or from subsea installations outside the safety zone on the continental shelf beyond 200 nm. This would amount to an interference with navigation and fisheries on the high seas above the shelf. Nothing in article 60 directly addresses that point. The safety zones are to be measured from the edge of the structures and must not exceed 500 metres 'except as authorised by generally accepted international standards or as recommended by the competent international organisation'.[152] The International Maritime Organization (IMO) has urged States to consider establishing fairways or routeing systems through exploration areas. It has also emphasized the need to disseminate information about activities connected to exploration and exploitation, including the location of pipelines.[153] One option for coastal States is to establish precautionary areas or areas to be avoided through the IMO.

A particular problem might arise when bottom fishing activity could risk damage to pipelines or other equipment that is part of the subsea installation. Within 200 nm, coastal States could prohibit bottom trawling in the vicinity of the installation without any legal difficulty. However, when the fishing is being conducted pursuant to high seas rights, it is far less clear whether the coastal State could require (as opposed to request) foreign vessels to refrain from bottom trawling in the vicinity beyond any safety zone that might be established. Any such measure would need to fulfil the requirement that it not infringe or unjustifiably interfere with high seas rights and freedoms as per article 78(2). The approach for evaluating this standard is discussed in Chapter 7. The obvious compromise is to designate areas in which advice is issued to mariners to either avoid the area or take particular care around infrastructure. One example is the Offshore Development Areas used by the United Kingdom.[154]

Some States are now exercising a right to create a safety zone around exploration vessels and not just installations. As discussed in Chapter 7, some governments have passed legislation restricting navigation in the vicinity of vessels conducting exploration or exploitation activities for the oil and gas industry. New Zealand prohibits entry of some vessels within 500 metres of a seismic survey vessel conducting a survey.[155]

C. Carbon Dioxide Sequestration

The right to sequester carbon dioxide in the seabed of the continental shelf is included in the sovereign rights of the coastal State. A coastal State has sovereign rights over the natural resources of the seabed and subsoil of the continental

---

[152] Article 60(5), LOSC.
[153] International Maritime Organization (IMO) Resolution A. 671(16) 'Safety Zones and Safety of Navigation around Offshore Installations and Structures'.
[154] http://www.hse.gov.uk/offshore/notices/on_42.htm.
[155] Sections 101B and 101C, Crown Minerals Act 1991 (New Zealand). See also Rule 43(a), Collision Regulations C.R.C., 1978, c. 1416 (Canada); Sections 1 and 8B, Marine Traffic Act No. 2 of 1981 (South Africa).

shelf.[156] Admittedly, the drafters of the LOSC would have considered non-living resources as those substances present in the shelf, and the right to deposit substances in the subsoil would not strictly fit with that interpretation. However, as Scott suggests, geological storage capacity would be considered an asset or resource in the present day.[157] State practice demonstrates that coastal States consider the right to sequester carbon in their continental shelf to be within their control.[158] Sovereign rights over the continental shelf include exclusive rights over the construction of installations and structures on the shelf, as well as any drilling into the shelf.[159] Therefore, the coastal State may regulate the physical activity of injecting the carbon into the sea floor.

One question that may arise is whether article 82 applies to any carbon sequestration. The relevant requirement is that the coastal State 'shall make payments or contributions in kind in respect of the exploitation of the non-living resources of the continental shelf beyond 200 nautical miles'. If, as explained above, the capacity for carbon storage can be viewed as a resource for the purposes of the LOSC, then it would presumably be a 'non-living resource'. This would make it potentially subject to article 82.[160]

However, article 82 was drafted with extractive resources in mind. It is very difficult to apply the article to carbon storage activities. For example, paragraph 2 refers to payment due 'with respect to all production'. It would be difficult to define the storage of carbon dioxide in the seabed as 'production'. Similarly, the ability for States to make 'contributions in kind' is difficult to apply to a process where a resource is not extracted, but is more a space into which something is injected. Paragraph 3 refers to mineral resources and would therefore not apply to carbon storage.

An argument could be constructed along the lines that a coastal State must attempt to fulfil the spirit of article 82 when carbon is stored under the extended continental shelf. The payments could be made on the basis of fees received by the coastal State rather than the production of hydrocarbons or minerals. A contribution in kind might be allowing the ISA to store carbon dioxide in the continental shelf, whether on its own behalf or for member States. However, this stretches the meaning of article 82 too far, and it is preferable to consider that article 82 would not apply to this activity. Simply put, carbon storage is not the sort of 'exploitation of the non-living resources' anticipated by the Convention.

D. Conclusion

This chapter has explored many of the questions arising from the exploitation of non-living resources on the continental shelf beyond 200 nm. Understandably, the

---

[156] Article 77, LOSC.
[157] Karen N Scott, 'The Day After Tomorrow: Ocean CO_2 Sequestration and the Future of Climate Change' *Georgetown International Environmental Law Review* 18 (2005–2006): pp. 57–108, p. 66.
[158] Ibid., p. 66.   [159] Articles 60, 80 and 81, LOSC.
[160] Scott, 'The Day After Tomorrow', p. 67.

key questions will revolve around the operation of article 82. This article is unique and many aspects of its operation remain open to question. However, the work of the ISA has drawn considerable attention to these processes. It remains to be seen how coastal States will work through the issues once exploitation commences on extended continental shelves. It can be expected that this will occur within a reasonably short space of time.

In Chapter 3, the status of article 82 under customary international law was discussed. This is relevant because at least one non-party to the LOSC, the United States, could be one of the first States to undertake exploitation activities on its extended continental shelf. If a traditional approach to the formation of customary international law is taken, it will be very difficult to argue that the obligations in article 82 are part of customary international law. The only reasonable argument is reliant on the fact that article 82 and article 76 were inextricably linked in the negotiations for the LOSC and so, if article 76 is customary law, then so too is article 82. The chances of this argument succeeding in an international tribunal are limited. Therefore, non-party States should be encouraged to commit to article 82's obligations, if not as a matter of customary international law, then as a matter of policy.

6
Marine Scientific Research

The principle of freedom of marine scientific research is well established in customary international law, and was implicitly referred to as a freedom 'recognised by the general principles of international law' in the 1958 Geneva Convention on the High Seas.[1] In contrast to the freedom of marine scientific research on the high seas, under the Continental Shelf Convention the consent of the coastal State was required for research concerning the continental shelf. This right was limited by the instruction that consent should not normally be withheld for pure scientific research.[2] This balance between the importance of encouraging marine scientific research and protecting the rights of the coastal State was the basis for the development of the Law of the Sea Convention (LOSC)[3] provisions on marine scientific research.[4]

The negotiation of the LOSC provisions on marine scientific research was characterized by the different negotiating positions of developed States and developing coastal States. Developing countries were concerned that marine scientific research in their exclusive economic zone (EEZ) would reveal information about resources that would be used to their detriment by other States.[5] They also had concerns that research activities might be used as a cover for military operations that would undermine their security.[6] The dramatic differences in marine scientific research capacity

---

[1] Convention on the High Seas (opened for signature 29 April 1958, entered into force 30 September 1962) 450 UNTS 82. Gorina-Ysern argues that the freedom of scientific research was not expressly included in the Convention owing to a debate about whether nuclear weapons testing was included as part of the freedom of scientific research. Monserrat Gorina-Ysern, *An International Regime for Marine Scientific Research* (New York: Transnational Publishers, 2003): pp. 246–51.

[2] Article 5(8), Convention on the Continental Shelf (opened for signature 29 April 1958, entered into force 10 June 1964) 499 UNTS 312 (Continental Shelf Convention).

[3] United Nations Convention on the Law of the Sea (opened for signature 10 December 1982, entered into force 16 November 1994) 1834 UNTS 397.

[4] Misunderstandings about the rights of coastal States over marine scientific research continue. Some States overstate their rights in relation to marine scientific research in their legislation as 'exclusive rights'—this is not the correct legal position under the LOSC. See D Nelson, D Ong and AG Oude Elferink, 'Preliminary Report', International Law Association, Committee on Legal Issues of the Outer Continental Shelf (15 January 2002): p. 15.

[5] Wesley S Scholz, 'Oceanic Research: International Law and National Legislation' *Marine Policy* 4 (1980): pp. 91–127, p. 95.

[6] Scholz, 'Oceanic Research: International Law and National Legislation', p. 95. See also Jon L Jacobson, 'Marine Scientific Research under Emerging Ocean Law' *Ocean Development and International Law* 9 (1981): pp. 187–218, p. 189.

between developed and developing States made many believe that research only benefited richer countries.[7] Therefore many developing coastal States preferred to see full coastal State control over research activities. Developed States with research capabilities, on the other hand, preferred to emphasize freedom of research on the basis that it would improve the common understanding of the oceans.[8]

Marine scientific research is not defined in the LOSC. However, article 240 sets out four general principles for the conduct of marine scientific research. These are that: the marine scientific research be conducted exclusively for peaceful purposes; marine scientific research shall be conducted with appropriate scientific methods and means compatible with the LOSC; research shall not unjustifiably interfere with the Convention; and that marine scientific research shall be conducted in compliance with all the relevant regulations that are adopted in conformity with the LOSC. According to article 241 of the LOSC, marine scientific research activities should not constitute the basis for any claim to any part of the marine environment or its resources.

Marine scientific research under the LOSC does not include all activities that might appear to have a research function. Activities designed to collect data about natural resources with the view to exploiting the resources will not be covered by article 246. Therefore, seismic surveying for the purpose of identifying commercial hydrocarbon resources would not fall into the category of marine scientific research in the Convention. Finally, many countries have argued that hydrographic and other military surveying does not amount to marine scientific research because it is treated as a separate category of activity from marine scientific research in the LOSC.[9] For example, article 21(1)(g) of the LOSC grants the coastal State the right to adopt laws for vessels in innocent passage in respect of 'marine scientific research and hydrographic surveys'.

This chapter first outlines the types of marine scientific research that may take place on or above the continental shelf beyond 200 nm in order to put the later discussion in context. It then provides a broad overview of coastal State jurisdiction over marine scientific research under the LOSC. Next, the particular issues arising in respect of the extended continental shelf are examined. The first significant question relates to which activities a coastal State has jurisdiction over. Drilling into the continental shelf would clearly be under coastal State jurisdiction, but what about seismic surveys for 'pure' scientific purposes? Where the research reveals information about the continental shelf resources but makes no physical contact with the shelf, does the coastal State have jurisdiction to refuse consent, or is it an exercise of high seas freedoms? A second important issue is

---

[7] Maureen N Franssen, 'Oceanic Research and the Developing Nation Perspective' in Warren S Wooster (ed), *Freedom of Oceanic Research* (Russak, New York: Crane, 1973): pp. 179–200.

[8] See Herman T Franssen, 'Developing Country Views of Marine Science and Law' in Warren S Wooster (ed), *Freedom of Oceanic Research* (Russak, New York: Crane, 1973): pp. 137–78, p. 153.

[9] Sam Bateman, 'Hydrographic Surveying in the EEZ: Differences and Overlaps with Marine Scientific Research' *Marine Policy* 29 (2005): pp. 163–74; J Ashley Roach, 'Marine Scientific Research and the Law of the Sea' *Ocean Development and International Law* 27 (1996): pp. 59–72.

how article 246(6), which restricts coastal State jurisdiction in relation to marine scientific research on the extended continental shelf, will be interpreted. It is argued that coastal States will be able to identify conservation areas such as marine protected areas in which the coastal State can refuse consent for research despite the provisions of article 246(6). Finally, the chapter discusses the sorts of conditions that States may wish to impose on marine scientific research on the continental shelf beyond 200 nm.

A. Types of Marine Scientific Research

The ocean and its seabed is the subject of study by a variety of scientific disciplines broadly fitting into the area of oceanography.[10] Oceanography can encompass a number of scientific disciplines, including 'biology, biotechnology, geology, chemistry, physics, geophysics, hydrography, physical oceanography, and ocean drilling and coring, which are dedicated to the study of oceans, marine flora, fauna, and physical boundaries with the solid earth and atmosphere'.[11]

Marine scientific research has an extensive history.[12] The famous *HMS Challenger* expedition was a four-year journey, begun in 1872, that first attempted to examine the ocean's chemical, physical and biological properties.[13] Since that time, marine scientific research has underpinned mankind's understanding of the earth and climate, as well as the expansion of activities such as fishing and hydrocarbon exploitation at sea. Marine scientific research provides information essential to the protection of the environment, safer navigation and military uses of the oceans.[14] Because of the importance of marine scientific research, it has traditionally been viewed as one of the freedoms of the seas. It was only with the 1958 Continental Shelf Convention that restrictions began to be placed on marine scientific research conducted beyond the territorial sea.

Biological oceanography is the study of living organisms in the sea. It considers the nature of their interactions, often referred to as marine food webs, and the exterior physical or chemical influences on those interactions. Biological oceanography consists of sub-disciplines such as marine planktonology, marine ecology, marine ichthyology, marine pharmacology and marine molecular biology.[15] In order to

---

[10] Oceanography can also be used more narrowly to describe interdisciplinary approaches to understanding the marine environment. Florian H Wegelein, *Marine Scientific Research: The Operation and Status of Research Vessels and Other Platforms in International Law* (Leiden: Brill, 2005): p. 16.

[11] Marko Pavliha and Norman A Martinez Gutierrez 'Marine Scientific Research and the 1982 United Nations Convention on the Law of the Sea' *Ocean and Coastal Law Journal* 16 (2011): pp. 115–33, p. 115.

[12] For an excellent discussion of the history of marine scientific research, see Wegelein, *Marine Scientific Research*, pp. 22–35.

[13] Gorina-Ysern, *An International Regime for Marine Scientific Research*, p. xxi.

[14] Pavliha and Gutierrez, 'Marine Scientific Research', p. 115.

[15] M Affholder and F Valiron, *Descriptive Physical Oceanography* (Boca Raton: CRC Press, 2001): p. 1.

Types of Marine Scientific Research

study benthic organisms (those found on the sea floor) scientists may use a range of methods, from simple hand-operated corers to complicated bottom dredgers, sledges, trawls, grabs and deeply penetrating geological cores. Also used are a wide assortment of nets, sampling bottles and traps.[16]

Marine chemistry is the study of the chemical properties of sea water. Chemical oceanography comprises of sub-disciplines such as marine physical chemistry, marine organic chemistry and marine nuclear chemistry. It is the study of salts, nutrients and particles in solution, dynamics of chemical reactions and action of gases in sea water.[17] Marine chemists will often focus on results obtained with water samples and instrument probes.[18]

Physical oceanography is concerned with the physical characteristics of sea water, including currents, temperature and pressure. Floating and moored instruments can be used to measure the temperature, rate and direction of water flow.[19]

Hydrography supports navigation at sea. This discipline is very important to seafarers as it includes mapping of the sea floor, depth soundings, wreck search and determining tide schedules.[20] Hydrographic surveying using sonar is a common method for hydrographers to describe and map the sea floor.

Marine geology studies the tectonic situation of the sea floor and the formation of sediments and rocks. It derives information from seismic and drilling studies of the lithosphere, from geomorphological studies of dredge and grab samples from the sea floor, and from studies of magnetic anomalies.[21]

A variety of the methods of research described can have an impact on the sea floor. Some make direct contact with the seabed. Dredging and trawling involve dragging equipment along the seabed with the goal of capturing samples of living organisms and sedimentary deposits.[22] Grab samplers collect the top layers of the seabed by bringing two steel shells together and cutting the soil.[23] Some research requires drilling into the seabed, for example extracting core samples for biological, geological and geophysical surveys.[24]

Other types of research can gather information without making physical contact with the sea floor apart from sound waves. Seismic surveys can gain information about the sediments beneath the sea floor, which can be used to determine the possible existence of hydrocarbons. Sonar can also be used to map the sea floor and examine the resources there. In both types of survey an acoustic signal (high or low frequency) is sent towards the sea floor and receivers record the response.

---

[16] Wegelein, *Marine Scientific Research*, p. 13.
[17] Affholder and Valiron, *Descriptive Physical Oceanography*, p. 1.
[18] Wegelein, *Marine Scientific Research*, p. 13.   [19] Ibid., p. 14.   [20] Ibid., p. 15.
[21] Ibid., p. 14.
[22] Food and Agriculture Organization of the United Nations, 'Dredges' (2015) www.fao.org/fishery/geartype/104/en; 'Bottom trawls' (2015) www.fao.org/fishery/geartype/205/en.
[23] Bureau of Ocean Energy Management, *Types of Geological and Geophysical Surveys and Equipment* (US Department of the Interior, 2013): p. 6 www.boem.gov/G-and-G-Survey-Techniques-Information-Sheet
[24] As an example, see the International Ocean Drilling Programme http://www.iodp.org/.

B. Coastal State Jurisdiction Over Marine Scientific Research Under the Law of the Sea Convention

While freedom of scientific research is a freedom of the high seas,[25] the LOSC gives control of marine scientific research in the territorial sea, EEZ and continental shelf to coastal States. According to article 246(3) of the LOSC, under normal circumstances States are to grant their consent to marine scientific research projects in the EEZ or on the continental shelf that will be carried out in accordance with the LOSC, exclusively for peaceful purposes and in order to increase scientific knowledge of the marine environment for the benefit of all mankind. This requirement, together with the provisions of article 240 on general principles, means that the coastal State can reject research projects that are not carried out in accordance with the LOSC, are not for peaceful purposes or are not intended to increase scientific knowledge of the marine environment for the benefit of all mankind.[26]

Even if a proposed marine scientific project meets the requirements of article 246(3), the LOSC empowers a coastal State to withhold consent for certain types of marine scientific research projects.[27] The first type of project, described in paragraph 246(5)(a), is those that are of direct significance for the exploration and exploitation of natural resources, whether living or non-living. This respects the fact that coastal States have sovereign rights over resources found in the EEZ and on the continental shelf. Other reasons for withholding consent include where the research involves drilling into the shelf or the introduction of harmful substances, the research involves the use of structures regulated by articles 60 and 80 or the researchers have provided inaccurate data or have outstanding obligations from prior projects.

When the research will take place on the extended continental shelf, a coastal State is limited in its right to refuse consent on the basis that the project is of direct significance for the exploration and exploitation of resources. Article 246(6) states:

> Notwithstanding the provisions of paragraph 5, coastal States may not exercise their discretion to withhold consent under subparagraph (a) of that paragraph in respect of marine scientific research projects to be undertaken in accordance with the provisions of this Part on the continental shelf, beyond 200 nautical miles from the baselines from which the breadth of the territorial sea is measured, outside those specific areas in which coastal States may at any time publicly designate as areas in which exploitation or detailed exploratory operations focused on those areas are occurring or will occur within a reasonable period of time. Coastal States shall give reasonable notice of the designation of such areas, as well as any modifications thereto, but shall not be obliged to give details of the operations therein.

---

[25] Articles 87(1) and 257, LOSC.
[26] Alfred HA Soons, *Marine Scientific Research and the Law of the Sea* (The Hague: TMC Asser Instituut/Kluwer, 1982): p. 165.
[27] Article 246(5), LOSC.

Article 246(7) provides that this limitation is 'without prejudice to the rights of coastal States over the continental shelf as established in article 77'. Article 246(8) stipulates that marine scientific research activities must not unjustifiably interfere with activities undertaken by coastal States in the exercise of their sovereign rights and jurisdiction.

This modification in respect of the continental shelf beyond 200 nm was the result of negotiations about whether coastal States should be entitled to exercise jurisdiction over the portion of their shelves that extend beyond 200 nm. In a similar way to article 82, article 246(6) was a compromise that restricted the sovereign rights of the coastal State in the extended shelf—it was a quid pro quo for the international community agreeing to coastal State jurisdiction over the continental shelf beyond 200 nm. Early proposed drafts suggested that consent should be implied in the case of marine scientific research projects on the extended shelf outside of designated areas.[28] Such a proposal would have significantly altered the general approach of article 246, which is that research in the EEZ and on the continental shelf should be subject to the consent of the coastal State. The final version of article 246(6), while limiting the right to refuse consent in some cases, still respects the general principle of coastal State consent.

The effect of article 246(6) is that, unless States have nominated a particular area as one in which exploitation will take place within a reasonable time, they must give consent for scientific research relating to mineral, hydrocarbon or living resources on the extended continental shelf upon request. There are still a number of reasons why the coastal State may refuse consent for research on the extended continental shelf. For example, the coastal State may withhold consent if the research is not exclusively for peaceful purposes and to increase scientific knowledge of the marine environment for all mankind. Exploratory activities designed to collect data about resources with a view to exploiting those resources fall outside this category and can be refused permission.[29] In addition, where the research involves one of the other three criteria in article 246(5)(b)–(d), the coastal State may refuse consent.

It is important to note at the outset of the discussion that the availability of the compulsory dispute settlement process is limited in respect of coastal State decisions to refuse consent for marine scientific research projects under article 246. Article 297(2)(a) of the LOSC provides that a coastal State is not required to accept the submission to compulsory dispute settlement of any dispute arising out of the exercise by the coastal State of a right or discretion in accordance with article 246. Article 297(2)(b) acknowledges that such disputes can be submitted to conciliation, but the conciliation commission 'shall not call in question the exercise by the coastal State of its discretion to designate specific areas as referred to in article 246, paragraph 6, or of its discretion to withhold consent in accordance with article 246(5)'. Thus, disputes that arise about the interpretation of important aspects of

---

[28] Myron H Nordquist (ed), *United Nations Convention on the Law of the Sea 1982: A Commentary, Volume IV* (The Hague: Martinus Nijhoff Publishers, 1991): p. 515.
[29] Soons, *Marine Scientific Research and the Law of the Sea*, pp. 170–1.

article 246 are excluded from compulsory settlement. This may give coastal States encouragement to interpret article 246 broadly in favour of coastal State rights. However, the exclusion in article 297(2) does not prevent disputes from arising and being pursued at a diplomatic level. In addition, if a coastal State exercises enforcement jurisdiction against research vessels while they are in the high seas above the extended continental shelf, this may raise issues of interference with the freedom of navigation, which is a permissible topic for compulsory dispute settlement.

A number of questions will arise for the coastal State when considering the provisions on marine scientific research. Some are practical questions arising from the implementation of article 246. Other problems arise because newer issues have emerged relating to the conservation and potential exploitation of genetic resources, which were not foreseen at the time of the LOSC negotiations.

C. The Scope of Coastal State Jurisdiction in Relation to Research Above the Extended Continental Shelf

The first important matter is to determine the activities over which the coastal State has jurisdiction. This raises two questions that are explored further in this section. First, what is the distinction between exploration and exploitation on the one hand, and marine scientific research on the other? Coastal States' rights to the resources of the continental shelf are regulated by article 77, and are exclusive, while rights to regulate marine scientific research are set out in article 246. Some activities that could generally be described as research actually amount to exploratory activities and are not treated as marine scientific research. The general practice of States is to distinguish between research for commercial purposes and research for the advancement of knowledge. This means that, on occasion, a vessel conducting exactly the same activities (for example a seismic survey) can be covered by a different legal regime, depending on the intent of the voyage. This raises obvious difficulties for the coastal State in deciding how to categorise any particular activity.

The second question is what types of marine scientific research may the coastal State exercise control over on the extended continental shelf? Different types of research can reveal information about the continental shelf. Some, such as seismic surveys or sonar scans, take place entirely in the water column and the only interface with the shelf itself is the impact of the sound waves from the sonar or airgun. Other research requires physical contact with the sea floor, and includes dredging, coring and trawling. Within 200 nm of the baseline, the distinction between these categories of research is largely irrelevant, as the coastal State has jurisdiction over research conducted in the water column, as well as on the seabed. Beyond 200 nm, however, the coastal State only has jurisdiction over research 'on the continental shelf'. This raises the question as to what research activities the coastal State may legitimately regulate when they take place over the extended continental shelf.

(1) Exploitation versus marine scientific research

The distinction between exploratory activities and marine scientific research is not always easy to determine. O'Connell has suggested that, if the aim of the research is to reveal the existence or otherwise of mineral resources, the activity must be considered to be exploratory in nature.[30] He claims that the coastal State has the right to exclusive knowledge about the existence of mineral resources.[31] Soons has defined exploration as 'data collecting activities (scientific research) concerning natural resources … conducted specifically in view of the exploitation … of those natural resources'.[32] The common approach of these two definitions is that research specifically intended to provide information about economically useful resources will be exploration, rather than marine scientific research.

The purpose of the research is often a criteria relied on to distinguish between bioprospecting and marine scientific research. As described in Chapter 4, many States rely on the purpose of the research to determine whether to deal with research into biological resources as exploitation or research.

However, it is clear that some scientific research is of relevance to resource exploitation, even when it does not amount to exploration.[33] One example was the research that discovered manganese nodules and subsequently estimated their extent and value.[34] These studies were focused on describing natural phenomena, which turned out later to have economic value. Economic exploitation was not the goal of the research. It seems that only once the existence and commercial viability of the resources is established will the possibility of exploration exist. Until that point it is marine scientific research. However, a deliberate search for commercially exploitable resources would not be. It has been suggested that voyages focused simply on collection of marine samples might be difficult to characterize; they could be marine scientific research with direct significance for the exploitation of living resources or they might be classed as exploitation, depending on the coastal State's view of the purpose of the research.[35]

It is very common for expensive scientific voyages to be funded by more than one source and carry researchers pursuing individual research objectives. This means that it can be difficult for the coastal State to adequately assess the true nature of the research voyage. In addition, the coastal State is reliant on the researchers providing accurate information about their purposes. However, this less than satisfactory situation is the result of the LOSC framework and how it has been applied by States. Coastal States will need to make an assessment in each case whether the activity is research or exploitation.

---

[30] DP O'Connell, *The International Law of the Sea, Volume 2* (Oxford: Clarendon Press, 1984): p. 1030
[31] Ibid., p. 1030.   [32] Soons, *Marine Scientific Research and the Law of the Sea*, p. 125.
[33] CH Allen, 'Protecting the Oceanic Gardens of Eden: International Law Issues in Deep-sea Vent Resource Conservation and Management' *Georgetown International Environmental Law Review* 13 (2001): pp. 563–660, p. 646.
[34] See HT Franssen, 'Developing Country Views of Marine Science and Law', p. 158.
[35] Allen, 'Protecting the Oceanic Gardens of Eden', pp. 646–7.

(2) Activities over which coastal States have jurisdiction on the extended continental shelf

A further issue is which research activities coastal States have jurisdiction over on the extended continental shelf. Research in relation to the continental shelf may take place by making physical contact with the shelf, for example by drilling into the seabed or by dredging. Research may also take place entirely in the water column, for example through seismic surveys or use of sonar and other detection equipment. Both activities involve collecting data about the seabed and, potentially, its resources. Within 200 nm of the coast, research that does not make physical contact with the shelf but is directed at obtaining information about the shelf and its resources will require the consent of the coastal State. This is because the research takes place above the continental shelf and within the EEZ, and both are zones over which the coastal State has jurisdiction in respect of marine scientific research.

In the case of the extended shelf, however, research may be conducted on the high seas, directed at the continental shelf, which does not make physical contact with the shelf. Does this fact mean that the coastal State will not have jurisdiction over research conducted in the water column that collects data about the resources of the continental shelf?

(a) The Continental Shelf Convention

The Continental Shelf Convention stated in article 5(8) that 'the consent of the coastal State shall be obtained in respect of any research concerning the continental shelf and undertaken there'. Many commentators and States considered that coastal State rights to consent to marine scientific research under the Continental Shelf Convention applied to research that concerned the continental shelf, regardless of where the research activities took place.[36] Authors have noted the ambiguous nature of the words of article 5(8) and relied on a purposive interpretation to conclude that the research applied to all research about the continental shelf.[37] O'Connell argued that, although seismic surveys do not physically make contact with the seabed, the principle of constructive presence could be relied on to argue that the seismic waves generated in the survey penetrated the geological structure of the seabed.[38] The fact that the jurisdiction would be exercised in relation to vessels operating in the high seas did not appear to cause problems for authors adopting this interpretation.

These views were not unanimous. Brown relied on the legislative history of article 5(8) of the Continental Shelf Convention to argue that the words 'and undertaken there' were meant to clarify that the article only referred to research that made

---

[36] Gorina-Ysern, *An International Regime for Marine Scientific Research*, p. 265; Soons, *Marine Scientific Research and the Law of the Sea*, p. 71. Rober Revelle, 'Scientific Research on the Sea-bed: International Cooperation in Scientific Research and Exploration of the Sea-bed' in Jerzy Sztucki, *Symposium on the International Regime of the Sea-bed*, cited in Ram Prakash Anand, *Legal Regime of the Sea-bed and the Developing Countries* (Leiden: Brill, 1976): p. 104.

[37] See, e.g. O'Connell, *The International Law of the Sea Volume 2*, p. 1032.   [38] Ibid.

physical contact.³⁹ However, he also acknowledged that this restrictive interpretation was unlikely to be adopted by many States.

State practice under the Continental Shelf Convention was mixed. Winner noted that some developed States required consent only when the researchers came into contact with the seabed. Others, often developing States, required consent for all research concerned with the shelf, even when there was no physical contact with the seabed.⁴⁰ However, State practice was not divided strictly along developed versus developing State lines.⁴¹ The United States and the USSR took the more restrictive view, not requiring consent unless the research came into contact with the seabed.⁴² In a survey of legislation from eleven States, Soons concluded that no uniform practice could be discerned, but eight of the eleven States appeared to claim control over scientific research concerning the continental shelf without making a distinction between the methods employed for the research.⁴³ His view was that the Convention covered all research concerning the continental shelf, however conducted.⁴⁴

The closest the International Court of Justice (ICJ) has come to addressing this issue was in the *Aegean Continental Shelf* case. Turkey authorized the conducting of seismic surveys in the waters above the continental shelf claimed by itself and Greece. Greece objected to the survey activities, arguing that they would reveal information about the resources of the sea floor that rightfully belonged to Greece. The ICJ ultimately decided that it did not have jurisdiction to hear the dispute.⁴⁵ At an earlier stage, Greece requested provisional measures to the effect that Turkey refrain from all actions that might prejudice any final decision on the basis that the exploration licences breached Greece's exclusive right to knowledge about the resources of the shelf.⁴⁶ The Court's order acknowledged that seismic exploration of the continental shelf without the consent of the coastal State could be a breach of the coastal State's exclusive right of exploitation.⁴⁷ However, the Court considered

---

³⁹ ED Brown, 'Freedom of Scientific Research and the Legal Regime of Hydrospace' *Indian Journal of International Law* 9 (1969): pp. 327–80, p. 356.
⁴⁰ Russ Winner, 'Science, Sovereignty, and the Third Law of the Sea Conference' *Ocean Development and International Law Journal* 4 (1977): pp. 297–342, pp. 322–3. See also PK Mukherjee, 'The Consent Regime of Oceanic Research in the New Law of the Sea' *Marine Policy* 5 (1981): pp. 98–113; Scholz, 'Oceanic Research: International Law and National Legislation'.
⁴¹ One developing State, Venezuela, considered that article 5(8) applies to 'any investigation relating to the continental shelf that is carried out there'. Directive No. MD-EMC-DI-004-79 of 1 February 1980 in Office for Ocean Affairs and the Law of the Sea, *The Law of the Sea: National Legislation, Regulations and Supplementary Documents on Marine Scientific Research in Areas Under National Jurisdiction* (New York: 1989): p. 285.
⁴² Michael Redford, 'Legal Framework for Oceanic Research' in Warren S Wooster (ed), *Freedom of Oceanic Research* (Russak, New York: Crane, 1973): pp. 41–96, pp. 54–55. However, Redford and Soons, *Marine Scientific Research and the Law of the Sea*, p. 70, note that the position of the United States was sometimes unclear.
⁴³ Soons, *Marine Scientific Research and the Law of the Sea*, p. 70. The eight States were: Australia, Denmark, Finland, France, the Netherlands, Norway, Spain and Sweden.
⁴⁴ Soons, *Marine Scientific Research and the Law of the Sea*, p. 216.
⁴⁵ *Aegean Sea Continental Shelf* Case *(Greece v Turkey)* (Judgment) [1978] ICJ Reports, p. 3.
⁴⁶ *Aegean Sea Continental Shelf* Case *(Greece v Turkey)* (Interim Protection) [1976] ICJ Reports, p. 3, para. 17.
⁴⁷ Ibid., para. 31.

that the possibility of the prejudice could be capable of reparation in a situation where Greece was successful on the merits.[48] The Court noted in its judgment that there was no allegation that the seismic surveys caused a risk of physical damage to the resources of the shelf.[49] It therefore declined to order interim measures. The issue was about seismic surveys for the purpose of exploration, rather than marine scientific research. However, it would have been interesting to learn what the Court thought about Greece's alleged right to exclusive right to information about the seabed.

(b) The Law of the Sea Convention

The wording in the LOSC is different to that in the Continental Shelf Convention. The right in article 246 of the LOSC is expressed as jurisdiction over marine scientific research 'in [States'] exclusive economic zone and on their continental shelf'. The use of the words 'on their continental shelf' could be viewed either as a grammatical tool or a requirement that the research physically takes place there.[50] Churchill and Lowe have suggested that, in relation to marine scientific research on the extended continental shelf, the words used in article 246(1), '*on* their continental shelf', implies that consent is required only for research that makes physical contact with the sea floor.[51] They argue that this interpretation is supported by article 257, which stipulates that: '[a]ll States, irrespective of their geographical location and competent international organizations have the right, in conformity with this Convention, to conduct marine scientific research in the water column beyond the limits of the exclusive economic zone'. The authors do not reach a conclusion about the correct interpretation, simply noting that in any case article 246(6) will limit potential restrictions on research that could otherwise be imposed by the coastal State.

It seems unlikely that article 257 was intended to deprive coastal States of any authority or jurisdiction. In fact, the article was a complement to article 256,[52] and was intended to clarify the difference between the treatment of marine scientific research in the Area, as opposed to in the water column.[53] Winner, commenting on the negotiating history of article 257, argued that the article seemed more likely to refer to research that has the water column as its object of study, rather than to research performed in the water column but directed at the continental shelf. Therefore, research concerned with the continental shelf or deep seabed would not be 'research in the waters of the high seas', whether it made contact with the sea

---

[48] Ibid., para. 33. [49] Ibid., para. 30.
[50] See Wegelein, *Marine Scientific Research*, p. 201; Soons, *Marine Scientific Research and the Law of the Sea*, p. 215.
[51] RR Churchill and AV Lowe, *The Law of the Sea*, 3rd edn (Manchester: Manchester University Press, 1999): p. 407.
[52] Article 256 confirms that all States and competent international organizations have the right to conduct marine scientific research in the Area in conformity with Part XI.
[53] Nordquist (ed), *United Nations Convention on the Law of the Sea 1982: A Commentary*, Volume IV, pp. 609–11.

floor or not.[54] Additionally, it could be argued that research conducted in the water column directed at collecting information about the resources of the continental shelf without the consent of the coastal State is not in conformity with the LOSC because it violates State rights.[55]

An arbitration between Guyana and Suriname[56] raised issues similar to that in the *Aegean Continental Shelf* case.[57] In an area claimed by both States, Guyana issued an oil exploration licence to a Canadian oil company. The Tribunal considered both parties' obligations under article 74(3) and 83(3) of the LOSC, which require States with overlapping claims to enter into provisional arrangements of a practical nature and make every effort not to jeopardize or hamper the reaching of a final agreement. In that context, the Tribunal found that the conduct of seismic surveys were not unilateral acts that caused a physical change to the marine environment and so could be undertaken by the parties without jeopardizing the final agreement.[58]

This decision does not imply that research activities conducted in the water above a continental shelf will fall outside the coastal State's jurisdiction. First, activities such as seismic testing were described by the Tribunal as 'exploratory' activities and not research activities.[59] Secondly, the arbitration was between two States with claims to the continental shelf and therefore exclusive rights to the resources. If a third State had conducted testing above the continental shelf, the outcome would likely have been different.

State legislation on marine scientific research generally does not clarify this matter. Commonly, the legislation mirrors the words in the LOSC. One exception is the German Federal Mining Law, in which research on the continental shelf is referred to as 'investigating the continental shelf *on the spot*'.[60] This wording implies that the research must physically take place on the shelf in order for the legislation to apply.

The scope of activities that the LOSC covers appears to be open to interpretation owing to the different possible meanings of 'on' the continental shelf. Soons argues that there was no evidence that the States negotiating the LOSC intended to provide for more limited coastal State jurisdiction than in the Continental Shelf Convention.[61] This would lead to a conclusion that the coastal State has jurisdiction over marine scientific research conducted in the water column using, for example, seismic surveys.[62]

---

[54] Winner, 'Science, Sovereignty, and the Third Law of the Sea Conference', p. 323.
[55] Soons, *Marine Scientific Research and the Law of the Sea*, p. 224.
[56] *Guyana v Suriname* (Arbitration) 139 ILR 566 (2007).   [57] See text above at Section C(2).
[58] *Guyana v Suriname*, para. 480.   [59] Ibid., para. 479.
[60] Federal Mining Law of 13 August 1980, article 132(1) in Office for Ocean Affairs and the Law of the Sea, *The Law of the Sea: National Legislation, Regulations and Supplementary Documents on Marine Scientific Research in Areas under National Jurisdiction* (New York: United Nations, 1989) (emphasis added).
[61] Soons, *Marine Scientific Research and the Law of the Sea*, p. 216.
[62] Wegelein agrees: see *Marine Scientific Research*, pp. 201–2.

D. Issues Arising from Article 246(6)

Paragraph 6 of article 246 restricts a coastal State's ability to refuse consent for marine scientific research projects on the extended shelf on the basis that the project is of direct significance for natural resources, unless the area has been designated as one in which detailed exploratory activities or exploitation is occurring, or will occur within a reasonable time. The provision raises a number of practical questions about how it can be implemented by States.

(1) Interpreting article 246(6)

The first question is what activities amount to 'detailed exploratory operations'. The plain meaning of the phrase indicates that something more than 'exploration' is required to allow a coastal State to create a designated area. During the negotiations for the LOSC, a proposal was made that the designated area be in respect of 'areas in which exploitation or exploratory operations, such as exploratory drilling, are occurring or are about to occur'.[63] Although the example of exploratory drilling was given, 'exploratory operations' could be broad enough to include general seismic surveys, for example. However, in the next session the proposed article referred to 'detailed exploratory operations'. The change in wording was highlighted by the representative of Norway, who said that he had 'been assured by various delegations that the term should be given a broad interpretation, in order to encompass a wide range of exploratory operations'.[64] During the following discussions, some of the proposed amendments focused on returning to the words in their previous formulation, omitting the word 'detailed'.[65] These proposals were not accepted and the Chairman's compromise proposal retained 'detailed exploratory operations'.[66]

The negotiating history of article 246(6) does not, therefore, conclusively settle the matter of what activities will amount to 'detailed exploratory activities'. Soons has argued that the wording suggests that reconnaissance surveys and broad prospecting would not be included,[67] and this seems reasonable in light of the ordinary meaning of the words of the paragraph.[68] However, at the point at which a coastal State issues licences for conducting detailed seismic surveys and the drilling of exploration wells in an area, it would be arguable that 'detailed exploratory

---

[63] The proposal was made by the chairman of the Third Committee during the Eighth Session in 1979. Division for Ocean Affairs and the Law of the Sea, *Marine Scientific Research: Legislative History of Article 246 of the United Nations Convention on the Law of the Sea* (New York: United Nations, 1994).
[64] Ibid., p. 159.
[65] Amendments that supported a broader 'exploratory operations' approach were made by Brazil, Malaysia, Uruguay, Yugoslavia, Canada, the Federal Republic of Germany, Norway and Spain.
[66] Division for Ocean Affairs and the Law of the Sea, *Legislative History of Article 246*, p. 164.
[67] Soons, *Marine Scientific Research and the Law of the Sea*, p. 218.
[68] See article 31, Vienna Convention on the Law of Treaties (opened for signature 23 May 1969, entered into force 27 January 1980) 1155 UNTS 331.

operations' will take place within a reasonable time. A coastal State could designate those areas as ones that meet the requirements of article 246(6).

In order for a coastal State to designate an area under article 246(6), the exploration or exploitation must be undertaken 'within a reasonable period of time'. Although reasonableness is a very flexible concept, it would not be acceptable to designate an area on the basis that it might be explored at some time in the unknown future. Arguably, a coastal State must have concrete plans to explore or exploit the area within the foreseeable future.

A further question is the permitted geographical extent of article 246(6) designated areas. The paragraph is silent about how much of the extended shelf can be declared a designated area. Rather, the entitlement to designate areas relies on the activities authorized by the coastal State. It might be possible, in the case of a small extended shelf, to declare the entire extended shelf a designated area under article 246(6). However, the legality of such a declaration will depend on whether the coastal State is in fact about to undertake exploration or exploitation.

Almost no States have, as yet, incorporated details of article 246(6) into their laws about marine scientific research. One significant exception is Russia, which states in respect of marine scientific research:[69]

> Permission may not be refused in the case of marine scientific research to be conducted on the continental shelf at a distance of more than 200 nm from the baselines from which the width of the territorial sea is measured in connection with resource research, with the exception of areas with respect to which the Government of the Russian Federation has announced that regional geological study of the continental shelf, prospecting, exploration or development of mineral resources, or harvesting of living resources is being or will be conducted. Information about such areas shall be published in the 'Notices to Navigators'.

There are several concerns with Russia's implementation of article 246(6). Russia's legislation does not refer to the requirement that the exploration or exploitation will occur 'within a reasonable period of time' when designating an area. It includes conducting a regional geological study and prospecting as an activity that can prompt the closure of an area, which may be inconsistent with the requirement for 'detailed exploratory operations'. However, this example does indicate that some States will be inclined to interpret article 246(6) very broadly to claim the maximum control over marine scientific research. It is interesting to note that Russia (as the USSR) was originally a State that argued for broad freedoms of scientific research under the Continental Shelf Convention.

(2) Application of article 246(6) to living resources

As established in earlier chapters, a coastal State has sovereign rights to explore and exploit the living resources of the continental shelf.[70] The coastal State also has

---

[69] Article 25, Federal Law on the Continental Shelf of the Russian Federation (adopted by the State Duma on 25 October 1995).
[70] Article 77(4), LOSC.

obligations under customary and treaty law to protect biodiversity and prevent harm to the marine environment. It is not entirely clear on the face of article 246(6) how these rights and obligations are to be reconciled with the requirement to identify areas in which exploitation will occur. At one level, if a coastal State is allowing the commercial fishing of sedentary species, then this would clearly form the basis for a designated area. However, because the extended continental shelf relates to the seabed beyond 200 nm from land, it is reasonably unlikely that commercial fishing will be the primary concern of coastal States. Instead, the coastal State is likely to be concerned with two issues: marine bioprospecting of sedentary species and conservation of vulnerable marine ecosystems.

In both cases, the coastal State will have a legitimate interest in possibly refusing consent for marine scientific research projects that have significance for living resources. First, there may be a concern that the means of the research may be harmful to the marine environment. For example, scientists may propose trawling or dredging the sea floor for samples of living organisms. This method of research, at face value, does not meet any of the other reasons for refusing consent in article 246(5) as it does not involve drilling into the continental shelf, the use of explosives, the introduction of harmful substances into the marine environment or the use of installations and structures. Would the coastal State be required to consent to such a research project on the extended continental shelf? It may be possible that a ground for refusal on the basis of general environmental harm could be read into article 246(5)(b).[71] This would be on the basis that the reasons for refusal in that paragraph are directly related to potential environmental threats.[72] A coastal State may be able to argue that this paragraph, together with the State's more general obligations to protect the marine environment, indicates that it may refuse consent where the research carries a risk of significant harm to the environment. However, as discussed below, there are other arguments on which the coastal State can rely to refuse consent.

A second reason that the coastal State may want to decline a marine scientific research project is on the basis that the research could lead to the identification of valuable genetic characteristics, which could lead to commercial production beyond the control of the coastal State. A researcher, in the course of pure environmental research, could observe valuable properties in an organism collected as part of the research. On publication of this research, a commercial entity may be able to obtain access to specimens from the original research trip, which are then used to synthesize the useful genetic compound and turn it into a commercial product. The point is that an initially non-commercial research project may lead to the commercial development of genetic material obtained from the continental shelf.[73]

---

[71] Marta Chantal Ribeiro, 'The "Rainbow": The First National Marine Protected Area Proposed Under the High Seas' *International Journal of Marine and Coastal Law* 25 (2010): pp. 183–207, p. 204.
[72] Soons, *Marine Scientific Research and the Law of the Sea*, p. 173.
[73] See, e.g. Tullio Scovazzi, 'Mining, Protection of the Environment, Scientific Research and Bioprospecting: Some Considerations on the Role of the International Sea-bed Authority' *International Journal of Marine and Coastal Law* 19 (2004): pp. 383–409, p. 403.

Therefore, the coastal State may wish to withhold consent for the initial project on the basis that it has direct significance for the exploitation of living organisms. However, it would be impossible for the coastal State to say that exploitation is about to commence, because this would not be the case. This is one example demonstrating that article 246(6) is not easily applied to situations involving research into living resources.

To illustrate this point further in relation to article 246(6), let us contrast the situation with regard to living resources and non-living resources. The consequence of article 246(6) is that, outside of the declared areas, consent should be granted to projects that have relevance to the exploitation of resources such as oil and gas. Therefore, researchers from other States may be able to obtain information about the geology of the shelf, which may indicate where mineral resources lie. However, obtaining this information does not grant them the right to explore or exploit the resource: those rights rest exclusively with the coastal State. In this case, although coastal States may be uncomfortable with the idea that information about the continental shelf is widely available, this would not ultimately impact on their sovereign rights.[74] In contrast, when the research involves living resources there is a possibility (however remote) that the research may lead to the commercialization of any genetic material obtained in the course of the research without the consent of the coastal State. Therefore, the consequences of allowing marine scientific research into living resources carries more risk of interference with the rights of the coastal States than research into non-living resources does.

Outside of article 246(5)(b)–(d), there are two ways in which a coastal State could refuse consent for marine scientific research with direct significance for the exploitation of living resources on the extended continental shelf.

First, the coastal State could refuse consent for the marine scientific research project on the basis that it would unjustifiably interfere with the coastal State's sovereign rights. Paragraph 7 indicates that the limitation in article 246(6) is without prejudice to the coastal State's rights in the continental shelf as established in article 77, which include the exclusive sovereign right to explore and exploit living and non-living resources.[75] Paragraph 8 makes it clear that marine scientific research must not 'unjustifiably interfere with activities undertaken by coastal States in the exercise of their sovereign rights and jurisdiction'.[76] Indeed, many of the delegations to the negotiations expressed acceptance of article 246(6) on the basis that it did not undermine the sovereign rights of the coastal States.[77] Soons has suggested

---

[74] Although it has been suggested that the availability of geological information about a State's continental shelf might disadvantage coastal States during the negotiation of exploration and exploitation leases and permits. See MN Franssen, 'Oceanic Research and the Developing Nation Perspective', p. 182.

[75] Article 77, LOSC.

[76] This is a more specific application of the principle in article 240(c) of the LOSC that 'marine scientific research shall not unjustifiably interfere with other legitimate uses of the sea compatible with this Convention and shall be duly respected in the course of such uses'.

[77] Division for Ocean Affairs and the Law of the Sea, *Marine Scientific Research: Legislative History of Article 246 of the United Nations Convention on the Law of the Sea* (New York: United Nations, 1994): pp. 153–9.

that research that will unjustifiably interfere with the coastal State's rights can be refused, because in that case the research would apparently not be carried out in accordance with the Convention, as required by article 246(3).[78]

A second approach would be to designate an area as a marine protected area under article 246(6).[79] There are three advantages to this approach. First, it is more easily dealt with in legislation or regulations. Secondly, the designation of areas clearly signals to researchers what the coastal State's policy is, rather than the more indeterminate approach of arguing on a case-by-case basis that the research constitutes an interference with coastal State rights. Thirdly, the exercise of the coastal State's discretion to designate areas under paragraph 246(6) is excluded from consideration in dispute settlement under the LOSC, even by a conciliation commission.[80]

The difficulty with using article 246(6) is that the requirement that 'detailed exploratory operations' take place 'within a reasonable period of time' is not really applicable to coastal State interests in living resources on the extended shelf. A coastal State may wish to conserve the marine environment rather than exploit it, both for broad conservation purposes and to protect any future intellectual property interests in the genetic resources.

It is suggested that a coastal State could establish a marine protected area on the extended shelf and simultaneously designate that area as one in which it reserves the right to refuse consent to marine research projects with significance for living resources. Although the reason for designation does not completely satisfy the provisions of article 246(6), it is certainly consistent with the purpose and intent of the Convention. As already explained, article 246(6) was a compromise between the broad margin coastal States who wished to retain control of marine scientific research and other States who favoured full freedom of scientific research on the extended continental shelf. The article recognizes that marine scientific research with significance for resources should not interfere with the exercise of the coastal State's sovereign rights (in the case of minerals this would be exploration and exploitation; in the case of living resources this may include conservation and future bioprospecting). Marine scientific research into minerals and hydrocarbons risks only the prospect that researchers from other States become aware of the likely presence of non-living resources in the shelf. The coastal State retains full control over the exploitation of those resources.

In contrast, allowing researchers to take samples of living resources risks the coastal State losing control over the use of those resources in future biotechnology applications. The parties to the LOSC would not have intended the consequence of a strict application of article 246(6) to living resources to be the undermining of coastal State rights. Therefore, despite the fact that living resources in a designated area may not be exploited for a long time, if ever, the coastal State would be justified in establishing the area under article 246(6).

---

[78] Soons, *Marine Scientific Research and the Law of the Sea*, pp. 176–7.
[79] Ribeiro, 'The "Rainbow": The First National Marine Protected Area Proposed Under the High Seas', p. 203.
[80] Article 297(2), LOSC.

An alternative argument would be that a decision by the coastal State to establish a marine protected area is actually 'exploitation' of living resources in that area. As the coastal State has exclusive rights of exploration and exploitation, a decision to conserve the resources is as valid an expression of coastal State rights as a decision to harvest them. Taking active steps to put in place legal protection over the living resources of a part of the shelf could be seen as equivalent to issuing prospecting licences in respect of non-living resources. This argument is reinforced by the fact that the coastal State has obligations under international law, specifically the Convention on Biological Diversity (CBD),[81] to take steps to conserve biological diversity and establish networks of marine protected areas.

It is important to remember that article 246(6) was a compromise between freedom of marine scientific research and the exercise of coastal State rights on the extended continental shelf. Therefore, the coastal State has an obligation to take into account the importance of marine scientific research and consider ways in which research can be facilitated, rather than simply rejected. The coastal State is also obliged to exercise its rights in good faith. These principles can be addressed in two ways. First, the coastal State should consider limiting the declaration of designated areas for living resources to areas that are reasonably considered to host unique or vulnerable marine ecosystems. Declaring large swathes of the extended continental shelf to be marine protected areas without any justification would run counter to the intentions of the LOSC. Secondly, the coastal State can consider using conditions to ensure that its interests are protected, while also allowing for marine scientific research. For example, the coastal State could require the researchers to obtain permission before sharing samples with researchers of other States. This would reduce the risk of the coastal State losing control of its resources once they have left its jurisdiction.

A further problem for the coastal State will be identifying when research relates to the continental shelf or not. Some research projects, especially those focused on the benthic marine environment, may involve study of living resources that are counted as sedentary species, and others that are not. The coastal State has an interest in this research in terms of ensuring that it has given its consent in relation to the research on sedentary species. The research activities in the water column may also be of interest to the coastal State if the activities have the potential to cause serious harm to the environment of the coastal State. However, in this case, the researcher is not required to gain the consent of the coastal State because the research is not directed at the continental shelf resources. This does not remove the obligation on flag States to ensure that activities under its jurisdiction do not cause harm to the coastal State's shelf environment. However, it may mean that the coastal State has less information or input about marine scientific research with the potential for environmental damage that may be taking place in the proximity of the extended shelf.

---

[81] Convention on Biological Diversity (opened for signature 5 June 1992, entered into force 29 December 1993) 760 UNTS 79.

Research into the genetic properties of marine organisms will be considered to be marine scientific research if it is not intended to lead to commercial products. If the coastal State gives consent to a research project, information that results from it must be shared with the coastal State.[82] A problem may arise where researchers wish to share the information resulting only from the research relating to the continental shelf but not the results of research in the water column. This requires a decision to be made about the distinction between the two types of research, which may be difficult when the research is conducted near the sea floor.

The coastal State only has rights in relation to sedentary species on the extended continental shelf. According to article 77(4) of the LOSC, sedentary species are 'organisms which, at the harvestable stage, either are immobile on or under the seabed or are unable to move except in constant physical contact with the seabed or the subsoil'. Who is to determine whether organisms subject to marine scientific research are sedentary or not? It has been difficult enough for States to agree on the classification of major commercial species, such as scallops or lobsters. When the research involves microorganisms or juveniles of particular species it will be even more difficult.[83]

In light of these difficulties, the coastal State may seek to obtain information in relation to all research conducted in the vicinity of the shelf including organisms that may be in the high seas. This would allow the coastal State to make independent determinations as to whether the research relates to shelf resources or high seas resources. One possibility is that the coastal State could require the broader information as part of the conditions for agreeing to marine scientific research on the shelf.

E. Conditions of Granting Consent for Marine Scientific Research Projects

The final question is whether the coastal State can impose conditions on the consent for marine scientific research to protect the environment. Article 249 sets out the obligation of researching States to comply with particular conditions. These include allowing the coastal State to participate in the project, providing the coastal State with results of the research, giving the coastal State access to the data, samples and research results, ensuring the research results are made internationally available and informing the coastal State of major changes to the research programme.[84] None of these conditions directly relate to protection of the environment. However, article 249(2) is clear that these obligations are without prejudice to conditions established by the laws of the coastal State in the exercise of its discretion to grant or withhold consent according to article 246(5). Soons distinguishes between research to which

---

[82] Article 249, LOSC.
[83] Allen, 'Protecting the Oceanic Gardens of Eden', pp. 619–28.
[84] Article 249, LOSC.

the coastal State is obliged to give consent and research projects to which the coastal State has discretion to refuse consent.[85] In the first case, he argues that the State is not able to impose additional conditions because it is required to grant consent in normal circumstances. In the second case, Soons suggests that the coastal State is able to impose any conditions it likes, 'since the discretionary power to withhold consent must be deemed also to include the power to grant consent under conditions to be established unilaterally by the coastal State'.[86]

State practice confirms that many States have imposed environmental conditions on foreign marine scientific research projects that go beyond the content of article 249(1).[87] Hubert cites the example of the Canadian Endeavour Hydrothermal Vents Marine Protected Area Regulations, which require all researchers, including foreign researchers, to comply with the environmental protection measures set out.[88] Particular parts of the marine protected area are set aside for research with different levels of impact on the marine environment.

However, if States are required to give consent to a marine scientific research project owing to the operation of article 246(6), are they permitted to impose conditions that are, for example, intended to protect the environment? Soons's analysis would suggest that they are not. However, it is arguable that imposing conditions to prevent undue harm to the marine environment is entirely consistent with States' general obligation to protect the marine environment found in the LOSC and customary international law.

F. Conclusion

This chapter has raised a series of issues about how article 246 will operate on the extended continental shelf. Of these, some of the most pressing are: distinguishing between exploratory and research activities; determining whether the coastal State has jurisdiction over research in the water column above the extended continental shelf where no physical contact is made with the seabed; and the application of article 246(6) to living resources.

It has been several decades since the conclusion of negotiations for the LOSC that sought to balance coastal State interests in an extended continental shelf with the rights of other States, among others, to conduct marine scientific research. Article 246(6) is a compromise that may come as an unwelcome surprise to coastal States used to exercising full discretion over marine scientific research activities that have significance for resource exploitation. However, it is important that coastal States give effect to the intention of the LOSC. Article 246(6) should not be applied in

---

[85] Soons, *Marine Scientific Research and the Law of the Sea*, p. 188.
[86] Ibid., p. 188. See also Churchill and Lowe, *The Law of the Sea*, p. 406.
[87] Gorina-Ysern, *An International Regime for Marine Scientific Research*, p. 18.
[88] Anna-Maria Hubert 'The New Paradox in Marine Scientific Research: Regulating the Potential Environmental Impacts of Conducting Ocean Science' *Ocean Development and International Law* 42 (2011): pp. 329–55, p. 343.

such a way as to attempt to circumvent the restrictions—for example, by designating areas that are not going to be explored or exploited within a reasonable time, or by designating unnecessarily large areas.

It is suggested that coastal States, when considering designation of particular areas, should consider the purpose of the designation and not impose greater restrictions than necessary. For example, if a coastal State has established a marine protected area on the continental shelf and designated it under article 246(6), it does not follow that research directed at non-living resources should always be refused. If the research would interfere with the purpose of the marine protected area by adversely impacting on the living resources, then it could be refused. However, marine scientific research, even that which has significance for the exploitation of living resources, can be an advantage to the coastal State, which may then obtain further information about its resources as a result of the research project. Finally, a coastal State's decision about marine scientific research on living resources must also consider the relationship between the LOSC and the CBD.[89] Processes that incorporate the requirements of the CBD, Nagoya Protocol and the LOSC need to be developed if none currently exist.

---

[89] See ch 4.

7
The Intersection Between Coastal State Rights and High Seas Freedoms

If sovereign rights belonging to coastal States in respect of the resources of the continental shelf are to have any meaning in international law, the coastal State must have the ability to control activities that affect the resources in that area. However, on the extended continental shelf, the problem of the intersection between the sovereign rights of the coastal State and the freedoms of the high seas becomes particularly problematic. This problem is illustrated by some simple examples.

In the first example, a coastal State has decided to establish a marine protected area (MPA) on its extended continental shelf in order to protect what it considers to be a vulnerable marine ecosystem. This is consistent with its obligations in relation to the protection of the marine environment, discussed in Chapter 4. However, the MPA is in an area that has recently become the site for bottom trawling by fishing vessels from other countries. The bottom trawlers are targeting fish that are not sedentary species, but live in close proximity to the sea floor. During the trawling, considerable damage is done to the organisms living on the sea floor by the rollers, weights and heavy nets that are dragged along the seabed. Corals, sponges and other species are caught as by-catch to the fishing effort. Although the fishing vessels are exercising the freedom of fishing on the high seas, the coastal State's ability to protect the marine environment on the continental shelf is seriously undermined. How should these two interests be balanced? Can the coastal State prohibit the foreign fishing vessels from trawling on its extended continental shelf?

In the second example, the coastal State has established an MPA but, this time, researchers decide to conduct experiments in the deep waters of the high seas above the extended shelf. They decide to pump large quantities of carbon dioxide into the ocean to observe whether the carbon dioxide is rapidly absorbed into the water column or whether it forms large 'lakes' of carbon dioxide owing to the pressure of the water at such depths. One possible side effect of the formation of 'lakes' of carbon dioxide is that they may affect any organisms on or near the sea floor. Can the coastal State require the researchers to seek permission for such activities based on its sovereign rights to the species of the sea floor, even though the researchers are primarily exercising their right to conduct marine scientific research on the high seas?

In a third example, how will coastal States deal with a situation where foreign researchers are conducting bioprospecting at a hydrothermal vent field? The coastal State will want to ensure that its rights in respect of sedentary species are respected but must do so without interfering with the high seas freedom in respect of other species. What would be the appropriate reach of the coastal State in trying to impose conditions on the researchers to, for example, obtain consent and provide information?

Finally, another possible issue for coastal States is where they extract mineral resources, not by means of a stationary ship or platform, but through the use of remotely operated vessels or other technology. These would not be artificial islands, installations or structures according to articles 60 and 80 of the Law of the Sea Convention (LOSC).[1] Can the coastal State impose restrictions on fishing in the areas around the exploitation areas when it cannot rely on articles 60 and 80 to establish safety zones?

If these examples occurred in the exclusive economic zone (EEZ) of the coastal State, there would not be an issue. In that case, the coastal State has rights over the fishing and scientific research by foreign vessels and could regulate the activity to ensure it did not cause harm to the environment protected by the MPA. Article 56(1) of the LOSC provides that, in the EEZ, the coastal State has sovereign rights regarding natural resources of the seabed and subsoil in addition to the waters above. Therefore, a coastal State has clear legislative and enforcement authority over resource-related activities in this area, although it must be exercised with due regard to the rights of other States.[2] Article 246(5) allows a coastal State to refuse consent for research projects in the EEZ or on the continental shelf that involve the introduction of harmful substances into the marine environment. However, on the continental shelf beyond 200 nm, the coastal State rights must coexist with high seas freedoms in the superjacent waters, and the two may sometimes be incompatible. For example, as in the second example above, researchers may claim to be exercising the freedom of marine scientific research on the high seas but, in doing so, may have a detrimental impact on the living resources of the coastal State.

Although there is no explicit obligation to give 'due regard' to the rights of coastal States in the continental shelf, this does not mean that flag States whose vessels are operating in the high seas above the extended continental shelf are free to infringe on the rights of the coastal State. Because coastal States have rights over both living and non-living resources, flag States have an obligation to ensure that their vessels do not undertake activities that infringe on those rights. This obligation may require flag States to restrict their exercise of high seas freedoms in certain situations.

---

[1] United Nations Convention on the Law of the Sea (opened for signature 10 December 1982, entered into force 16 November 1994) 1834 UNTS 397.

[2] Article 56(2), LOSC.

This chapter evaluates the LOSC provisions that relate to the intersection between high seas freedoms and coastal State rights. It focuses on the situation that arises if a coastal State wishes to take unilateral action that affects a State whose nationals are exercising high seas freedoms. Of course, in many cases, international organizations exist that regulate activities on the high seas—such as regional fisheries management organizations (RFMOs)—and it will usually be preferable for the coastal State to work cooperatively through these organizations to achieve its goals. The benefits and risks of working with international organizations are discussed further in Chapter 9. This chapter proposes a framework to assess whether a coastal State's claim to be able to interfere with high seas freedoms is justifiable. It then illustrates the framework with reference to two examples.

A. The Law of the Sea Convention and the Balancing of State Rights

Part VI of the LOSC is largely silent about how States should balance coastal State rights and high seas freedoms in relation to the continental shelf. There are two provisions that deal with this intersection.

The first is article 80, which applies article 60 (relating to artificial islands, installations and structures) to the continental shelf, *mutatis mutandis*. Article 60 is an example of a situation in which the drafters of the Convention considered in detail the balance between coastal State rights and high seas freedoms.[3] The construction of a platform to exploit oil or gas means that other States cannot navigate through the area where the platform is placed.[4] Fishing and navigating in the immediate vicinity of the platform may be incompatible with the operation and safety of the structure. Article 60 recognizes that States must have the right to construct such structures to pursue their rights to the resources of the shelf, and they are given exclusive jurisdiction over those structures. A number of conditions, such as notification to seafarers, limitation on placement and the maximum breadth of safety zones ensure that the interference with navigation rights is minimal. Therefore, in this situation, the LOSC dictates how the balance of rights between the coastal State and other States must be resolved.[5] Coastal State rights have been given preference

---

[3] Article 60 is based on article 5(2)–(7) of the Convention on the Continental Shelf (29 April 1958, entered into force 10 June 1964) 499 UNTS 312 (Continental Shelf Convention). Under this Convention, the waters surrounding installations outside the territorial sea would be the high seas.

[4] David Attard, *The Exclusive Economic Zone in International Law* (Oxford: Clarendon Press, 1987): p. 87.

[5] The ILC referred to the fact that installations were an interference with high seas freedoms during its considerations prior to the Continental Shelf Convention and accepted the coastal State's rights to build installations was a justifiable interference. 'Report of the International Law Commission to the General Assembly: Covering the work of its Eighth Session', A/3159, *Yearbook of the International Law Commission* 1956 Vol. II, pp. 253–302, p. 299.

over high seas freedoms to the extent necessary for the State to extract the resource, and no further.

The second relevant provision is article 78. The first paragraph of this provision establishes that the rights of the coastal State over the continental shelf do not affect the legal status of the superjacent waters or of the air space above those waters. This is a clear limitation on the exercise of jurisdiction by the coastal State. The second paragraph states: 'The exercise of the rights of the coastal State over the continental shelf must not infringe or result in any unjustifiable interference with navigation and other rights and freedoms of other States as provided for in this Convention'.

This prohibition on unjustifiable interference with the rights and freedoms of other States applies to all areas of the continental shelf, not just that part beyond 200 nm. However, as already mentioned, the provision will be of particular importance for the resolution of issues in relation to the extended shelf.

Article 87(2) sets out how States should consider the rights of other States on the high seas. It provides that 'these freedoms shall be exercised by all States with due regard for the interests of other States in their exercise of the freedom of the high seas' and also the rights in relation to the Area. No mention is made of giving due regard to the rights of coastal States.

The question is, in the absence of any other relevant provisions of the LOSC, how should States interpret the article 78 requirement not to infringe or unjustifiably interfere with high seas freedoms? From the wording of the article it is clear that there must be some accommodation—high seas freedoms will not necessarily trump coastal State rights. The provision implies that there may be justifiable interferences with high seas freedoms[6]—but what these might be are not identified. This is nicely illustrated by the bottom trawling example given above. Must the coastal State accept the destruction of the sedentary species (such as coral) that is the by-product of the bottom trawling on the basis that the targeted species is subject to high seas freedoms? Or, can no bottom trawling ever take place over an extended continental shelf on the basis that it might harm sedentary species or other coastal State interests?

(1) Drafting history of article 78

Two significant influences impacted on the negotiation of article 78 of the LOSC. The first was the Convention on the Continental Shelf,[7] which provided a starting

---

[6] Churchill and Lowe agree that it is inevitable that some interference with navigation will result, and does result, from coastal State activities, including where installations are constructed on the shelf. RR Churchill and AV Lowe, *The Law of the Sea*, 3rd edn (Manchester: Manchester University Press, 1999): p. 154.

[7] Convention on the Continental Shelf, (opened for signature 29 April 1958, entered into force 10 June 1964) 499 UNTS 312. Allot described the 1958 Geneva conventions, including the Continental Shelf Convention, as 'a brooding omnipresence that haunts many of the UNCLOS texts'. Philip Allott, 'Power Sharing in the Law of the Sea' *American Journal of International Law* 77 (1983): pp. 1–30, p. 12.

point for discussing the relative weight of rights and freedoms. The second was the extension of coastal State jurisdiction over the continental shelf beyond 200nm in some cases, and a concern to ensure that this jurisdiction did not interfere with high seas freedoms.

(a) The 1958 Convention on the Continental Shelf

As discussed in Chapter 3, many of the provisions of Part VI of the LOSC are based to some extent on the Continental Shelf Convention. This is true of the predecessor articles to article 78. Consideration of the commentary to the Continental Shelf Convention is therefore useful in interpreting article 78.

Article 3 of the Continental Shelf Convention stated that: 'The rights of the coastal State over the continental shelf do not affect the legal status of the superjacent waters as high seas, or that of the airspace above those waters'. Article 5(1) provided:

> The exploration of the continental shelf and the exploitation of its natural resources must not result in any unjustifiable interference with navigation, fishing or the conservation of the living resources of the sea, nor result in any interference with fundamental oceanographic or other scientific research carried out with the intention of open publication.

The rest of article 5 dealt with the right of coastal States to establish installations on their continental shelves, protection of the marine environment and marine scientific research.

It is interesting to compare the provisions of the Continental Shelf Convention and article 78 of the LOSC, as set out in Table 7.1. The influence of the Continental Shelf Convention is striking.

Although they are not identically worded and the extent of the shelf is defined differently in each convention, the substantive nature of the provisions is similar. One

Table 7.1 A comparison between articles 3 and 5 of the Continental Shelf Convention and article 78 of the LOSC

Convention on the Continental Shelf	Law of the Sea Convention
Article 3 The rights of the coastal State over the continental shelf do not affect the legal status of the superjacent waters as high seas, or that of the airspace above those waters.	Article 78(1) The rights of the coastal State over the continental shelf do not affect the legal status of the superjacent waters or of the air space above those waters.
Article 5(1) The exploration of the continental shelf and the exploitation of its natural resources must not result in any unjustifiable interference with navigation, fishing or the conservation of the living resources of the sea, nor result in any interference with fundamental oceanographic or other scientific research carried out with the intention of open publication.	Article 78(2) The exercise of the rights of the coastal State over the continental shelf must not infringe or result in any unjustifiable interference with navigation and other rights and freedoms of other States as provided for in this Convention.

potentially significant point is that article 78(2) adds the word 'infringe' to the prohibitions on the coastal State. This will be discussed below.

In its commentary on the draft of what would become article 3 of the Continental Shelf Convention, the International Law Commission (ILC) made the following comments:[8]

> Article [3] is intended to ensure respect for the freedom of the seas in face of the sovereign rights of the coastal State over the continental shelf. ... A claim to sovereign rights in the continental shelf can only extend to the seabed and subsoil and not to the superjacent waters; such a claim cannot confer any jurisdiction or exclusive right over the superjacent waters, which are and remain a part of the high seas. The articles on the continental shelf are intended as laying down the regime of the continental shelf, only as subject to and within the orbit or the paramount principle of the freedom of the seas and of the airspace above them. No modification of or exceptions to that principle are admissible unless expressly provided for in the various articles.

Although these comments sound as though the ILC was ruling out *any* interference with high seas freedoms, in other parts of the commentary it is accepted that some interference would be inevitable. Earlier in the report, the ILC noted:[9]

> In the view of the Commission, the coastal State, when exercising its exclusive rights, must also respect the existing rights of nationals of other States. Any interference with such rights, when unavoidably necessitated by the requirements of exploration and exploitation of natural resources, is subject to the rules of international law concerning respect for the rights of aliens.

Finally, in relation to the draft of article 5(1), the ILC stated that it applied the basic principle of article 3 to the main manifestation of the freedom of the seas: navigation and fishing. The discussion on this point is worth repeating in full.[10]

> It will be noted, however, that what the article prohibits is not any kind of interference, but only unjustifiable interference. The manner and the significance of that qualification were the subject of prolonged discussion in the Commission. The progressive development of international law, which takes place against the background of established rules, must often result in the modification of those rules by reference to new interests or needs. The extent of that modification must be determined by the relative importance of the needs and interests involved. *To lay down, therefore, that the exploration and exploitation of the continental shelf must never result in any interference whatsoever with navigation and fishing might result in many cases in rendering somewhat nominal both the sovereign rights of exploration and exploitation and the very purpose of the articles as adopted.* The case is clearly one of assessment of the relative importance of the interests involved. Interference, even if substantial, with navigation and fishing might, in some cases, be justified. On the other hand, interferences even on an insignificant scale would be unjustified if unrelated to reasonably conceived requirements of exploration and exploitation of the continental shelf.

---

[8] 'Report of the International Law Commission to the General Assembly: Covering the work of its Eighth Session', A/3159, *Yearbook of the International Law Commission* 1956 Vol. II, pp. 253–302, p. 298.

[9] ILC Report 1956, p. 298.   [10] Ibid., p. 299, emphasis added.

The ILC concluded its remarks by commenting that the coastal State must be the judge of the reasonableness of the justification of the measures in the first instance but that States could have recourse to dispute settlement processes if there is conflict over those measures.[11]

The main subsequent change to these two draft articles discussed in the commentary was that no interference with oceanography and scientific research was permitted in article 5(1). However, this does not change the relevance of the commentary of the ILC quoted above. In all other respects the draft articles were adopted as set out in the 1956 report.

In summary, the ILC approach to articles 3 and 5(1) of the Continental Shelf Convention was as follows. First, the rights over the continental shelf do not limit the rights and freedoms in the water column or airspace above. Indeed, the coastal State was to have no jurisdiction in relation to activities such as navigation or fishing, which remained unchanged. Secondly, some interference with high seas freedoms may be necessary in order for the coastal State to exercise its rights of exploitation in relation to the natural resources of the continental shelf. It is not possible to predict completely what these might be in the future. Thirdly, whether an interference is unjustifiable, or not, will depend on an assessment of the 'relative importance of the interests involved'. Finally, the assessment in the first instance is conducted by the coastal State although affected States naturally may object to this assessment.

This approach was echoed by scholars writing about the Continental Shelf Convention who considered that the coastal State had to be permitted some interference with high seas rights to pursue its interests, although not all agreed on the extent of such a possibility. Young, for example, anticipated that coastal States could impose a ban on trawling over oyster beds.[12] He considered that it was obvious that including sedentary fisheries in the continental shelf regime 'increases the potential control which a coastal State can exert over other uses of the high seas lying above its shelf'.[13] In that case, Young was considering a situation where the sedentary species are put at risk by the high seas activity, either because they are themselves the target of the activity or because they are inevitably affected by the activity. In contrast, McDougal and Burke argued that: 'it seems doubtful if the Convention ought to be construed as permitting consequential interference, except perhaps when the sedentary resources are of very great importance to the coastal State concerned'.[14] The authors considered interference with activities that had an indirect impact on the continental shelf resources to be problematic.

---

[11] For discussion of the ILC debate, see also MS McDougal and WT Burke, *The Public Order of the Oceans: A Contemporary International Law of the Sea* (New Haven: New Haven Press, 1987): pp. 706–10.

[12] Richard Young, 'Sedentary Fisheries and the Convention on the Continental Shelf' *American Journal of International Law* 55 (1961): pp. 359–73, p. 372.

[13] Ibid. [14] McDougal and Burke, *Public Order of the Oceans*, p. 721.

(b) The Law of the Sea Convention

The important achievement in Part VI of the LOSC was the extension of coastal State jurisdiction in relation to the continental shelf. The EEZ had been agreed on at an early stage, and considerable debate was taking place about whether States could exercise control over the continental shelf beyond 200 nautical miles. Although debate about article 78 was not extensive, several proposals that cast light on the negotiators' thinking were rejected.

In 1974 the Informal Group of Juridical Experts proposed that coastal States should exercise their rights 'without undue interference with navigation, fishing, and other legitimate uses of the sea'.[15] The words 'undue interference' imply that the interference is unwarranted because it is disproportionate.[16] This would appear to give coastal States some leeway to determine what proportionate interference would be. A draft from this group the following year returned to the concept of 'unjustifiable' which, it can be argued, imposes a higher standard on coastal States.[17]

The wording of the drafts continued to change. The 1975 draft from the Informal Group of Juridical Experts included a requirement for States and coastal States to have due regard to the rights and duties of each other in relation to the continental shelf.[18] This reflected an intention to balance the rights of coastal States and other users of the ocean.

For the following three years the draft text did not refer to 'unjustifiable interference' by the coastal State. In 1979, the USSR proposed to reintroduce the requirement. Its proposal read that coastal States must not '*infringe* or result in any unjustifiable interference' with the rights and freedoms of other States. This wording was picked up by the chairman of the negotiating group in his compromise text and is what prevails in the final text.[19] It is considered that this change was an attempt to reinforce that the extension of coastal State jurisdiction on the continental shelf would not interfere with high seas freedoms.[20]

Kwiatkowska has commented in relation to article 78 that the joining of the two principles formerly contained in articles 3 and 5(1) of the 1958 Geneva Convention made sense. She argued that the conjunction of these principles in article 78 showed 'more clearly that the coastal State's rights inevitably require some control of the superjacent waters'.[21] In rejecting both 'undue interference' and 'due

---

[15] Myron H Nordquist (ed), *United Nations Convention on the Law of the Sea 1982: A Commentary, Volume II* (The Hague: Martinus Nijhoff Publishers, 1993): p. 903.

[16] Based on the Concise Oxford Dictionary (1999) definition of 'undue'. See Judy Pearsall (ed), *The Concise Oxford Dictionary* 10th edn (Oxford: Oxford University Press, 1999).

[17] See also the proposal by the United States in 1974 that 'unjustifiable interference' should be used in relation to the EEZ. A/CONF.62/C.2/L.47, draft article 8.

[18] Nordquist (ed), *United Nations Convention on the Law of the Sea 1982: A Commentary, Volume II*, p. 904. Compare articles 56(3) and 58(3), LOSC.

[19] A/CONF.62/L.37 (1979).

[20] Nordquist (ed), *United Nations Convention on the Law of the Sea 1982: A Commentary, Volume II*, p. 905.

[21] Barbara Kwiatkowska, 'Creeping Jurisdiction Beyond 200 Miles in the Light of the 1982 Law of the Sea Convention and State Practice' *Ocean Development and International Law* 22 (1991): pp. 153–87, p. 161.

regard' in earlier drafts, the negotiators seemed to imply a different standard for balancing rights with respect to the continental shelf compared with the EEZ, which uses 'due regard'.

It should be noted that the words 'unjustifiable interference' are used in one other place in the LOSC. Article 194(4) requires States, when taking measures to control pollution of the marine environment, to 'refrain from unjustifiable interference with activities carried out by other States' in the exercise of their rights and duties under the Convention. Early drafts of this paragraph required States to 'have due regard to the legitimate uses of the marine environment and [to] refrain from unjustifiable interference with such uses'.[22] By 1976, the reference to due regard was dropped in favour of mentioning 'unjustifiable interference with activities', as in the final text.[23] The authors of the *Commentaries* suggest that this reflects a general rule of international law in its application to the subject matter of Part XI.[24] They state that there is no indication of what is meant by 'unjustifiable interference' in the text.

(2) Other balancing provisions in the Law of the Sea Convention

The entire LOSC is an exercise in balancing between the rights and interests of different States. On the one hand, States are interested in protecting access to the seas for purposes of navigation and other activities, such as fishing and marine scientific research. On the other hand, coastal States desire to exercise control over activities in their maritime zones in order to protect their interests. The Convention as a whole strikes a careful balance between these perspectives. It is not surprising, then, that other provisions in the LOSC address how the intersection of rights should be dealt with. Article 78 sits between two contrasting approaches: the prohibition on interference and the requirement to give due regard.

(a) Prohibition on interference

The primary example of a prohibition on the coastal State from interfering with the rights of other States is found in article 24 of the LOSC. Coastal States must not 'hamper' the innocent passage of foreign ships. This is a clear instruction that coastal States may not interfere with foreign vessels in innocent passage unless permitted to do so by the LOSC. Similar instructions are made in relation to straits[25] and, to a certain extent, archipelagic waters.[26] In these cases, the balance of rights not expressly provided for in the Convention falls in favour of the foreign vessels exercising the right of navigation in those areas. The presumption must be that any exercise of right by the coastal State must be supported by explicit provisions of the LOSC or other international law instruments.

---

[22] Myron H Nordquist (ed), *United Nations Convention on the Law of the Sea 1982: A Commentary, Volume IV* (The Hague: Martinus Nijhoff Publishers, 1993): p. 57.
[23] Ibid., p. 63.   [24] Ibid., p. 67.   [25] See articles 42(2) and 44, LOSC.
[26] Articles 52 and 53(2), LOSC.

(b) 'Due regard'

A second approach to balancing the rights of States is found in the obligations for States to have 'due regard' to the rights of other States while exercising their own rights. This is found in several places in the LOSC in different contexts.

The first category is in relation to maritime zones in which States' rights have to coexist. In the EEZ, coastal States must have due regard to the rights and duties of other States and must act in a manner compatible with the LOSC.[27] Other States exercising their rights in a coastal State's EEZ must have due regard to the rights and duties of the coastal State and must also comply with the laws and regulations adopted by the coastal State that are in accordance with the LOSC.[28] On the high seas, all States must exercise their freedoms with due regard for the freedoms and interests of other States.[29] Allott describes the high seas and the EEZ as 'horizontally shared' areas in which a variety of States have rights that must coexist.[30]

A second category in which due regard is used in the Convention is where activities near the boundaries of maritime zones could lead to the undermining of a State's rights. Due regard to States' rights or interests is referred to in article 142, which requires activities in the Area with respect to transboundary deposits to be undertaken with due regard to the rights and legitimate interests of any coastal State across whose jurisdiction such deposits lie.

Thirdly, some provisions of the LOSC also require 'due regard' for certain interests that do not directly relate to respective State rights. For example, in Part VI, article 79 requires States to have due regard to existing cables and pipelines when laying their own cables and pipelines. When removing installations and structures, due regard must be given to fishing and the protection of the marine environment.[31] The involvement of developing States in activities in the Area must be promoted, 'having due regard to their special interests and needs'.[32] Article 267 requires States that are cooperating to promote the transfer of marine technology to have due regard for all legitimate interests including those of holders, suppliers and recipients of marine technology.

There is no guidance as to the meaning of 'due regard' in the Convention. During the negotiations for article 56, the issue of how to balance coastal State and other States' rights in the EEZ was contentious. However, debate largely focused on the elaboration of coastal State rights, rather than on the due regard provision, which was settled upon by 1975. Interestingly, an early proposal suggested that the rights of the coastal State in the EEZ be exercised 'without undue interference with other legitimate uses of the seas', reflecting an approach more similar to article 78.[33] This demonstrates that the negotiators could have taken a similar approach in both articles, but did not.

---

[27] Article 56(2), LOSC.   [28] Article 58(3), LOSC.   [29] Article 87(2), LOSC.
[30] Philip Allott, 'Power Sharing in the Law of the Sea' *American Journal of International Law* 77 (1983): pp. 1–30, pp. 15–16.
[31] Article 60(3), LOSC.   [32] Article 148, LOSC. See also articles 160(2)(d) and 163(4).
[33] A/CONF.62/l.4 (1974), Nordquist (ed), *United Nations Convention on the Law of the Sea 1982: A Commentary, Volume II*, p. 526.

Commentators have come to a variety of conclusions about how the obligation to give due regard affects States. For example, the obligation to give due regard for others has been suggested to impose an obligation on States to exercise their power of regulation in a reasonable manner.[34] This could also be considered as an obligation on other States not to interfere unreasonably with the coastal State's interests.[35]

Most commentators would agree that the LOSC does not ascribe any priority between the States exercising rights relating to 'due regard', and so each case has to be determined on its own merits.[36] In relation to the EEZ, the LOSC provides some guidance where there is a conflict between States as to how to resolve disputes by requiring States to consider equity and the relevant circumstances, including the respective interests of the parties and the international community.[37]

The idea of the 'due regard' obligation as involving a balancing exercise was reinforced by the decision of the Arbitral Tribunal in the *Chagos Marine Protected Area* arbitration.[38] In that case, the United Kingdom established a MPA around the Chagos Islands, which Mauritius claims are subject to its sovereignty. The United Kingdom had acknowledged the rights of Mauritian fishermen to fish in the area for many years. The Tribunal considered whether, by failing to consult adequately with Mauritius before declaring the MPA, the United Kingdom had breached article 56(2). The Tribunal concluded that the article did not impose a universal rule of conduct.[39] Nor did the obligation favour one party's interests over the other. Instead:[40]

[t]he extent of the regard required by the Convention will depend on the nature of the rights held by Mauritius, their importance, the extent of the anticipated impairment, the nature and importance of the activities contemplated by the United Kingdom, and the availability of alternative approaches.

The Tribunal went on to conclude that 'in the majority of cases, this assessment will necessarily involve at least some consultation with the rights-holding State'.[41] This conclusion was in the context of a bilateral dispute between a coastal State and another State that claimed sovereignty over the island. In that case, it is unarguable that consultation should have taken place with the specially affected State.

---

[34] JC Phillips, 'The Exclusive Economic Zone as a Concept in International Law' *International and Comparative Law Quarterly* 26 (1977): pp. 585–618, p. 604.

[35] Moritaka Hayashi, 'Military and Intelligence Gathering Activities in the EEZ: Definition of Key Terms' *Marine Policy* 29 (2005): pp. 123–37, p. 133.

[36] Boleslaw Adam Boczek, 'Peacetime Military Activities in the Exclusive Economic Zone of Third Countries' *Ocean Development and International Law* 19 (1988): pp. 445–468, p. 455; Mark Valencia and Kazumine Akimoto, 'Guidelines for Navigation and Overflight in the Exclusive Economic Zone' *Marine Policy* 30 (2006): pp. 704–11, p. 705; Horace B Robertson, Jr, 'Navigation in the Exclusive Economic Zone' *Virginia Journal of International Law* 24 (1983–1984): pp. 865–915, p. 882–3. For a contrasting view, see Dr Ren Xiaofeng and Cheng Xizhong, 'A Chinese Perspective' *Marine Policy* 29 (2005): pp. 139–46, p. 145, where the authors argue that rights of States in the EEZ are not evenly balanced and therefore due regard requires priority to be given to the coastal State.

[37] See article 59, LOSC.

[38] Chagos Marine Protected Area *(Mauritius v United Kingdom)* PCA Case No. 2011-3, 18 March 2015.

[39] Ibid., para. 519.   [40] Ibid.   [41] Ibid.

However, it is unlikely that article 56(2) will always require consultation, especially in a situation where there is no easily identifiable, specially affected State. If a particular State's interests will be affected by the decision, then consultation may be important. Nevertheless, the decision is a reminder that coastal States, when making decisions affecting the rights of other States in the EEZ, should consult with these States.

(c) Comparison of balancing provisions in the Law of the Sea Convention

All of the prohibitive provisions in relation to the territorial sea relate to the freedom of navigation. This reflects the high priority that States put on protecting navigation from coastal State interference. This also can be explained by the fact that the rights of coastal States in respect of navigation through the territorial sea, straits and archipelagic waters was codified in great detail in the LOSC. Therefore, it could be argued that unless the Convention expressly permitted interference with navigation in these zones, the coastal State should refrain from interference.

'Due regard' makes sense in the context of the EEZ. Navigation was more tightly regulated than rights such as fishing and marine scientific research. The EEZ was a compromise that established a complex series of rights and freedoms.[42] Therefore, it was likely that unforeseen problems would arise, and that the rights of the States would have to be balanced. Thus it was appropriate to give 'due regard' to each other's rights. This was also appropriate on the high seas where States are equal and no State's rights are greater than another's.

The negotiators rejected a 'due regard' approach when it came to the continental shelf. Although the 'due regard' approach was raised as a possibility during the negotiations, ultimately the delegates settled on a more restrictive approach. This makes sense because the intersection between high seas and continental shelf rights is a different question to the balancing of rights within a zone. The continental shelf is not an 'area of legal mutuality' in the same way as the EEZ and the high seas.[43] States were concerned to limit the impact of coastal State jurisdiction beyond that granted by the LOSC. At the same time, coastal States are not completely forbidden from interference. The 'infringe/unjustifiable interference' approach lies on the spectrum between the prohibitive and the 'due regard' approaches to balancing State rights found in the Convention. However, it has many similarities to 'due regard'.

Although it is not strictly within the scope of this book, it can be noted that the similarities between the standards of 'due regard' and 'infringe/unjustifiable interference' can cause confusion when dealing with issues arising within 200 nm. The EEZ and continental shelf regimes were developed autonomously within the LOSC, but there is a connection between them owing to the fact that article 56(1)

---

[42] Allott described the EEZ as a 'horizontally shared zone' between the coastal States and other States, and the high seas as a 'horizontally shared area'. Allott, 'Power Sharing in the Law of the Sea', pp. 15–16.

[43] Allott, 'Power Sharing in the Law of the Sea', p. 16.

refers to the seabed and subsoil. Nevertheless, article 56(3) states that the rights in respect of the seabed and subsoil are to be exercised in accordance with Part VI. In a case when the coastal State is protecting its rights in respect of the continental shelf within 200 nm, is the appropriate standard 'due regard' or 'infringe/unjustifiable interference'?

An example of the confusion that this can cause is highlighted by the arbitral decision in the *Arctic Sunrise* case.[44] In that case, Russia arrested protesters who had boarded an oil rig on its continental shelf within 200 nm from the coast. In discussing Russia's rights, the Tribunal referred frequently to the EEZ, when in fact, Russia's rights were derived from its sovereign rights to the resources of the continental shelf.[45] The Tribunal referred a number of times to the obligation to have due regard to the rights of other States in the EEZ.[46] Only once did it refer to the standard in article 78.[47]

Attard has suggested that the introduction of the EEZ has tilted the balance of interests further in favour of the coastal State: 'Because EEZ rights also cover shelf resources, a coastal State, in defending measures to exploit its shelf resources, can now also be expected to quote its EEZ rights as a further justification to interfere with the said freedoms'.[48] It is possible that the perceived overlap between the EEZ and continental shelf rights within 200 nm means that the 'infringe /unjustifiable interference' standard will be interpreted as very similar to 'due regard' in disputes arising within 200 nm from the coast. However, this would not be consistent with the drafting of the LOSC. In any case, beyond 200 nm, the two standards must be considered to be different, at least in theory.

(3) Interpreting article 78(2)

A starting point for interpretation are the comments by the Law Commission in relation to the Continental Shelf Convention.[49] According to those comments, some interference with high seas freedoms must be allowed in order to protect coastal State rights to the resources of the continental shelf. The appropriate interference can be determined by a consideration of the relative importance of the interests involved, which is a matter for the coastal State in the first instance. These conclusions are a useful beginning for the interpretation of article 78.

There has been one significant change in wording between article 5(1) of the Continental Shelf Convention ('must not result in any unjustifiable interference') and article 78(2) of the LOSC ('must not *infringe or* result in any unjustifiable interference'). The question must be asked whether the addition of the words

---

[44] Arctic Sunrise *(Netherlands v Russian Federation)* (Merits) PCA Case 2014-02, 14 August 2015.
[45] The Tribunal's recital of the facts shows that Russia based its establishment of a safety zone on the continental shelf regime rather than the EEZ regime. See paras. 217–19. Contrast this with the Tribunal's finding at para. 225 that the case was governed by Part V of the Convention relating to the EEZ.
[46] See, e.g. *Arctic Sunrise* (Merits), paras. 166, 228 and 230. [47] Ibid., para. 331.
[48] Attard, *The Exclusive Economic Zone in International Law*, p. 144.
[49] See above at Section A(1).

'infringe or' has changed the approach from that considered by the ILC in 1956 to be appropriate. As discussed above, the original intention of the USSR in proposing them was to emphasize that coastal State rights must not limit freedoms of the high seas. Importantly, there is no qualifier to the word 'infringe' as there is in relation to 'interference'. To infringe a law is to violate it.[50]

It is also important to note that, as well as adding the word 'infringe', the negotiators did not remove the words 'unjustifiable interference'. Therefore, it is reasonable to suppose that the negotiators wished to retain the general approach of article 5(1) with a reinforcement in relation to the rights of non-coastal States.[51] There is no reason to suppose that the intention was to remove any right to protect coastal State rights from activities on the high seas—otherwise, more prohibitive language could have been used, such as that the coastal State must not 'hamper' other States' rights and freedoms. Therefore it is clear that flag States must accept some level of reasonable interference with their high seas activities if they interfere with the coastal State's exercise of sovereign rights in relation to the continental shelf beyond 200 nm. The key question is to ensure that the interference by the coastal State is as minimal as possible.

In the *Chagos Marine Protected Area* case, the Tribunal considered the meaning of 'unjustifiable interference with activities' in article 194(4).[52] It concluded:

> The Tribunal considers the requirement that the United Kingdom 'refrain from unjustifiable interference' to be functionally equivalent to the obligation to give 'due regard', set out in Article 56(2), or the obligation of good faith that follows from article 2(3). Like these provisions, Article 194(4) requires a balancing act between competing rights, based upon an evaluation of the extent of the interference, the availability of alternatives, and the importance of the rights and policies at issue.

The Tribunal did point out that, unlike article 56(2), article 194(4) applies to activities carried out by other States pursuant to their rights, rather than the rights themselves. For the purposes of that case, this meant that the obligation did not apply to rights that had not yet vested in Mauritius.[53]

Despite the Tribunal's conclusion in relation to article 194(4), it can be argued that there is in fact a difference between the obligations to have due regard to another State's rights (as in articles 56(2) and 58(2)) and the obligation to refrain from unjustifiable interference with that State's rights (as in article 78(2)). First, article 78(2) differs from article 194(4). As the Tribunal pointed out, article 194(4) refers to interference with activities, as opposed to rights and freedoms as in article 78(2). Article 78(2) provides that the exercise of coastal States' rights must not '*infringe* or result in any unjustifiable interference' with other rights and freedoms. Secondly, the negotiating history of article 78(2) indicates that States must have

---

[50] Judy Pearsall (ed), *The Concise Oxford Dictionary*, p. 728.
[51] According to Kovalev, the change 'materially expanded' the 'fundamental provision' of article 5(1) of the Continental Shelf Convention. AA Kovalev, *Contemporary Issues of the Law of the Sea: Modern Russian Approaches*, translated by WE Butler (Utrecht: Eleven International Publishing, 2004): p. 95.
[52] *Chagos Marine Protected Area* arbitration, para. 540.   [53] Ibid.

intended a meaning different from 'due regard' or else they would have retained that language when it was proposed for the continental shelf.

The Arbitral Tribunal in the *Arctic Sunrise* case made a passing reference to article 78 of the LOSC.[54] The Tribunal was considering the Russian arrest of a Greenpeace vessel in the Russian EEZ following protests at an oil platform.[55] The Tribunal found that, at the time of the arrest, the *Arctic Sunrise* was not interfering with Russia's sovereign rights to explore and exploit the continental shelf and was exercising freedom of navigation.[56] The Tribunal noted that if Russia *had* been defending its sovereign rights, the boarding 'would have infringed and unjustifiably interfered with the navigation and other rights and freedoms of the Netherlands'.[57] It was not the exercise of enforcement jurisdiction per se that the Tribunal was referring to because, earlier in the judgment, the Tribunal accepted that a coastal State had jurisdiction to arrest vessels interfering with the coastal State's exercise of sovereign rights.[58] Rather, the Tribunal's reasoning appeared to be that, *in the circumstances* (where the *Arctic Sunrise* had retreated from the platform and showed no intention to return), it was an unjustifiable interference with the vessel's rights of navigation. However, these remarks do not substantially add to the general approach to the interpretation of article 78(2).

The following conclusions can be drawn as to the appropriate interpretation of article 78(2). First, a coastal State may need to interfere to some extent with high seas rights and freedoms in order to exercise its sovereign rights in relation to the resources of the extended continental shelf. Secondly, this interference must be as minimal as possible and only to the extent strictly necessary for the coastal State to exercise its rights. To determine this, the relative importance of the interests must be assessed but high seas freedoms must be protected as far as possible. Thirdly, this assessment is first made by the coastal State but non-coastal States are entitled to challenge the assessment. Commentators have noted that this standard is vague and lacks certainty.[59] However, this is not dissimilar to other parts of the Convention, notably the obligation to have due regard to the rights of other States.

B. A Framework for Balancing Rights on the Extended Continental Shelf

The types of activities that a coastal State may wish to prevent range from activities which in themselves breach their sovereign rights (e.g. drilling for oil or the catching of sedentary species on the extended shelf) through to activities that are an

---

[54] *Arctic Sunrise* (Merits), para. 331.
[55] James Harrison, 'The *Arctic Sunrise* Arbitration (Netherlands v Russia)' *International Journal of Marine and Coastal Law* 31 (2016): pp. 145–59. Russia did not accept the jurisdiction of the Tribunal.
[56] *Arctic Sunrise* (Merits), para. 330. [57] Ibid., para. 331. [58] Ibid., para. 284.
[59] McDougal and Burke found the formulation in the Continental Shelf Convention to be unhelpful to States trying to determine what is reasonable in light of all the factors. McDougal and Burke, *The Public Order of the Oceans*, p. 721.

exercise of high seas freedoms but which have an indirect impact on the resources of the extended shelf. For example, the coastal State could prohibit trawling for sedentary species on the seabed, which would be a direct violation of the State's sovereign rights if undertaken by another State. However, can the State prohibit trawling on the sea floor that is not targeting sedentary species, but rather fish technically living in the high seas? What if the trawling is done in mid-water but above a highly sensitive ecosystem, which might be detrimentally affected by lost gear or accidental impact?

The impact on the resources of the extended continental shelf will vary depending on the activity undertaken, so it is impossible in advance to set out many concrete conclusions about what activities can and cannot be challenged by the coastal State. In fact, any purported regulation of an activity that is not the deliberate targeting of continental shelf resources is one that must be carefully considered by the coastal State. In most cases the foreign actor will be exercising high seas rights or freedoms, and so an assessment must be made whether any coastal State action infringes or unjustifiably interferes with the activity. As with many balancing exercises, the extent of the coastal State rights will depend on the nature of the rights in the circumstances. This, of course, is not particularly satisfactory for States and users who desire more detailed guidance as to how their respective rights are to be determined. Therefore, it is useful to establish a practical process for determining whether their regulation of activities on the extended continental shelf is consistent with the LOSC and the requirement not to infringe or unjustifiably interfere with high seas freedoms.

O'Connell, recognizing that a concept of reasonable use underpins the balancing of rights and duties of States in the law of the sea, nevertheless recognized it as 'relativistic and hence capable of subjective evaluation'.[60] He noted that the application of such a principle can lead to 'diametrically opposite inferences' and concluded that the concept was not capable of resolving specific questions. Instead, it excludes the automatic resolution according to rigid rules and requires resolution to be based on 'appraisal as distinct from mandate'.[61] Nevertheless, based on the previous evaluation of the history and purpose of article 78, it is still possible to identify factors that are appropriate for States to consider in making that appraisal.

The following discussion proposes a framework for assessing whether a particular coastal State action will infringe or unjustifiably interfere with a high seas activity. The proposed guidelines are based in part on the factors identified by authors commenting on the Continental Shelf Convention provisions, in part on judicial commentary and in part on a logical application of principles in the LOSC and international law generally.[62] To illustrate this process, the framework will then be

---

[60] DP O'Connell, *The International Law of the Sea: Volume I*, edited by Ivan Shearer (Oxford: Clarendon Press, 1982): p. 58.
[61] Ibid.
[62] These guidelines were previously outlined in Joanna Mossop, 'Beyond Delimitation: Interaction Between the Outer Continental Shelf and High Seas Regimes' in Clive Schofield, Seokwoo Lee and Moon-Sang Kwon (eds), *The Limits of Maritime Jurisdiction* (Leiden: Martinus Nijhoff, 2014): pp. 753–68. See also Joanna Mossop, 'Regulating Uses of Marine Biodiversity on the Outer Continental

A Framework for Balancing Rights on the Extended Continental Shelf

applied to examples of coastal State regulation that potentially interfere with high seas freedoms.

(1) Principles and approach to balancing rights under article 78(2)

The obligation on coastal States not to infringe or cause unjustifiable interference with other States' rights requires them to act with caution. Ultimately, this will involve a balancing of respective rights in order to determine the relative importance of the rights involved. In the *Chagos Marine Protected Area* arbitration, the Tribunal identified the factors relevant to a 'due regard' or 'unjustifiable interference' consideration: these included the extent of the interference, the availability of alternatives and the importance of the rights and policies at issue.[63] The Tribunal in the *Arctic Sunrise* arbitration found that measures taken by coastal States had to be reasonable, necessary and proportionate.[64]

The following factors incorporate these, and other considerations, to evaluate whether coastal State measures to protect sovereign rights on the extended continental shelf infringe or result in unjustifiable interference with the rights and freedoms of other States.

(a) Likelihood of interference with continental shelf resources and level of harm

First, a coastal State must establish that a high seas activity will have a detrimental impact on its exercise of sovereign rights over the resources of the continental shelf. This will require the coastal State to consider the evidence that the regulated activity is interfering, or could interfere, with the coastal State's rights over shelf resources. There should be a real probability that a particular activity could impact on the resource before a coastal State would have a right to regulate such an activity. This is consistent with the Arbitral Tribunal's finding in the *Arctic Sunrise* case that, even if the *Arctic Sunrise* had interfered with Russia's sovereign rights, the Russian arrest of the vessel would have been inconsistent with article 78, because at the time of the arrest the vessel did not actually pose a threat to activities at the platform.[65] This factor does not require evidence of *actual* harm; rather, there must be a real probability that if the activity takes place there will be harm to the resources. This is consistent with a precautionary approach to environmental management.

A second consideration is the nature and level of harm or interference for the shelf resources as a result of the activity to be regulated. Is the potential or actual damage minimal, or more serious? This may require consideration not just of the extent of any physical harm, but the impact on the coastal State's interest in the

---

Shelf' in Davor Vidas (ed), *Law, Technology and Science for Oceans in Globalisation* (Leiden: Brill, 2010): pp. 319–37.

[63] *Chagos Marine Protected Area* arbitration, para. 540. See also para. 519.
[64] *Arctic Sunrise* (Merits), para. 326.   [65] Ibid., para. 331.

resource. For example, bottom trawling has been demonstrated to have a serious impact on the benthic ecosystem. On the other hand, the impact of longlining appears to be more limited and could be mitigated by the way the longlining is conducted (e.g. by ensuring that the line does not descend far enough to touch the sea floor). However, if there is a highly vulnerable ecosystem at stake, even longlining above the shelf may pose some risks.

These factors contribute to a conclusion that the proposed coastal State regulation is necessary to protect an interest of the coastal State.

(b) Relative importance of the interests

The next step is to compare the impact on the interests of the coastal State as opposed to the interests of other States operating on the high seas. This was a factor that was considered to be important by the ILC in relation to the Continental Shelf Convention.[66]

This question requires consideration of the relative importance of the various interests potentially affected, and the numbers of actors affected by the regulation.[67] This may require a comparison of the relative economic values. In evaluating a clash between oil exploitation and navigation, Attard suggested comparing the value of a resource deposit to the cost, for example, of changing shipping routes.[68] McDougal and Burke considered that, in a conflict between oil and fishing interests, decision-makers could take into account the importance of the productivity of the area for both fisheries and mineral exploration and the importance of the competing objectives including the contribution of the disrupted fisheries to the economies and food supply of the States concerned.[69]

In a modern context, it is also likely that coastal States will include conservation values in their interests in the extended continental shelf resources. Indeed, coastal States have a number of obligations to protect the marine environment, which includes the extended continental shelf.[70] Therefore, if there is an area that a coastal State considers to have high biodiversity value because it contains endemic species or is a vulnerable marine ecosystem, this should be factored into the decision making.

It can be difficult for policymakers to measure economic interests directly against conservation or social objectives. Significant work has been done on attempting to put an economic value on biodiversity or ecosystem services.[71] However, economic

---

[66] 'Report of the International Law Commission to the General Assembly: Covering the work of its Eighth Session', A/3159, *Yearbook of the International Law Commission* 1956 Vol II, pp. 253–302, p. 299. See above at Section A(1).
[67] See McDougal and Burke, *Public Order of the Oceans*, p. 721.
[68] Attard, *The Exclusive Economic Zone in International Law*, p. 144.
[69] McDougal and Burke, *Public Order of the Oceans*, p. 721.   [70] See ch 4.
[71] For example, see Patricio Bernal et al., 'Chapter 55: Overall Value of the Oceans to Humans' in *First Global Integrated Marine Assessment* (New York: United Nations, 2015) http://www.un.org/Depts/los/global_reporting/WOA_RegProcess.htm; NJ Beaumont et al., 'Economic Valuation for the Conservation of Marine Biodiversity' *Marine Pollution Bulletin* 56 (2008): pp. 386–96; Robert Costanza, 'The Ecological, Economic, and Social Importance of the Oceans' *Ecological Economics* 31

values do not take all interests into account, or may simply be inappropriate in some cases. The deep-sea environment is one of the least studied from an ecosystems services viewpoint,[72] and so the economic value of deep-sea marine ecosystems may not be readily available to policy-makers.

There is an interesting question as to whether coastal States, in determining whether their actions infringe or unjustifiably interfere with high seas freedoms, should consider the interests of the international community as a whole and not just one or two other States that may be immediately affected. In the case of conflicts between rights of States operating in the EEZ, article 59 requires consideration of the interests of the international community. Although article 78 contains no similar exhortation, a coastal State could not ignore the interests of the international community in assessing the relative importance of the rights involved. Because article 78 is designed to protect high seas freedoms as much as possible, the broader goals of the LOSC in protecting high seas freedoms should be taken into consideration. This will be particularly important if the immediate impact on particular States is low but the proposed coastal State action may have more general implications for the international community and its interests. For example, this might be the case if a coastal State purported to exclude activities in an area of the extended continental shelf where current activities were not taking place but it was arguable that States would regard the restriction as an unjustifiable interference with navigation or fishing.

(c) Minimal interference

The next consideration reflects the conclusion that coastal State interference with high seas freedoms must be as minimal as possible. In other words, it must be reasonable and proportionate. The coastal State must assess whether the proposed interference with the high seas rights is as minimal as possible to achieve the coastal State's objectives, or whether a less restrictive option is available.

Considerations include whether the restriction is procedural in nature (e.g. a requirement for conducting environmental impact assessments or reporting on activities) or is it a substantive restriction (e.g. a prohibition on activities covered by the freedoms of the high seas)? If there is a less restrictive method of achieving the coastal State's goals it is arguable that the more restrictive approach will be unjustifiable. For example, a coastal State wishing to protect seabed ecosystems that prohibits all bottom fishing above its extended continental shelf may be found to be unjustifiably interfering with the freedom of fishing, but a State that only restricts bottom fisheries in the vicinity of vulnerable marine ecosystems may not

---

(1999): pp. 199–213; Robert Costanza et al., 'The Value of the World's Ecosystem Services and Natural Capital' *Nature* 387 (15 May 1997): pp. 253–60; David Pearce and Dominic Moran, *The Economic Value of Biodiversity* (London: Earthscan, 1994).

[72] Camino Liquete, 'Current Status and Future Prospects for the Assessment of Marine and Coastal Ecosystem Services: A Systematic Review' *PLOS One* 8(7) (3 July 2013) DOI: 10.1371/journal.pone.0067737.

be. McDougal and Burke have suggested that the ability of the activities to be managed so they coexist should be explored.[73]

(d) Role of international or regional institutions and soft law instruments

Fourthly, a coastal State may not ignore the international legal framework. It may be possible that the issues facing the coastal State have been addressed by international organizations such as the International Maritime Organization (IMO), RFMOs or regional seas organizations. One possibility is that the international organization may create particular rules or impose duties on flag States when operating above an extended continental shelf. In the absence of any binding obligations, the relevant international or regional organizations may have established guidelines or principles that are relevant to the coastal State's considerations. Although it is most likely that any conflicts will arise between a coastal State and another State whose nations are operating in the vicinity of the continental shelf, there is a possibility that the problems may be discussed in a multilateral forum, such as the IMO or an RFMO. The results of discussion may simply be soft law instruments such as guidelines or a decision of a regional organization, and the practice of other States may be reflected in those instruments. States should consider those statements when weighing up their options.[74]

(e) Consultation with affected States

Where certain States are particularly affected by the proposed measure, it will be important for the coastal State to consult with those States prior to implementing any measure.[75] Where a State is likely to have vessels exercising high seas freedoms in the water above the extended shelf, and those rights would be impeded in some way, the coastal State should consult with that State. It is argued that this should not necessarily extend to all States that may be affected in some way, such as flag States whose vessels simply transit through the area. It would be very difficult for a coastal State to meet such a consultation requirement. Instead, the coastal State should consult with specially affected States.

Of course, one option for a coastal State is to work with any of the relevant international institutions mentioned above. For example, if the issue is about the impact on fisheries, then working through the relevant RFMO would be a useful way of consulting with many of the States affected. In many RFMOs, areas have been closed to bottom trawling to protect a vulnerable marine ecosystem and there is no reason why this approach could not apply to protect a part of a coastal State's extended continental shelf.

---

[73] McDougal and Burke, *Public Order of the Oceans*, p. 721.
[74] The possibilities for international organizations to address these issues are discussed in ch 9.
[75] See the *Chagos Marine Protected Area* arbitration, para. 519, where the Tribunal found that the obligation to have due regard would, 'in the majority of cases', involve consultation with the 'rights-holding State'.

A Framework for Balancing Rights on the Extended Continental Shelf 193

This final consideration is a procedural rather than a normative consideration. However, the conduct of the coastal State will support an argument that it has acted reasonably in considering its options.

The weighing of these factors in different contexts is likely to lead to different outcomes. As indicated by the ILC, some forms of substantial interference with high seas freedoms may be justifiable. Other forms of interference may be insignificant, yet unjustifiable.[76]

It is important to emphasize that the analysis is made on the same basis, whether a foreign vessel is within the EEZ or above the extended continental shelf. As discussed above, Article 78 applies to the entire continental shelf. The due regard obligation is relevant to the EEZ and article 56(3) provides that the rights in relation to the seabed and subsoil must be exercised in accordance with Part VI. Therefore, any issue that balances the rights of the coastal State to the resources of the seabed (no matter where those resources are on the seabed) must be analysed according to article 78 and the 'unjustifiable interference' standard. The proposed framework can also be used in a 'due regard' analysis, but the ultimate balancing and conclusion may be slightly different.

(2) Application of the framework

The factors outlined above are nicely illustrated by article 60 of the LOSC, dealing with installations and structures. In this case, the balance between the rights of coastal States and others was expressly considered by the negotiators and an elaborate system set out to manage the balance. The resulting provision reflects a consideration of the importance of the interests affected by the creation of an artificial island, installation or structure, and the need to minimize the interference with high seas freedoms.

It is clear that the existence of an oil platform is inconsistent with navigation or fishing in the area taken up by the platform, but the value of the interest to the coastal State in the oil justifies an interference with those rights. Additionally, the security of the platform is strengthened by the establishment of a safety zone around it in which navigation may be limited. However, the rights of the coastal State are limited. The platform may not be established where it would interfere with sea lanes essential to international navigation, and the size of the safety zone is limited to 500 metres. So, in most circumstances the coastal State may build platforms, but not where essential navigational routes will be disturbed.

A similar analysis can underpin a coastal State's proposed interference with high seas freedoms to protect the resources of the extended continental shelf. To illustrate this, two examples are considered. The first is a hypothetical situation where a State may wish to prohibit bottom trawling above its extended continental shelf in order to protect vulnerable marine ecosystems, which include sedentary species. In the second example based on existing State practice, the framework is applied to

---

[76] Attard, *The Exclusive Economic Zone in International Law*, p. 144.

a law prohibiting navigation within 500 metres of vessels conducting seismic and other surveys.

(a) Example one: bottom trawling for high seas resources

To take the example of bottom trawling, there is indisputable evidence that this practice leads to significant harm to the sea floor environment, including sedentary species.[77] A coastal State would be able to assume that, if bottom trawling takes place on its extended continental shelf, significant environmental damage would result.

When weighing up the relative value of the interests, there are several to be taken into account. The first is the clear right of the fishing State to conduct fishing on the high seas, as set out in Part VII of the LOSC. Deep-sea fishing is a valuable economic activity and bottom trawling is dependent on location, as many deep-sea fish species aggregate above particular seamounts.[78] Therefore, a prohibition on bottom trawling in a specific location may directly impact on the fishing activity. The level of economic harm will differ, depending on the location and the history of the fishery.

The second interest is the rights of the coastal State in the resources of the shelf. It is possible that there may be an interest in fishing for the sedentary species, but it is more likely that the interests of the coastal State will be in the conservation value of the species, or in protecting them for future potential bioprospecting activities. It can be difficult to balance a clear economic value (in the fish) against a conservation or future-use interest (in the sedentary species). Relevant considerations could include the endemism of the species found in the area, the extent to which other areas are protected from bottom trawling and the ease with which the fishery can relocate.

States must consider their obligations to conserve the marine environment when considering the strength of relevant interests.[79] Although Part VI of the LOSC does not contain an express obligation to conserve sedentary species,[80] the Convention does contain a general obligation to protect and preserve the marine environment.[81] Also important is the requirement in article 118 for States fishing in the high seas to cooperate with other States to conserve living resources. Other conservation obligations may also apply, including those found in the Convention on Biological Diversity.[82]

---

[77] See ch 2.
[78] Malcolm Clark, 'Deep-sea Seamount Fisheries: A Review of Global Status and Future Prospects' *Latin American Journal of Aquatic Research* 37(3) (2009): pp. 501–12, p. 507.
[79] See ch 4.
[80] Compare this with article 61 of the LOSC, which requires the conservation of the living resources of the EEZ.
[81] Article 192, LOSC.
[82] AC de Fontaubert, DR Downes and TS Agardy, 'Biodiversity in the Seas: Implementing the Convention on Biological Diversity in Marine and Coastal Habitats' *Georgetown International Environmental Law Review* 10 (1998): pp. 753–854; DK Anton, 'Law for the Sea's Biodiversity' *Columbia Journal of Transnational Law* 36 (1997): pp. 341–71.

The next step is to consider whether the interference with the rights of the fishing State is as minimal as possible. As mentioned above, a coastal State that closes its entire continental shelf to fishing may be accused of unjustifiable interference, whereas targeted closures may be more acceptable. One of the problems with targeting closures or restrictions to particular areas is that a vast majority of continental shelf ecosystems have not been adequately sampled or studied. A coastal State may wish to take precautionary measures until the science is more fully understood, but this does risk challenge from other States.

Fourthly, where the species targeted by the fishing State is subject to conservation and management by an RFMO, it is possible that the organization may have promulgated measures related to bottom trawling.[83] Where regulations are adopted by such organizations and a coastal State adopts consistent regulations, this may assist in arguing that the restrictions are not an unjustifiable interference with high seas freedoms. This will be particularly useful when dealing with foreign fishing vessels from States not party to the RFMO.

Finally, the coastal State must consider whether there are any specially affected States that must be consulted about the measure. These might include, for example, States whose vessels currently conduct bottom trawling on the extended shelf.

It is clear from the assessment that some forms of restrictions on bottom trawling may be appropriate in some circumstances but blanket bans on trawling would probably be seen to be unjustifiable, particularly where the coastal State has not adequately considered the rights of other States. The coastal State's case will be improved if there is scientific evidence that a particular site will be adversely affected by the bottom trawling.

(b) Example two: restrictions on navigation around survey vessels operating above the continental shelf

Some governments have passed legislation restricting navigation in the vicinity of vessels conducting exploration or exploitation activities for the oil and gas industry.[84] These restrictions have been prompted by potential or actual protests against oil exploration and exploitation activity. For example, Canada's Collision Regulations create a safety zone around a vessel in Canadian waters 'that is in position for the purpose of exploring or exploiting the non-living natural resources of the sea bed …'.[85] The safety zone extends from the outer extremities of the vessel for 500 metres, although the relevant minister may establish a larger safety zone if it is necessary to ensure navigational safety. No vessel may navigate within a safety zone

---

[83] Glen Wright et al., 'Advancing Marine Biodiversity Protection through Regional Fisheries Management: A Review of Bottom Fisheries Closures in Areas Beyond National Jurisdiction' *Marine Policy* 61 (2015): pp. 134–48. See also ch 9.
[84] This analysis previously appeared in Joanna Mossop, 'Protests Against Oil Exploration at Sea: Lessons from the *Arctic Sunrise* Arbitration' *International Journal of Marine and Coastal Law* 31 (2016): pp. 60–87, pp. 80–6.
[85] Rule 43(a), Collision Regulations C.R.C., 1978, c. 1416.

unless it is supporting the vessel or is in distress.[86] In South Africa, no vessel may enter a safety zone of 500 metres around an offshore installation, which is defined to include any exploration or production vessel used in prospecting for or the mining of any substance.[87] The only exceptions are for emergencies or for permitted servicing vessels. The legislation expressly applies to installations above the continental shelf, as well as in the EEZ.[88]

The issue was recently debated in New Zealand, which in 2013 introduced similar legislation.[89] The legislation was prompted by protests against seismic surveys being undertaken in New Zealand's EEZ.[90] One protest against oil exploration in 2011 took the form of small vessels obstructing the path of a seismic survey vessel, Petrobras' *Orient Explorer*, forcing it to divert its course to avoid a collision. This created extra expense for the petroleum company because it required additional time at sea to complete the survey. The New Zealand Government deployed the defence force and the New Zealand police to the area to monitor the protests and prevent interference. A New Zealand citizen was subsequently charged with offences against domestic legislation for his actions in the protest.[91]

The new legislation provides for the creation of a non-interference zone around structures[92] and vessels undertaking prospecting, mining or exploration activity in the territorial sea, the EEZ or above the continental shelf. The zone may extend up to 500 metres from the outer edge of the vessel or any attached equipment and be in effect for up to three months.[93] It is an offence for any vessel or person to enter the non-interference zone 'without reasonable excuse'.[94] The legislation does not define what a 'reasonable excuse' is. Enforcement officers may stop, board or detain a vessel, or remove or arrest a person on board the ship.[95] The fact that the application of this legislation against foreign vessels may be controversial is reflected in the requirement for the Attorney-General's consent prior to prosecution of a person on

---

[86] Rule 43(c) provides that the restriction on navigation does not apply to a vessel that is in distress, is attempting to save life or provide assistance to a vessel in distress, is operated by or on behalf of the State having jurisdiction over the exploration or exploitation operations, or has received permission from the person in charge of the exploration or exploitation vessel to enter the safety zone around that vessel.

[87] Sections 1 and 8B, Marine Traffic Act No. 2 of 1981, South Africa.

[88] Section 1, Marine Traffic Act No. 2 of 1981, South Africa.

[89] Section 55, Crown Minerals Amendment Act 2013 (NZ). New Zealand legislation is available at www.legislation.govt.nz.

[90] The protests derived from opposition to the prospect of deep-sea oil and gas exploitation on New Zealand's continental shelf. The protests took place in the EEZ. 'Police Issue Notices to Oil Company Protesters' *Dominion Post*, 12 April 2011 http://www.stuff.co.nz/national/politics/4874546/Police-issue-notices-to-oil-company-protesters.

[91] *Teddy v NZ Police* [2014] NZCA 422.

[92] 'Structures' are defined as meaning any fixed, moveable or floating structure or installation and includes petroleum pipelines, pumping stations, tank stations or valves. See section 101A, Crown Minerals Act 1991 (NZ). This definition goes well beyond article 60 of the LOSC.

[93] Section 101B(6) and (7), Crown Minerals Act 1991 (NZ). It seems likely that the 500 metre distance was chosen to echo article 60(5) of the LOSC.

[94] Section 101B(2), Crown Minerals Act 1991 (NZ).

[95] Section 101C(1), Crown Minerals Act 1991 (NZ).

board a foreign ship.[96] A number of notices to mariners have been issued, notifying vessels of non-interference zones. At least one has been issued in respect of seismic surveying in an area of the continental shelf beyond 200 nm.[97]

Any restriction on navigating within 500 metres of a vessel undertaking a seismic survey is not legitimized by any explicit provision in the LOSC. In fact, it is a limitation on a foreign vessel's right of navigation in the EEZ and above the continental shelf akin to that imposed by installations and structures on the continental shelf. Because the right to impose restrictions on navigation around vessels is not a right conferred by the LOSC, the question is whether the legislation infringes or unjustifiably interferes with the freedom of navigation.

The first step is to identify the impact of navigation within 500 metres of a survey vessel on the coastal State's interests. A survey vessel conducts operations to determine the extent of the hydrocarbon resources in the continental shelf—a matter within the State's sovereign rights to explore and exploit the natural resources of the shelf. It is common practice for States to license private companies to undertake the exploration and exploitation of oil and gas. Regulation of this activity is an expression of the coastal State's sovereign rights.

The effects of shipping on seismic surveys can be significant. The survey is carefully designed to build a composite picture of an area of the sea floor by following predetermined grids or tracks over the area.[98] Interruption of the path of the vessel can cause expensive delays while the vessel aborts the data acquisition and returns to repeat the track.[99] A survey vessel has limited manoeuvrability because it tows long streamers carrying equipment.[100] There can be safety concerns if a vessel becomes entangled in the streamers, as well as financial implications for the operator of the vessel.[101]

If the interruption to the survey is deliberate, as a result of protest action, the cost to the State in taking enforcement action can also be significant.[102] There may also be a concern that protest action might deter other companies from applying to prospect in the coastal State's waters.

The second step is to compare the interests of the coastal State with the interests of foreign vessels that are inhibited from exercising their high seas freedoms. In this

---

[96] Section 101B(9), Crown Minerals Act 1991 (NZ). This requirement appears in other New Zealand legislation where there may be controversy about the exercise of jurisdiction over a foreign national. For example, see section 6, Antarctica (Environmental Protection) Act 1994 (NZ).

[97] Notice to Mariners NZ 48(T)/16. The notice applies to an area covering approximately 191,000 km² of the continental shelf beyond 200 nm.

[98] International Association of Oil and Gas Producers, *An Overview of Marine Seismic Operations*, Report 448, April 2011, p. 7.

[99] Ibid., p. 42.

[100] Ibid., p. 12. Streamers can be as much as twelve kilometres long, and turning the vessel to run another track can take as long as three hours. See also p. 41.

[101] New Zealand Government, *Offshore Seismic Surveying* http://www.nzpam.govt.nz/cms/iwi-communities/government-role/doc-library/seismic-surveying-factsheet.pdf.

[102] New Zealand's response to the 2011 protests against the *Orient Explorer* by police and the defence force was estimated to cost NZ$1.7 million. Isaac Davison, '$1.7m bill for oil protest on high seas' *New Zealand Herald*, 14 August 2013.

case, the inhibition relates to the freedom of navigation within 500 metres of a vessel conducting a seismic survey.

Freedom of navigation is not an absolute right to take whatever route a master prefers. Vessels are subject to rules that restrict their navigation when in close proximity with one another. In particular, all vessels must keep out of the way of other vessels restricted in their ability to manoeuvre, which includes vessels engaged in surveys or towing gear.[103] In practice, survey vessels often use chase vessels to warn other vessels of the hazard posed by the survey vessel and direct them away from the path of the main vessel. Therefore, some alteration of their course may be necessary in the normal course of navigation on the high seas. If the ship has early warning of the survey vessel's presence, the necessary evasive action could be factored in and cause little delay to the navigating vessel. However, the creation of an exclusion or safety zone around a vessel and its streamers could lead to a considerable detour in some cases.

An additional problem is determining where the survey ship may be. Unlike installations and structures, survey vessels move constantly and cannot be marked on charts. Although it is possible to issue a maritime notice that a survey vessel is operating in a particular area, the precise location of the survey vessel at any particular time will not be known until another vessel comes upon it. Large commercial ships may have equipment that gives prior warning of the proximity of other vessels, but not all vessels will be so equipped.

In some cases, the vessel exercising the right of navigation may be undertaking a protest—as seen in New Zealand and the United States in recent times. The protesters intend to disrupt the activities of the survey vessel to draw public attention to their cause. In the *Arctic Sunrise* case, the Tribunal found that protest at sea is 'an internationally lawful use of the sea related to the freedom of navigation'.[104] These vessels also exercise freedom of navigation but are legally required to observe the rules relating to the collision regulations mentioned above.[105] The protesters argue that they are exercising rights of freedom of expression or other human rights.[106] The freedom of expression is rarely absolute in domestic legal systems, and is often subject to lawful, proportionate limitations.[107] These may include the protection

---

[103] Rule 18, Convention on the International Regulations for Preventing Collisions at Sea (opened for signature 10 October 1972, entered into force 15 July 1977) 1050 UNTS 18 (COLREGs).
[104] *Arctic Sunrise* arbitration (Merits), para. 227.
[105] See Richard Caddell, 'Platforms, Protesters and Provisional Measures: The Arctic Sunrise Dispute and Environmental Activism at Sea' in M Ambrus and RA Wessel (eds), *Netherlands Yearbook of International Law 2014* (The Hague: TMC Asser Press): pp. 359–84.
[106] See, e.g. Jasper Teulings, 'Peaceful Protests against Whaling on the High Seas: A Human Rights-based Approach' in Clive R Symmons (ed), *Selected Contemporary Issues in the Law of the Sea* (Leiden: Martinus Nijhoff, 2011): pp. 221–49; Duncan Currie, 'Proposed Amendments to Crown Minerals (Permitting and Crown Land) Bill Under International Law' (5 April 2013) www.greenpeace.org/new-zealand/en/reports/Legal-opinion-on-the-proposed-Crown-Minerals-Act-amendments/.
[107] Matthew McMenamin, 'Protest at Sea: An Analysis of the Crown Minerals Amendment Act 2013' *Human Rights Research Journal* 8 (2013) www.victoria.ac.nz/law/centres/nzcpl/publications/human-rights-research-journal/publications/vol-8/MATTHEW-MCMENAMIN-HRR-2013.pdf.

A Framework for Balancing Rights on the Extended Continental Shelf 199

of national security or public order, or of public health or morals.[108] Inevitably, the considerations involve a balancing of interests.[109]

The right to protest has been acknowledged by the IMO's Maritime Safety Committee (MSC), which also had serious concerns about the safety implications of protest action. The MSC called on governments to discourage protesters from intentionally endangering safety or the marine environment and to take action in respect of protests that breached international regulations for the safety of navigation.[110]

With this in mind, a coastal State wishing to restrict protests at sea should consider the human right to freedom of expression, although this must be balanced against the safety of navigation. In one interesting example, the United States has prohibited protesters in the territorial sea from navigating within 500 yards of transiting vessels associated with Shell's exploration activities in the Arctic. At the same time, the regulation designated a 'Voluntary First Amendment Area' for protesters in which navigation speeds would be reduced to minimise the risk of collisions.[111]

In summary, both the coastal State and the flag State have legitimate interests that they are seeking to protect. In economic terms the coastal State interest outweighs that of other States impacted by the restriction on navigation. The safety of the survey vessel and the other vessel must also be considered. However, other interests must be borne in mind, including not unduly limiting navigation and the importance of respecting freedom of expression.

The third question to be considered is whether the proposed interference with the high seas rights is as minimal as possible to achieve the coastal State's objectives, or whether a less restrictive option is available. One argument is that the coastal State should not impose greater restrictions than those available under international law. Therefore, the coastal State could choose to respond only to breaches of the COLREGs, or the Convention for the Suppression of Unlawful Acts that Endanger the Safety of Navigation.[112] However, as set out above, this may not achieve the aim of the coastal State, which is to minimize the interference of other vessels with the seismic survey.

The question, therefore, is whether the attempted regulation is as minimal as possible. For example, is it possible to impose a smaller exclusion zone around the

---

[108] Article 19, International Covenant on Civil and Political Rights (opened for signature on 19 December 1966, entered into force 28 March 1979) 999 UNTS 171.

[109] Teulings discusses a range of international and national court decisions on non-violent protests at sea. Teulings, 'Peaceful Protests against Whaling on the High Seas'.

[110] Resolution MSC.303(87) 'Assuring Safety During Demonstrations, Protests or Confrontations on the High Seas' (adopted on 17 May 2010) www.maritimenz.govt.nz/adygil/imo-resolution.pdf.

[111] 33 CFR 165 'Safety Zones and Regulated Navigation Area: Shell Arctic Drilling/Exploration Vessels and Associated Voluntary First Amendment Area, Puget Sound, WA' www.gpo.gov/fdsys/pkg/FR-2015-04-28/pdf/2015-09858.pdf. For background, see Victoria Cavaliere, 'Shell Oil Rig Arrives in Seattle Waters Amid Protests, Permit Controversy', *Reuters*, 14 May 2015 www.reuters.com/article/2015/05/15/us-usa-shell-arctic-idUSKBN0NZ2LD20150515.

[112] (Opened for signature 10 March 1988, entered into force 1 March 1989) 1988 UNTS 1992 (SUA Convention).

survey vessel, one that would still protect the safety and navigation interests of the survey vessel, while also allowing for greater freedom of navigation? It should be noted that, under the New Zealand legislation, the exclusion zone may be *up to* 500 metres, allowing for some discretion on the part of the minister.[113] This allows for a smaller zone to be declared if that would be possible. However, in practice the non-interference zone is usually set at 500 metres in New Zealand. The South African legislation also provides for a smaller safety zone if the minister so decides. In contrast, 500 metres has been criticized as being insufficient to protect installations from navigational collisions or terrorist attacks. There were concerns about this during the negotiation of article 60, but it was decided to leave the distance at 500 metres but to allow for law-making by international organizations.[114] Some discussion has been held in the IMO but no agreement on extending safety zones beyond 500 metres has been reached.[115] The concern that has been expressed is that 500 metres is insufficient to prevent intentional attacks on installations. A vessel approaching at a speed of 25 knots would pass from the outer edge of a safety zone to the installation in approximately 39 seconds.[116] In light of this fact, it is arguable that 500 metres is a reasonable distance to protect the safety of navigation.

A second consideration is whether there is any allowance for inadvertent or necessary entry into the safety zone. Both South Africa and Canada provide exceptions when a vessel is in a distress or responding to an emergency. However, no allowance is made for situations where a vessel enters the zone through error. New Zealand's legislation is more vaguely worded: entry into the zone is not permitted unless the vessel has a 'reasonable excuse', which is not defined. It is arguable that New Zealand's legislation could be interpreted to cover inadvertent entry into the zone as an incident of ordinary navigation. So far, no judicial interpretation has cast light on this wording.

Finally, the coastal State should consider any guidance that may exist at a regional or international level that would cast light on whether its actions would be unjustifiable. The IMO issued a resolution on 'Assuring Safety During Demonstrations, Protests or Confrontations on the High Seas' in 2010.[117] While this was primarily in response to the actions of the Sea Shepherd organization against Japanese whaling vessels in the Southern Ocean and applies to the high seas with no reference to the continental shelf, it does assist in the present case. The resolution called on governments to urge persons under their jurisdiction to refrain from actions that intentionally imperil human life, the marine environment or property during demonstrations, protests or confrontations at sea. Governments should take action over any breaches of international collision or other applicable regulations[118] and

---

[113] Section 101B(7) and (8), Crown Minerals Act 1991 (NZ).
[114] O'Connell, *The International Law of the Sea: Volume I*, p. 502.
[115] Assef Harel, 'Preventing Terrorist Attacks on Offshore Platforms: Do States Have Sufficient Legal Tools?' *Harvard National Security Journal* 4 (2012): pp. 131–84, p. 149.
[116] Stuart Kaye, 'International Measures to Protect Oil Platforms, Pipelines, and Submarine Cables from Attack' *Tulane Maritime Law Journal* 31 (2006–2007): pp. 377–424, p. 405.
[117] Resolution MSC 303.87 (17 May 2010) www.maritimenz.govt.nz/adygil/imo-resolution.pdf.
[118] Including breaches of the SUA Convention.

cooperate to address any such action. Although this resolution did not address the particular situation addressed by the non-interference zones, it reinforces the importance of protecting safety during protests. There is a real danger that protesters deliberately obstructing a moving vessel will put safety of life and property at risk.

Overall, a moving vessel is of a very different character from an installation that is fixed and which can be noted on charts. The question is whether requiring vessels to remain 500 metres from a moving survey vessel is an unjustifiable interference with the freedom of navigation, which includes the freedom to protest. Given that protest vessels must comply with the law of the sea, including collision regulations, it does not seem unreasonable to require them to keep their distance from a vessel conducting a seismic survey. The use of chase vessels should help to prevent inadvertent breaches by the usual commercial traffic. Provision should be made so that inadvertent but safe entry to the safety zone as a result of navigation will not be the subject of coastal State jurisdiction.

This conclusion is reinforced by the Tribunal in the *Arctic Sunrise* arbitration. The Tribunal found that coastal States faced with protest action beyond the safety zone of structures would be justified in preventing dangerous situations or delays in essential operations.[119] However, the coastal State should tolerate some level of nuisance through civilian protest, 'so long as it does not amount to an "interference with the exercise of its sovereign rights"'.[120] Any measures by the coastal State must be reasonable, necessary and proportionate.[121]

As with any balancing of interests, how reasonable the action is will depend on the circumstances. For example, if the seismic survey is being conducted in areas of heavy shipping it may be unreasonable to impose a 500 metre non-interference zone. In more isolated waters this may not be such an issue.

C. Conclusion

The intersection between the high seas regime and the rights of coastal States in their extended continental shelves is potentially a source of conflict where States disagree about their respective rights and interests. Any attempt by the coastal State to regulate activities taking place on the high seas risks creating discord. This chapter has considered the appropriate meaning of the coastal State's responsibility not to allow their activities to 'infringe or result in unjustifiable interference' with the rights and freedoms of other States in the waters above the extended continental shelf as set out in article 78(2). It is clear from an analysis of the negotiating history of the Continental Shelf Convention and the LOSC that neither the coastal State's rights, nor the rights of the State exercising high seas freedoms, will automatically

---

[119] *Arctic Sunrise* (Merits), para. 327.
[120] Ibid., para. 328. It should be noted that the Tribunal was discussing protest in the EEZ, not above the extended continental shelf.
[121] Ibid., para. 326.

prevail over the other. Rather, the correct approach is that the coastal State should only interfere with the rights of flag States where it is essential to protect important coastal State interests.

Five important factors should be taken into account by the coastal State in evaluating whether a proposed measure is permissible under article 78. First, the activity of the flag State's vessels must be likely to harm the coastal State's rights in the extended continental shelf. Secondly, the rights of the coastal State and the flag State must be evaluated and compared. This assessment should also consider any relevant interests of the international community such as the importance of freedom of navigation. Thirdly, in order that the measure be reasonable, it must interfere with the flag State's rights as little as possible. If there is an effective alternative measure that is less intrusive, the coastal State should prefer the less intrusive approach. Fourthly, the guidance or rules of relevant international institutions may need to be considered. Finally, in order to act reasonably, the coastal State should ensure it consults with specially affected States about the proposed measure if there is reason to believe their interests will be detrimentally impacted.

In many cases, the best course will be for the coastal State to work through multilateral forums to negotiate acceptance of the measures, as discussed in Chapter 9. However, this is not strictly required under article 78(2). As noted by the Law Commission in relation to the Continental Shelf Convention, the determination of the reasonableness of the measure is in the hands of the coastal State. This does not, of course, prevent other States from objecting or resorting to dispute settlement to challenge the coastal State's decision.[122]

---

[122] Dispute settlement under the LOSC is set out in Part XV. See Natalie Klein, *Dispute Settlement in the UN Convention on the Law of the Sea* (Cambridge: Cambridge University Press, 2005).

8

Enforcement Powers of Coastal States in Relation to the Continental Shelf Beyond 200 Nautical Miles

A significant question for coastal States is the extent of their right to enforce their laws and regulations against foreign vessels operating above the extended continental shelf. If a vessel on the high seas above the extended continental shelf is suspected of bottom trawling for sedentary species or researchers are taking samples of sedentary species, can a coastal State board and inspect the vessel, or even detain the vessel and its crew for violations of coastal State law? It is well recognized in international law that enforcement jurisdiction may not always coexist with prescriptive jurisdiction.[1]

This chapter explores the arguments for and against a right of visit and/or enforcement by a coastal State against a foreign vessel. It is concluded that there are good grounds for arguing there is a right to exercise enforcement jurisdiction, including detention, where a foreign vessel breaches lawful coastal State regulations above the extended continental shelf.[2]

A. Arguments Against an Enforcement Right

The first and most obvious obstacle to concluding that the Law of the Sea Convention (LOSC)[3] confers enforcement jurisdiction on the coastal State is that there is no explicit recognition of this right. In respect of other zones, the LOSC contains express recognition of a right to exercise jurisdiction over foreign vessels.

In relation to the territorial sea, article 25 provides that coastal States may 'take the necessary steps' to 'prevent passage which is not innocent'. Article 33 confers on

---

[1] See, e.g. Vaughan Lowe, 'Jurisdiction' in Malcolm Evans (ed), *International Law* (Oxford: Oxford University Press, 2003): pp. 329–55, p. 332; Malcolm N Shaw, *International Law*, 6th edn (Cambridge: Cambridge University Press, 2008): pp. 645–6.
[2] This argument was first explored in Joanna Mossop, 'Regulating Uses of Marine Biodiversity on the Outer Continental Shelf' in Davor Vidas (ed), *Law, Technology and Science for Oceans in Globalisation* (Leiden: Brill, 2010): pp. 319–37.
[3] United Nations Convention on the Law of the Sea (opened for signature 10 December 1982, entered into force 16 November 1994) 1834 UNTS 397.

the coastal State the right to 'exercise the control necessary' to prevent infringement of its national laws on certain matters, and to punish infringements committed in the territorial sea and territory of the State. Article 73 provides for coastal State jurisdiction over foreign ships in its exclusive economic zone (EEZ) to protect its sovereign rights over the living resources in the EEZ, whilst also imposing restrictions on the coastal State such as prohibiting imprisonment as a penalty and providing for prompt release of foreign vessels. Article 73 is limited, however: it does not confer enforcement jurisdiction over all activities in the EEZ over which a State has jurisdiction under article 56. For example, it does not apply to non-living resources; nor does it apply to a State's jurisdiction over installations or to marine scientific research.[4] Article 220 provides for enforcement by coastal States in respect of pollution offences by foreign vessels. It might be expected that matters contained in Part VI regarding the continental shelf would also be covered by an enforcement clause, but this is not the case.

The history of negotiation of article 73 indicates an intention that enforcement rights for the sovereign rights contained in article 56(1) of the LOSC would be dealt with in the relevant parts of the Convention. Some delegations did suggest that article 73 should confer a general enforcement right for all activities in the EEZ.[5] However, most drafts of article 73 focused on enforcement in respect of living resources only. The authors of the Virginia Commentaries argue that:[6]

[t]he extensive enforcement powers set out in article 73 do not relate to the sovereign rights of the coastal State over the nonliving resources of the exclusive economic zone ... which are governed by Part VI, or to the matters over which the coastal State has 'jurisdiction' or 'other rights' under article 56, paragraphs 1(b) and (c), respectively. For example, the enforcement of the coastal State's laws and regulations for the protection and preservation of the marine environment is dealt with in article 220.

No explicit enforcement right is found in Part VI of the LOSC. Instead, there is confirmation that the waters above the extended continental shelf are high seas, and that coastal States may not unjustifiably interfere with the rights of the high seas, including freedom of navigation.[7] The freedom of navigation is supported by exclusive flag State jurisdiction, and on the high seas there are only limited circumstances in which another State may board and inspect, let alone detain, a foreign vessel.[8] It could not be said that a vessel operating on the high seas but interfering with continental shelf resources fits into any of the limited exceptions to exclusive flag State jurisdiction on the high seas in article 110.

In the application for provisional measures in the *Arctic Sunrise* case, Judges Wolfrum and Kelly expressed a view in a joint separate opinion that the enforcement jurisdiction of the coastal State is limited if it is not legitimized by one of

---

[4] Myron H Nordquist (ed), *United Nations Convention on the Law of the Sea 1982: A Commentary Volume II* (The Hague: Martinus Nijhoff Publishers, 1993): p. 794.
[5] Ibid., p. 793.   [6] Ibid., p. 794.   [7] Article 78, LOSC.
[8] Articles 92 and 110, LOSC.

the express provisions allowing for the exercise of jurisdiction such as article 73.[9] This view represents the arguments against an enforcement right in relation to the resources of the continental shelf.

B. Arguments in Favour of an Enforcement Right

The LOSC does imply that a coastal State may exercise enforcement jurisdiction over vessels in the waters above the continental shelf in some circumstances. First, article 80 provides that article 60 applies *mutatis mutandis* (with the necessary changes) to installations on the continental shelf—and article 60 provides for exclusive jurisdiction on installations and the creation of safety zones, within which the coastal State may exercise control to ensure the safety of the installation and navigation. Secondly, article 111(2) provides that a right of hot pursuit may apply to violations in the EEZ 'or on the continental shelf, including safety zones around continental shelf installations'. There is no distinction in article 111(2) between the inner and extended continental shelf, implying that a coastal State would be permitted to exercise jurisdiction over vessels breaching regulations applicable to the extended continental shelf, even if the vessel is on the high seas above the shelf at the time of the suspected offence. It must be acknowledged, of course, that the right applies *mutatis mutandis*, which could involve some diminution of States' rights in those zones.

The most convincing argument relates to the understanding of the Convention on the Continental Shelf[10] and enforcement jurisdiction. Part VI of the LOSC is heavily influenced by the Continental Shelf Convention. Articles 77 and 78(1) are substantially the same as articles 2 and 3 of the Continental Shelf Convention. Article 78(2), which provides that the exercise of the rights of the coastal State over the continental shelf must not 'infringe or result in any unjustifiable interference' with navigation and other rights under the LOSC, is very similar to article 5(1) of the Continental Shelf Convention. As with the LOSC, the Continental Shelf Convention did not provide for any express right of enforcement by States protecting their rights over the continental shelf in the high seas. However, the International Law Commission (ILC) commentary to the Convention states that:[11]

[t]he text as now adopted leaves no doubt that the rights conferred upon the coastal State cover all rights necessary for and connected with the exploration and exploitation of the

---

[9] The *Arctic Sunrise* case *(Kingdom of the Netherlands v Russian Federation), Provisional Measures*, ITLOS, Order of 22 November 2013, Separate Joint Opinion of Judge Wolfrum and Judge Kelly, para. 13. For further discussion of this case, see below at footnote 55.
[10] Convention on the Continental Shelf (opened for signature 29 April 1958, entered into force 10 June 1964) 499 UNTS 312.
[11] Commentary to the Articles Concerning the Law of the Sea, *Yearbook of the International Law Commission* [1956] II: pp. 265–91, p. 297. See also ED Brown, *The Legal Regime of Hydrospace* (London: Stevens & Sons, 1971): p. 91; MS McDougal and WT Burke, *The Public Order of the Oceans: A Contemporary International Law of the Sea* (New Haven: New Haven Press, 1987): p. 698.

natural resources of the continental shelf. Such rights include jurisdiction in connexion with the prevention and punishment of violations of the law.

This brief comment suggests that the ILC considered enforcement jurisdiction as inextricably linked with the 'sovereign rights' to explore and exploit the natural resources of the continental shelf.[12] The limited practice of States from that period also supports the existence of an enforcement right.

A final argument is that many States with rights to the extended continental shelf under the LOSC have enforcement powers in their legislation. These are discussed below.

(1) Academic views of the Continental Shelf Convention and enforcement jurisdiction

The existence of enforcement jurisdiction seems to have raised some questions for States and jurists at the time. Brown considered that 'the question is not free from doubt', although he concluded that a coastal State did have some degree of jurisdiction and control over foreign vessels in the high seas above the continental shelf.[13] Oda came to a similar conclusion, but also proposed that the right be explicitly set out in an amended Continental Shelf Convention.[14]

Brown and Oda relied for their conclusions on a functional interpretation of the Continental Shelf Convention. Brown argued that, where a vessel operating on the high seas came into physical contact with the continental shelf, the coastal State would have jurisdiction to exercise control over the interference with its rights: '[I]t would be unduly optimistic on the part of the intruder to think that a distinction could be made between, for example, interference with the dredge on the seafloor and interference with the towing vessel on the surface'.[15] Brown likened the exercise of jurisdiction over vessels on the high seas for offences committed on the continental shelf to the exercise of jurisdiction in the contiguous zone for offences committed, or about to be committed, in the territorial sea.[16] Oda, like the ILC, considered the enforcement jurisdiction to be a logical extension of the existence of sovereign rights over the resources of the continental shelf. He gave the example of foreign vessels acting against installations constructed on the continental shelf by the coastal State, and argued it would be unrealistic to expect the coastal State to postpone seizure and punishment until the offending vessels entered the territorial sea.[17]

---

[12] Similar comments were contained in earlier reports, including in 1953. See McDougal and Burke, *The Public Order of the Oceans*, p. 698.
[13] ED Brown, *The Legal Regime of Hydrospace* (London: Stevens & Sons, 1971): p. 94.
[14] Shigeru Oda, 'Proposals for Revising the Convention on the Continental Shelf' *Columbia Journal of Transnational Law* 7 (1968): pp. 1–31, pp. 19–20. Oda's argument in favour of coastal State jurisdiction was made in the context of a proposal to include a clear statement of that jurisdiction in a revised treaty, and so could be read as a suggestion for reform rather than as a discussion of existing rights; however, his comments do not appear to be intended to be interpreted in that light.
[15] Brown, *Legal Regime of Hydrospace*, p. 92.   [16] Ibid., p. 94.
[17] Oda, 'Revising the Convention on the Continental Shelf', pp. 19–20.

Logic would require that the coastal State be competent to prevent intrusion on the superjacent waters of the continental shelf by foreign vessels which intend to explore the continental shelf or to exploit its resources without first securing permission of the coastal State. Coastal States similarly should be able to prevent infringement by foreign vessels of their regulations regarding exploration or exploitation and to punish violators.

Academic writers were not fully in support of the coastal State right of enforcement, however. Slouka has argued that, at least by 1958, there was no acceptance of jurisdiction by a coastal State to take enforcement action against foreign vessels in the high seas above the continental shelf.[18] However, Slouka's comments in this regard were in the context of coastal State duties to maintain reasonable standards of order and safety on the high seas above the continental shelf, rather than directly addressing enforcement jurisdiction relating to resource exploitation, over which he suggested there was a 'question'.

O'Connell, in considering whether there was a right of hot pursuit under the Continental Shelf Convention, claimed that there was a difference between hot pursuit from the waters above the continental shelf and hot pursuit from a fishery zone:[19]

[i]n as much as in the former case international law authorizes the exhibition of the coastal State's power only with respect to the seabed, and only for the purposes of its exploration and exploitation; whereas in the latter case it authorizes it with respect to a species of act done in the sea itself.

O'Connell's view was that pursuit of a vessel from waters above the continental shelf required some nexus—physical or constructive contact—between the vessel's unlawful acts and the seabed. It follows from this conclusion that he would consider that the right to enforce the coastal State's continental shelf laws would similarly require such a nexus. The important question is what circumstances establish a sufficient nexus between ship and continental shelf. However, he also stated that the LOSC 'seeks to overcome the difficulties described by applying the doctrine of hot pursuit to violations on the continental shelf, including safety zones'.[20] The implication of this comment is that the existence of an enforcement right above the continental shelf is made clearer by article 111 of the LOSC.

In his discussion of whether hot pursuit could commence above the continental shelf under the Continental Shelf Convention, Poulantzas emphasized the ILC discussions before concluding that the jurisdiction of the coastal State included 'preventative and repressive measures against violations of its laws in force on the continental shelf'.[21] In particular, Poulantzas approved of hot pursuit applying to safety zones around installations and artificial islands.

---

[18] See Zdenek J Slouka, *International Custom and the Continental Shelf* (The Hague: Martinus Nijhoff, 1969), p. 168. See also the discussion in Slouka (p. 120), where he describes the enforcement jurisdiction issue as 'a point of considerable legal uncertainty'.
[19] DP O'Connell, *The International Law of the Sea: Volume II*, edited by Ivan Shearer (Oxford: Clarendon Press, 1984): p. 1087.
[20] Ibid., p. 1087.
[21] NM Poulantzas, *The Right of Hot Pursuit in International Law* (Leiden: AW Sijthoff, 1969): p. 170.

(2) State practice under the Continental Shelf Convention

Contemporary State practice reflected a belief by significant continental shelf States that they had enforcement jurisdiction over activities directed at living resources of the continental shelf. The United States included in its legislation a right to enforce domestic regulations relating to sedentary species against foreign vessels from the mid-1960s.[22] However, in practice, the United States appeared to take a conservative approach to actual enforcement action against foreign vessels, preferring instead to enter into bilateral arrangements with States whose nationals traditionally harvested continental shelf species.[23] Fidell reported that, by 1974, no foreign vessel had been seized and no foreign master had been prosecuted under the legislative regime.[24] This may partly have been related to potential disagreements over the inclusion of species such as king crabs in the list of continental shelf resources under the United States legislation.[25]

The Soviet Union and Australia also included enforcement powers in the legislation implementing their rights over continental shelf resources. The Australian Pearl Fisheries Act (No. 2) 1953 expressly applied to foreign vessels in proclaimed waters, including areas beyond the territorial sea.[26] Powers in the Act included boarding, inspection and arrest.[27] Goldie approved of the powers included in the Act, noting that the legislation assumed powers 'whose exercise is conducted on the surface of the sea'.[28] He asserted that those powers had a 'necessary connection' with the control of sedentary species, and were therefore appropriate.[29] In 1968, Australia continued this practice in its Continental Shelf (Living Natural Resources) Act. Enforcement officers were granted considerable powers to board and inspect vessels above the continental shelf, and arrest persons suspected of violating the Act.[30] The Soviet Union also prohibited foreign nationals from exploiting the resources of the continental shelf without permission, and provided for the exercise of jurisdiction on the high seas above the continental shelf regarding unauthorized activities.[31] Brown noted with some disapproval that the jurisdiction was claimed over vessels engaged in activities relating to the continental shelf but which may not come into physical contact with the shelf, such as surveying of the shelf.[32] It is interesting to note that the more recent Russian legislation also contains enforcement powers in

---

[22] Public Law 88-308, 88th Congress, S.1988, 20 May 1964, as cited by Brown, *The Legal Regime of Hydrospace*, p. 93. See also United States Regulations for Continental Shelf Fisheries Resources, Part 295.6 of CFR Title 50; 41 Federal Register No 123, 24 June 1976.

[23] AW Rovine, 'Contemporary Practice of the United States Relating to International Law' (1975) *American Journal of International Law* 69 (1975): pp. 141–53, p. 149.

[24] ER Fidell, 'Ten Years Under the Bartlett Act: A Status Report on the Prohibition on Foreign Fishing' *Boston University Law Review* 54 (1974): pp. 703–56, p. 725.

[25] Ibid., p. 710.

[26] See the discussion in LFE Goldie, 'Australia's Continental Shelf: Legislation and Proclamations' *International and Comparative Law Quarterly* 3 (1954): pp. 535–75.

[27] Ibid., p. 542.   [28] Ibid., p. 551.   [29] Ibid., p. 552.

[30] Section 14, Continental Shelf (Living Natural Resources) Act 1968 (Australian Commonwealth).

[31] Edict of the Presidium of the USSR Supreme Soviet concerning the Continental Shelf, 6 February 1968 (1968) 7 ILM 392, cited in Brown, *The Legal Regime of Hydrospace*, p. 93.

[32] Brown, *The Legal Regime of Hydrospace*, p. 94.

respect of the continental shelf as a whole, including the extended shelf.[33] Finally, the United Kingdom legislation assumed enforcement jurisdiction in respect of installations and safety zones on the continental shelf.[34]

In the reports of disputes that arose between States regarding the exploitation of the living resources on the continental shelf, it appears that the disputes tended to focus on whether particular species fell into the category of sedentary species, rather than on the legality of any exercise of jurisdiction connected to the sovereign rights. Two prominent disputes involved the arrest by the coastal State of vessels targeting sedentary species in the high seas above the continental shelf. In 1963, Brazil arrested three French vessels harvesting lobsters and an ongoing dispute arose, which included the dispatch of a French warship to protect its nationals.[35] In arguing that Brazil had no legal authority to exercise jurisdiction over the French vessels, France appeared to rely on the argument that lobsters were not sedentary species and therefore Brazil could not interfere with the high seas freedom to fish.[36] Reports of the dispute do not suggest that the exercise of enforcement jurisdiction per se was at issue.

A similar dispute arose between the United States and Canada.[37] Two United States vessels were arrested by Canada on 26 July 1994 approximately ten miles beyond the 200 nm EEZ.[38] The vessels, the *Warrior* and *Alpha & Omega 2*, were dredging for scallops on Canada's continental shelf when they were detained.[39] According to Canada's Coastal Fisheries Protection Act 1990, the vessels had committed an offence by fishing for Icelandic scallops on the continental shelf without a licence.[40] The vessels were escorted to St John's, Newfoundland, where the captains of the vessels were charged but ultimately released on bond.

United States officials immediately protested against the detention.[41] The main reason for the protest centred on the US position that Icelandic scallops were not

---

[33] Russian Federal Law on the Continental Shelf of the Russian Federation, adopted by the State Duma on 25 October 1995 (1996) 35 ILM 1500.
[34] United Kingdom Continental Shelf Act 1964, cited in Brown, *The Legal Regime of Hydrospace*, pp. 89–90 and p. 92. See also Oda, 'Revising the Convention on the Continental Shelf, p. 23.
[35] Issam Azzam, 'The Dispute Between France and Brazil over Lobster Fishing in the Atlantic' *International and Comparative Law Quarterly* 13 (1964): pp. 1453–9.
[36] Ibid., p. 1454. See also LFE Goldie, 'The Ocean's Resources and International Law: Possible Developments in Regional Fisheries Management' *Columbia Journal of Transnational Law* 8 (1969): pp. 1–53, p. 13.
[37] Michael Matza, 'US, Canada Go At It Again Over Their Fishing Rights' *Philadelphia Inquirer* (13 August 1994), p. A.1. See also JM Van Dyke, 'Modifying the 1982 Law of the Sea Convention: New Initiatives on Governance of High Seas Fisheries Resources: the Straddling Stocks Negotiations' *International Journal of Marine & Coastal Law* 10 (1995): pp. 219–27, pp. 221–2.
[38] At the time of the dispute, both Canada and the United States were party to the Continental Shelf Convention but not the LOSC. Canada's claim was based on the fact that the Continental Shelf Convention allows for a State's continental shelf to extend to 'where the depth of the superjacent waters admits of the exploitation of the natural resources of the said areas'. See article 1.
[39] James H Andrews, 'Behind US–Canada Fishing Flap' *The Christian Science Monitor* (8 August 1994), p. 8.
[40] Colin Nickerson, 'Canada Seizes Mass. Scallopers' *Boston Globe* (27 July 1994), p. 1.
[41] Jose Martinez, 'State Department protests seizure of scallop draggers' *Bangor Daily News* (28 July 1994).

sedentary. The State Department maintained that Icelandic scallops were capable of moving by rapidly clapping their bivalve shells together.[42] Therefore, the argument was that Canada had no rights in relation to the scallops because they did not meet the definition of sedentary species in article 77(4).[43]

The interesting aspect of this dispute is that the debate focused on whether Icelandic scallops were sedentary species, rather than on the arrest of the vessels on the high seas. Von Zharen notes that, at the time, both States had enforcement powers in their domestic legislation to protect sedentary species.[44] The absence of official protest on this point indicates that both States considered that enforcement jurisdiction was a consequence if the scallops were indeed a sedentary species.

Another incident involving Canadian and American fishermen arose in 2001, when Canada detained a fisherman collecting snow crab on the continental shelf twenty-four nm beyond the EEZ. In the subsequent prosecution of the fisherman, the defendant argued both that snow crab was not a sedentary species and that Canada could not exercise enforcement jurisdiction because the precise extent of the continental shelf was unknown.[45] The Court found that Canada had asserted a claim to the continental shelf and that the fishing took place far from any area that may be contestable. Therefore, the prosecution could proceed. The defendant at no point argued that there was no enforcement right on the continental shelf beyond 200 nm.

(3) Developments under the Law of the Sea Convention

(a) State practice

A number of States with extended continental shelves have legislation providing for the enforcement of laws relating to the extended continental shelf. The usual approach is to state that the relevant act applies to the continental shelf, while making it clear that the continental shelf includes those areas beyond the EEZ.

Australian enforcement officials may exercise considerable powers against vessels taking sedentary species, including boarding, inspection, detention and arrest.[46] The powers apply to areas of the continental shelf that are notified by regulation. The continental shelf is defined consistently with article 76 of the LOSC.[47]

In New Zealand, an enforcement officer may inspect structures on the continental shelf and a vessel in the waters above the extended continental shelf to determine

---

[42] 'US Fishing Boats Seized; Scallop Debate Ensues' *The Wall Street Journal* (28 July 1994).
[43] The United States subsequently accepted that the Icelandic scallops were sedentary species. 'Offshore Shell Game' *Boston Globe* (5 December 1994), p. 12.
[44] WM von Zharen, 'An Ecopolicy Perspective for Sustaining Living Marine Species' *Ocean Development & International Law* 30 (1999): pp. 1–42, p. 4.
[45] *R v Perry*, 2003 Carswell Nfld 23, 222 Nfld and PEIR 313, 663 APR 313, 59 WCB (2d) 92. Section 17 of the Oceans Act stated that the continental shelf consists of the seabed and subsoil of the shelf throughout the natural prolongation of the land territory of Canada. The outer edge of the continental shelf is the 'outer edge of the continental margin'.
[46] Section 14, Continental Shelf (Living Natural Resources) Act 1968 (Australia).
[47] Ibid., Section 5.

whether offences have been committed against the environmental protection measures established under the Act.[48] Additionally, if a vessel is found to have breached a safety zone around survey vessels operating above the continental shelf, including the continental shelf beyond 200 nm, an enforcement officer may board the vessel and arrest a person.[49]

The United Kingdom gives broad powers to enforcement officers in relation to oil and gas licensing matters in the United Kingdom marine area, which includes the continental shelf beyond 200 miles.[50] However, the legislation does limit United Kingdom jurisdiction over foreign vessels to matters that are consistent with international law.[51] This does not settle over which matters the United Kingdom considers it has enforcement jurisdiction on the extended continental shelf.

(b) International decisions

The functional approach to interpreting the content of sovereign rights has been approved by the International Tribunal for the Law of the Sea (ITLOS). In the *M/V Virginia G* case, the ITLOS considered whether Guinea-Bissau's regulation of bunkering of fishing vessels in its EEZ was part of its sovereign rights under the LOSC.[52] Referring to the 'sovereign rights' of a State in article 56, the Tribunal stated that: '[t]he term "sovereign rights" in the view of the Tribunal encompasses all rights necessary for and connected with the exploration, exploitation, conservation and management of the natural resources, *including the right to take the necessary enforcement measures*'.[53] In including enforcement as part of sovereign rights under the LOSC, the Tribunal relied on the above-mentioned ILC comments in relation to sovereign rights on the continental shelf.[54]

The issue of a coastal State's right to exercise enforcement jurisdiction in relation to activities in its EEZ and above the continental shelf arose in the *Arctic Sunrise* case. The Netherlands brought a case under the dispute settlement provisions of the LOSC regarding Russia's arrest of a Greenpeace protest vessel. Although some rigid-hull inflatable boats had entered the safety zone around the oil platform, the *Arctic Sunrise* itself did not. Russian forces boarded the vessel and arrested the crew in the EEZ outside the safety zone two days later. The Netherlands applied for

---

[48] Section 141, Exclusive Economic Zone and Continental Shelf (Environmental Effects) Act 2012 (New Zealand).
[49] Sections 101A–101C, Crown Minerals Act 1991 (New Zealand).
[50] Section 42 of the Marine and Coastal Access Act 2009 (United Kingdom) provides that the UK marine area includes 'the area of sea within the limits of the UK sector of the continental shelf' that is not included in the EEZ. See also sections 236 and 240, Marine and Coastal Access Act 2009 (UK).
[51] See section 42(2), which states that the area of sea in the sector of the continental shelf beyond the EEZ 'is to be treated as part of the UK marine area for any purpose only to the extent that such treatment for that purpose does not contravene any international obligation binding on the UK'. See also section 237 (in relation to non-EU vessels contravening nature conservation legislation), Marine and Coastal Access Act 2009 (UK).
[52] M/V 'Virginia G'*(Panama v Guinea-Bissau)* (Judgment) (2014) 53 ILM 1164.
[53] Ibid., para. 211, emphasis added.   [54] See above at Section B.

provisional measures from the ITLOS, including the release of the vessel and crew, pending the constitution of an arbitral tribunal.

At the provisional measures stage, the ITLOS found that the Arbitral Tribunal would have jurisdiction over the dispute and ordered that Russia release the vessel and the crew upon the posting of a bond of €3,600,000.[55] Russia did not appear at the hearing as it argued that the Tribunal did not have jurisdiction, based on a declaration under article 298(1)(b) that it made when it ratified LOSC.[56] The Tribunal recorded the Netherlands' submission that the Russian Federation could only exercise jurisdiction within the safety zone around installations. Outside that area, the Netherlands argued that only the flag State had jurisdiction. This was based on the view that 'any exceptions to the general rule [of exclusive flag State jurisdiction] over foreign vessels are explicit and cannot be implied'.[57]

The ITLOS did not give any indication of its view on this point in the order for provisional measures. However, two judges provided a separate joint opinion essentially adopting the Netherlands' position. Judge Wolfrum and Judge Kelly stated that:[58]

[a] coastal State has only limited enforcement jurisdiction in its exclusive economic zone. These are amongst others the competences set out in articles 73, 110, 111, 220, 221 and 226 of the Convention … As far as enforcement actions in the exclusive economic zone are concerned the enforcement jurisdiction of the coastal State is limited if it is not legitimized by one of the exceptions mentioned above.

In contrast Judge Golitsyn, in his dissenting opinion, argued that:[59]

Laws and regulations enacted by the coastal State in furtherance of its exclusive jurisdiction under article 60, paragraph 2, of the Convention would be meaningless if the coastal State did not have the authority to ensure their enforcement. Consequently, it follows from article 60, paragraph 2, of the Convention that the coastal State has the right to enforce such laws and regulations governing activities on artificial islands, installations and structures.

The Arbitral Tribunal that was convened to hear the merits of the dispute arrived at a position that is somewhat between these two views.[60] First, the Tribunal found that the Netherlands, as flag State, had exclusive jurisdiction over its vessel while it was in the Russian EEZ.[61] Although there are exceptions to this principle contained in the LOSC, Russia did not have the right to enforce violations of its laws relating to safety zones beyond the 500 metre safety zone unless it had met the criteria of hot pursuit.[62] The conclusion was based on the fact that the LOSC expressly refers to the 500 metre safety zone within which the coastal State may take 'appropriate measures'.[63]

---

[55] *Arctic Sunrise (Netherlands v Russian Federation)* (Provisional Measures) (2014) 53 ILM 607.
[56] Ibid., para. 40.   [57] Ibid., para. 63.
[58] Ibid. Separate Joint Opinion of Judge Wolfrum and Judge Kelly, paras. 12–13.
[59] Ibid. Dissenting Opinion of Judge Golitsyn, para. 23.
[60] *Arctic Sunrise (Netherlands v Russian Federation)* (Merits) PCA Case 2014-02, 14 August 2015. Russia did not appear in this case.
[61] Ibid., para. 231.   [62] Ibid., para. 244.
[63] Article 60(4), LOSC. See *Arctic Sunrise* (Merits), para. 229.

This approach is supported by academic commentators, who point to the fact that article 111 of the LOSC explicitly refers to hot pursuit from safety zones around continental shelf installations.[64] The point is that there would have been no need for inclusion of a reference to safety zones if the enforcement jurisdiction in relation to breaches extended throughout the EEZ or on the continental shelf. Oude Elferink suggested that this is consistent with the fact that article 60 creates special rights for coastal States and it is appropriate that its enforcement jurisdiction in relation to violation of safety zones is limited to the safety zones themselves.[65] Tho Pesch has argued that reasonableness requires that enforcement jurisdiction in safety zones be limited to those acts ensuring the safety of the installation and navigation. Enforcement should only be used if the vessel fails to comply with the request to leave the zone.[66]

Secondly, the Tribunal confirmed that coastal States do have the right to enforce their laws relating to non-living resources in the EEZ, although it declined to expand on the extent of the right.[67] The Tribunal noted the fact that article 73 only refers to living resources,[68] but that the Continental Shelf Convention was based on draft articles by the ILC, which considered enforcement to be an integral part of sovereign rights.[69] It concluded that there was no evidence that the *Arctic Sunrise* had breached any Russian laws in relation to exploration and exploitation activities on non-living resources in the EEZ, apart from the violation in relation to the safety zone.[70]

Thirdly, the Tribunal found that the types of situations in which it would be reasonable for a State to take preventative action included:[71]

(i) violations of its laws adopted in conformity with the Convention; (ii) dangerous situations that can result in injuries to persons and damage to equipment and installations; (iii) negative environmental consequences ... and (iv) delay or interruption in essential operations.

Although the Tribunal was focused on the incidents within the EEZ, its reasoning can be extended to areas above the extended continental shelf. This is because the rights discussed by the Tribunal affect the rights of navigation that apply equally in the EEZ and above the extended continental shelf. The rights of navigation must be exercised in light of the coastal State's sovereign rights above the continental shelf both within and outside 200nm.

---

[64] A Oude Elferink, 'The *Arctic Sunrise* Incident: A Multi-faceted Law of the Sea Case With a Human Rights Dimension' *International Journal of Marine and Coastal Law* 29 (2014): pp. 244–89, p. 258.
[65] Oude Elferink, 'Arctic Sunrise Incident', p. 259.
[66] Sebastian Tho Pesch, 'Coastal State Jurisdiction around Installations: Safety Zones in the Law of the Sea' *International Journal of Marine and Coastal Law* 30 (2015): pp. 512–32.
[67] *Arctic Sunrise* (Merits), para. 284. See also para. 324, where the Tribunal found that: 'A coastal State has the right to take measures to prevent interference with its sovereign rights for the exploration and exploitation of its non-living resources'.
[68] Ibid., para. 281.   [69] Ibid., para. 283.   [70] Ibid., para. 284.
[71] Ibid., para. 327.

Fourthly, the Tribunal specified that measures taken by the coastal State must be reasonable, necessary and proportionate.[72] The right of enforcement must consider the obligation to have due regard for the rights of other States, and so some protest action should be tolerated as long as it does not amount to an interference with the coastal State's sovereign rights.[73]

Finally, the Tribunal referred to the requirement in article 78 that the exercise of the coastal State's rights over the continental shelf must not infringe or result in any unjustifiable interference with navigation.[74] The Tribunal emphasized that the involvement of the *Arctic Sunrise* in protesting at the oil platform had come to an end.[75] It concluded that, because there was no evidence that at the time of the detention the *Arctic Sunrise* was interfering with the operation of the platform, the detention would have amounted to an unjustifiable interference, even if it was purported to be conducted pursuant to Russia's rights over the continental shelf. This seems to imply that the detention must coincide with the breach of the rights and any delay in detaining the vessel might mean that it becomes unjustifiable.

This latter observation imposes a significant limitation on the ability of the coastal State to exercise enforcement rights against vessels that have violated its legitimate laws in place to protect its rights over the continental shelf. The Tribunal appears to take the view that the detention must occur at the time of the violation and not at a later time. The *Arctic Sunrise* was detained a day and a half after the protest, but at the time it had returned to the oil platform and was circling it at a distance of four miles.[76] Although no evidence was presented by Russia, the Tribunal seemed to discount any potential concern that the vessel might resume its protest activity. It is therefore left uncertain whether a coastal State has a right to take measures to *prevent* breaches of its legitimate regulations unless the vessel is undertaking the activity or it is clearly imminent.

It is noteworthy that if the Tribunal's decision is that enforcement jurisdiction has to be exercised while there is still a threat of an interference with coastal State rights, this is inconsistent with the explicit enforcement provisions in the LOSC. For example, in article 73, the coastal State is not limited to enforcing its laws against foreign vessels only when they are committing an offence. Restricting the coastal State's enforcement rights in such a way would be a serious limitation on the ability of a coastal State to protect its rights and does not reflect the reality of enforcement in remote parts of the oceans. A better view would be that the coastal State has the right to enforce violations of its laws protecting its rights to exploration and exploitation even if the violation occurred in the past. The exception might be the enforcement of safety zones around installations, for the reasons expressed above. Of course, all enforcement actions will need to be conducted reasonably and proportionately.

---

[72] Ibid., para. 326.　[73] Ibid., para. 328.　[74] Ibid., para. 331.
[75] Ibid., para. 330.　[76] Ibid., paras. 85–100.

C. Conclusion

On the basis of the above evidence it appears reasonable to argue that, following the Continental Shelf Convention, the ILC, States and academic commentators assumed that 'sovereign rights' for the exploitation and exploration of the resources of the continental shelf included a right to enforce domestic regulations directed at protecting those rights. That enforcement right would include the right to board, inspect and detain vessels suspected of violating legitimate coastal State regulation regarding the living resources of the continental shelf. It is useful to recall that, because the EEZ did not yet exist, the right was exercised in the high seas. This is very similar to the situation in respect of the continental shelf beyond 200 nm.

The reliance on the Continental Shelf Convention explains the absence of an explicit enforcement right in Part VI of the LOSC. The inclusion of the continental shelf in article 111 on hot pursuit reinforces the conclusion that enforcement rights were inextricably part of the concept of 'sovereign rights' over the resources of the continental shelf.

The *Arctic Sunrise* decision has suggested that there are limits on the exercise of the enforcement jurisdiction. In particular, where the vessel is not arrested at the time of the interruption of coastal State activities, it may amount to an unjustifiable interference with navigation to detain the vessel once the interruption has ceased. In such a situation the coastal State would need to apply to the flag State to pursue measures against the vessel. This conclusion is not entirely satisfactory when compared to other enforcement rights in the LOSC, which are not limited to the time of the violation. It is clear that the coastal State must act reasonably in all circumstances.

9

Cooperative Approaches to Regulating Activities on the Continental Shelf Beyond 200 Nautical Miles

In previous chapters, the focus has been on the rights that may be exercised by coastal States unilaterally. However, there are clear limitations on taking a unilateral approach to regulating activities on the continental shelf beyond 200 nm. In particular, in situations where there is no, or little, existing State practice it is likely that disputes will arise between coastal and flag States when the interpretation of the Law of the Sea Convention (LOSC)[1] is unclear. There will be considerable difficulty in monitoring and enforcing activities beyond 200 nm, and so the cooperation of flag States will be important.

Where the coastal State wishes to regulate activities that are, or could be, under the mandate of regional or global organizations it makes sense for the coastal State to seek cooperation through such organizations. In a case where there is no relevant organization, then the coastal State should (and may in fact have a duty to) consult with affected States and others in the region.[2] Ideally, these States should reach agreement about how activities that impact on the extended continental shelf will be managed. Such agreement will encourage compliance with any measures, which is important when the areas involved will be very remote from land, and in deep water.[3]

This chapter discusses two examples of how coastal States have sought cooperative solutions to protect their sovereign rights to the extended continental shelf. The first is in the North-east Atlantic, where Portugal, and more recently the United Kingdom, have worked with regional organizations to establish protected areas on their extended shelf. The second is a bilateral treaty established by Mauritius and the Seychelles. The chapter then considers the possibilities for coastal States to cooperate with a range of regional and international organizations in relation to activities over the extended continental shelf. Much of the discussion in the chapter

---

[1] United Nations Convention on the Law of the Sea (concluded 10 December 1982, entered into force 16 November 1994) 1834 UNTS 397.
[2] See ch 7.
[3] Marta Chantal Ribeiro, 'The "Rainbow": The First National Marine Protected Area Proposed under the High Seas' *International Journal of Marine and Coastal Law* 25 (2010): pp. 183–207, p. 198.

is related to how coastal States can use cooperative arrangements to achieve protection for environmentally sensitive areas on the continental shelf beyond 200nm. But as the Mauritius–Seychelles example shows, bilateral cooperation is also a possible approach to dealing with potential conflicts over non-living resources.

Coastal States have been very protective of their rights in respect of continental shelves beyond 200 nm. Coastal States naturally want to avoid international organizations limiting their choices in protecting their sovereign rights. For this reason, it is not uncommon for the rules of organizations to exclude extended continental shelves from their jurisdiction.[4] However, these matters should not prevent coastal States from cooperating with these organizations and requesting management measures that protect the resources of the shelf. This was the approach taken by Portugal in respect of its extended continental shelf, which has been very successful.

It will be necessary to consider the issue of third-party States. Even where measures are agreed through a regional organization, it is possible, even likely, that vessels will be operating in the area that are flagged to non-party States. This means that a coastal State may also need to rely on unilateral rights. However, the interpretation of such rights may be strengthened if there is a regional agreement on the appropriate balancing of rights in certain situations.

A. Example 1: Multilateral Regional Cooperation in the North-east Atlantic

(1) Portugal in the North-east Atlantic

The Rainbow hydrothermal vent field was discovered in 1997. It is 1.5 km² and located 235 nm from Portugal's coastline.[5] It is between 2279 and 2320 metres below sea level and is the largest known hydrothermal vent field in the Mid-Atlantic ridge area.[6] It is of scientific interest for several reasons. The vents are in a geologically unusual site, the species found at the site differ from other, shallower, vent fields in the region and the vents are spatially and temporally dynamic.[7] Because of the interest in the Rainbow field, many scientific expeditions were sent to the area and some tourism has even occurred at the site.[8]

---

[4] For example, article I(4) Convention on Future Multilateral Cooperation in the North-west Atlantic Fisheries (opened for signature 24 October 1978, entered into force 1 January 1979) 1135 UNTS 369. See also article 1(f) of the Amendment (open for signature 28 September 2007) NAFO/GC Doc 07/4. In contrast, NEAFC can include areas under national jurisdiction so long as the coastal State agrees. See Convention on Future Multilateral Cooperation in North-east Atlantic Fisheries (opened for signature 18 November 1980, entered into force 17 March 1982, as amended in 2006) 1285 UNTS 129.

[5] Ribeiro, 'The Rainbow', p. 186.

[6] H Calado et al., 'Introducing a Legal Management Instrument for Offshore Marine Protected Areas in the Azores: The Azores Marine Park' *Environmental Science and Policy* 14 (2011): pp. 1175–87, p. 1178.

[7] Ibid., p. 1178.   [8] Ribeiro, 'The Rainbow', p. 187.

The OSPAR Commission was established by the 1992 OSPAR Convention[9] to protect the marine environment of the North-east Atlantic.[10] It has fifteen European State parties and is a member of the United Nations Environment Programme (UNEP) Regional Seas Agreements.[11] OSPAR's work programme includes issues related to the protection and conservation of ecosystems and biological diversity, hazardous and radioactive substances, eutrophication and environmental goals for offshore activities. One of the organization's goals is to establish marine protected areas (MPAs) and, in 2003, OSPAR agreed to establish a network of MPAs by 2010.[12]

In 2005, the World Wide Fund for Nature (WWF) lobbied OSPAR parties to have the Rainbow field nominated as an MPA in the high seas. Initially, the parties thought that the vent field was beyond national jurisdiction in the Area.[13] However, when Portugal established its Task Group for the Extension of the Portuguese Continental Shelf (EMEPC) in 2005, early indications were that the Rainbow vent field was on Portugal's extended continental shelf.[14]

When Portugal discovered the vent field was within its jurisdiction it agreed to proceed with the MPA on the basis that Portugal had jurisdiction over the seabed.[15] It made a formal proposal to OSPAR that the Rainbow vent field be included in the OSPAR network of MPAs. Portugal put in place management measures in respect of the Rainbow vent field but no measures were implemented by OSPAR in respect of the superjacent waters. It is significant that the members of OSPAR recognized Portugal's jurisdiction even though Portugal was still working on its submissions to the Committee on the Limits of the Continental Shelf (CLCS).[16]

Portugal has promoted the protection of four other areas on its continental shelf beyond 200 nm: Altair Seamount, Antialtair Seamount, Josephine Seamount and a series of seamounts on the North Atlantic Ridge.[17] Unlike the Rainbow vent field, OSPAR declared high seas MPAs over these seamounts in 2010. Thus, the measures are complementary based on jurisdiction. Portugal is responsible for measures

---

[9] Convention for the Protection of the Marine Environment of the North-East Atlantic (opened for signature 22 September 1992, entered into force 25 March 1998) 32 ILM 1069. (OSPAR Convention).

[10] The OSPAR Convention merged two Conventions: the Convention for the Prevention of Marine Pollution by Dumping from Ships and Aircraft 1972 (the Oslo Convention); and the Convention for the Prevention of Marine Pollution from Land-based Sources 1974 (the Paris Convention).

[11] State parties to OSPAR are: Belgium, Denmark, Finland, France, Germany, Iceland, Ireland, Luxembourg, the Netherlands, Norway, Portugal, Spain, Sweden, Switzerland, the United Kingdom and the European Union. Iceland, Norway and Switzerland are not members of the European Union. For a discussion of regional seas agreements, see Julien Rochette et al., 'Regional Oceans Governance Mechanisms: A Review' *Marine Policy* 60 (2015): pp. 9–19.

[12] Bremen Ministerial Statement, First Joint Ministerial Meeting of the Helsinki and OSPAR Commissions in Bremen, Germany (2003).

[13] Ribeiro, 'The Rainbow', p. 187.

[14] Ibid., p. 190. Resolution of the Council of Ministers, Portugal, 9/2005, of 17 January 2005. https://dre.pt/application/file/626140.

[15] Ibid., p. 195. [16] Ibid., p. 196.

[17] Marta Chantal Ribeiro, 'Marine Protected Areas: The Case of the Extended Continental Shelf' in Marta Chantal Ribeiro (ed), *30 Years After the Signature of the United Nations Convention on the Law of the Sea: The Protection of the Environment and the Future of the Law of the Sea* (Coimbra: Coimbra Editora, 2014): pp. 179–209, p. 195.

relating to the protection of the sea floor while OSPAR is responsible for measures in respect of the high seas.

OSPAR's jurisdiction is limited. It has competency over activities such as marine scientific research, cable laying, dumping, construction of installations and tourism. It is unable to exercise any jurisdiction over fishing, mining or navigation.[18] A variety of other organizations have competence for some form of human activity in the water column or on the seabed of the OSPAR area. These organizations include the North-east Atlantic Fisheries Commission (NEAFC), the International Commission for the Conservation of Atlantic Tunas (ICCAT), the North Atlantic Salmon Conservation Organization (NASCO), the North Atlantic Marine Mammal Commission (NAMMCO), the International Whaling Commission (IWC) and the International Maritime Organization (IMO).[19] OSPAR has pursued cooperation with these organizations. Memoranda of Understanding (MOUs) were signed with the International Council for the Exploration of the Seas (ICES) in 2006, the NEAFC in 2008 and the International Seabed Authority (ISA) in 2011, among others.[20] The MOU with NEAFC recognizes that the organizations have 'complementary competences and responsibilities for fisheries management and environmental protection respectively'.[21] It undertakes to ensure a free flow of mutually useful information, to have joint discussions over areas of mutual interest and cooperate on marine spatial planning and area management. Representatives from one organization are allowed to participate in the meetings of the other.[22] A 'Collective Arrangement' was concluded in 2014 between OSPAR and the NEAFC, with the goal of widening cooperation and coordination to other international organizations, including the IMO.[23]

The NEAFC is a regional fisheries management organization (RFMO) with competence over a range of fisheries in the North-east Atlantic.[24] Through the

---

[18] Julien Rochette et al., 'The Regional Approach to the Conservation and Sustainable Use of Marine Biodiversity in Areas Beyond National Jurisdiction' *Marine Policy* 49 (2014): pp. 109–17, p. 113.

[19] Tullio Scovazzi, 'Marine Protected Areas in Waters Beyond National Jurisdiction' in Marta Chantal Ribeiro (ed), *30 Years after the Signature of the United Nations Convention on the Law of the Sea: the Protection of the Environment and the Future of the Law of the Sea* (Coimbra: Coimbra Editora, 2014): pp. 209–38, p. 234. Clearly, not all of these organizations have competence in respect of seabed activities.

[20] For details of all OSPAR's MOUs, see www.ospar.org/about/international-cooperation/memoranda-of-understanding.

[21] Memorandum of Understanding between the NEAFC and the OSPAR Commission, http://www.ospar.org/site/assets/files/1357/mou_neafc_ospar.pdf.

[22] Stefán Ásmundsson and Emily Corcoran, '*Information Paper on the Process of Forming a Cooperative Mechanism between NEAFC and OSPAR: From the First Contact to a Formal Collective Arrangement*' (2016) www.neafc.org.

[23] 'Collective Arrangement between Competent International Organizations on Cooperation and Coordination regarding Selected Areas in Areas beyond National jurisdiction in the North-East Atlantic' OSPAR Agreement 2014-9. In addition to cooperation, the intention is to allow the compilation of comprehensive information on area management in the North-east Atlantic in annexes to the Arrangement. See Ásmundsson and Corcoran, '*Information Paper*', p. 17.

[24] Member States are Denmark (in respect of the Faroe Islands and Greenland), the European Union, Iceland, Norway and the Russian Federation.

2000s the organization began to introduce closed areas in which bottom fishing would be prohibited.[25] In 2006, the NEAFC Convention was amended to clarify that the organization had the legal competence to create conservation and management measures to protect marine ecosystems and preserve marine biodiversity.[26] By 2008, the NEAFC had introduced a rule restricting bottom fishing to areas that had already been exploited.[27] Fishing in new areas was subject to restrictions for exploratory fisheries, including the conducting of prior assessment of impacts on the benthic environment, as well as monitoring. In 2009, the NEAFC agreed to close areas to bottom trawling that largely mirrored the OSPAR proposals for MPAs.[28] A new rule was introduced in 2014, which lists thirteen areas in which bottom fishing is prohibited.[29] Under the NEAFC there is protection for the Altair and Antialtair seamounts and parts of the North Atlantic Ridge on Portugal's extended continental shelf.

Although Portugal has been successful in convincing NEAFC member States to close certain areas on its continental shelf to bottom fishing, it has encountered some resistance. For example, it has struggled to obtain acceptance from the NEAFC for the closure of Josephine Seamount as a conservation and management measure. Portugal has banned its fishing vessels from bottom fishing or gillnet fishing on the Josephine seamount and others on its extended continental shelf, but this has not been supported by NEAFC measures.[30]

At the national level, Portugal has taken steps to incorporate measures in support of the MPAs in domestic law.[31] In addition to establishing MPAs, legislation has established a legal framework for activities in Portugal's maritime zones, including the extended continental shelf.[32] Portugal has laws applicable to the extended shelf for the access and use of biological resources for scientific

[25] For a background to NEAFC's bottom fishing measures, see Stefán Ásmundsson, *The Development of NEAFC's Protection of Vulnerable Marine Ecosystems* (2015) www.neafc.org.
[26] Article 5, NEAFC Recommendation 19: 2014 'Protection of VMEs in NEAFC Regulatory Areas as Amended by Recommendation 9: 2015'.
[27] NEAFC Recommendation XVI: 2008 www.neafc.org/system/files/16-rec_bottom_fishing_em_2008.pdf.
[28] Rochette et al., 'The Regional Approach', p. 113.
[29] NEAFC Recommendation 19: 2014 'Protection of VMEs in NEAFC Regulatory Areas as Amended by Recommendation 9: 2015'.
[30] http://eu.savethehighseas.org/north-east-atlantic-fisheries-commission-makes-limited-progress-to-protect-deep-sea-species-and-habitats/. The Josephine Seamount is not listed in the current list of areas closed to bottom fishing. Article 5, NEAFC Recommendation 19: 2014. It seems that the EU has not always accepted Portugal's proposals or goals in respect of the extended continental shelf area, which has led to some internal tensions within the EU.
[31] For example, Law 114/2014, 28 May 2014, established the conditions applicable to Portuguese fishing vessels. See https://dre.pt/application/file/25346153.
[32] For a description of the broader Azores Marine Park initiative and its implementation, see Calado, 'Azores Marine Park' and Rita C Abecasis et al., 'Marine Conservation in the Azores: Evaluating Marine Protected Area Development in a Remote Island Context' *Frontiers in Marine Science* 2 (2015): Article 104, doi: 10.3389/fmars.2015.00104. See also Vasco Becker-Weinberg, 'Portugal's Legal Regime on Marine Spatial Planning and Management of the National Maritime Space' *Marine Policy* 61 (2015): pp. 46–53. Some tensions have arisen regarding the respective capabilities of the Azores regional government and the Portuguese government and the process of establishing the legal governance framework is continuing.

purposes and bioprospecting as well as exploration and exploitation of mineral resources.³³ In the past some criticism was aimed at MPAs in Portugal's waters because they lacked management plans and enforcement.³⁴ It is important that management of MPAs is operationalized to avoid 'paper parks' with no practical result.

One reason why Portugal has been so progressive in seeking to protect its extended shelf is that Portugal's EMEPC has been well resourced and the government has put significant effort into building public awareness of the existence of the extended continental shelf area. This initiative can be seen in the context of Portugal's large maritime economy and substantial investment in marine science and policy from the mid-1990s into the 2000s.³⁵ Portugal has conducted considerable scientific research into the sea floor ecosystems on its continental shelf, which has enabled it to make persuasive arguments at a regional level. The MPA initiatives have been used to foster a sense of ownership and pride in the extended continental shelf.

The issue of protected areas on extended continental shelves has arisen for other States in the OSPAR region. The United Kingdom has followed Portugal's lead, and has nominated three sites on its extended continental shelf for OSPAR MPA status, as well as European Union recognition.³⁶ The sites are on part of the UK extended continental shelf for which a submission has been made to the CLCS. Hatton Bank is a rocky pinnacle home to corals and deep-sea sponges. Hatton-Rockall Basin is an area of muddy sediment, which is a significant habitat for deep-water sponges. North-west Rockall is a bank which hosts a variety of reefs.³⁷ The seabed is protected under UK legislation, although the water column is not protected by an OSPAR MPA. In 2015, the NEAFC closed areas on the Hatton Bank, Hatton-Rockall Basin and South-west Rockall Bank.³⁸ North-west Rockall Bank is not included in the current closed areas.

Matters were more difficult in relation to a proposed MPA for the Charlie-Gibbs Fracture Zone. Part of the proposed MPA lay above the extended continental shelf claimed by Iceland. At the time, Iceland was not willing to commit to the establishment of an MPA on its continental shelf beyond 200 nm.³⁹ The MPA was split into two parts. The Southern Charlie-Gibbs MPA was established in respect of the water column and the seabed. The OSPAR parties then decided to implement an MPA above Iceland's continental shelf in the water column that was without prejudice to the outcome of the CLCS submissions.⁴⁰

---

³³ Ribeiro, 'Marine Protected Areas: The Case of the Extended Continental Shelf', pp. 202–3.
³⁴ Abecasis et al., 'Marine Conservation in the Azores', p. 10.   ³⁵ Ibid., p. 6.
³⁶ OSPAR Commission, '2014 Status Report on the OSPAR Network of Marine Protected Areas' (2015): p. 19.
³⁷ http://jncc.defra.gov.uk/page-6895.
³⁸ Article 5, NEAFC Recommendation 9: 2015.
³⁹ BC O'Leary et al., 'The First Network of Marine Protected Areas (MPAs) in the High Seas: The Process, the Challenges and Where Next' *Marine Policy* 36 (2012): pp. 598–605, p. 602.
⁴⁰ Ibid., p. 602.

(2) Lessons from the North-east Atlantic experience

Ribeiro notes that the coexistence of the continental shelf and the high seas regime 'requires a real cooperative effort, even if the coastal State has competence to adopt, unilaterally, restrictive protective measures.[41] The advantage of a cooperative approach includes the ability to persuade third-party States of the reasonableness of its actions and encourage them to observe the measures in practice.[42] The following section explores some lessons from the North-east Atlantic example.

(a) Importance of the existence and competence of the international organizations

The success of the OSPAR/NEAFC example is supported by the fact that the membership of the organization is drawn from European nations with similar interests. Portugal is a member of the European Union, which has a strong environmental protection agenda. Over time, a number of directives have required EU Member States, among other things, to protect wildlife habitats and create MPAs.[43] The EU has been pivotal in ensuring that Member States identify potential protected areas, for example through the Habitats Directive and the Natura 2000 project.[44] Thus, there is significant legal and institutional pressure on European coastal States to create MPAs. In addition, there is a strong overlap between the membership of OSPAR and the NEAFC. Only Russia is a party to the NEAFC but not a party to OSPAR.

There is a problem in assuming that coastal States will be able to replicate the OSPAR/NEAFC model in other places. First, only four regional seas organizations exercise any jurisdiction over the high seas: these apply to the Southern Ocean; the North-east Atlantic; the Mediterranean; and the central Pacific. This means that in many regions States will have to look for another forum for the negotiations, and they may struggle to find an organization that accepts any mandate for the creation of an MPA. Even Portugal has struggled with this, because it has identified areas for protection on its extended shelf that are south of the OSPAR area and outside its competence. One option is to seek to extend OSPAR's region, but the lack of existing institutional structures adds a significant hurdle to achieving cooperative measures.

The Sargasso Sea Alliance is another model for protecting biodiversity on the high seas.[45] In that case there is no regional seas agreement that covers the high seas

---

[41] Ribeiro, 'The Rainbow', p. 198.   [42] Ibid.
[43] These include the EU Habitats Directive 92/43/EEC (1992) and the European Marine Strategy Framework Directive 2008/56/EC (2008). See Calado et al., 'The Azores Marine Park', p. 1177.
[44] Abecasis et al., 'Marine Conservation in the Azores', p. 4. See also http://ec.europa.eu/environment/nature/natura2000/. Abecasis et al., emphasize that Portugal was the first EU member State to apply some EU directives to deep-water areas in the Azores before other member States did the same.
[45] The Sargasso Sea is a 2 million square mile ecosystem in the North Atlantic that is primarily outside the jurisdiction of States. David Freestone et al., 'Can Existing Institutions Protect Biodiversity in Areas Beyond National Jurisdiction? Experiences from Two Ongoing Processes' *Marine Policy* 49 (2014): pp. 167–75, p. 168; Karen N Scott, 'Conservation on the High Seas: Developing the Concept

area of the Sargasso Sea, and there are gaps in coverage by RFMOs.[46] The Sargasso Sea Alliance was created in 2011 as a partnership between Bermuda, NGOs, scientists and private donors. One aim of the alliance is to use existing regional, sectoral and international organizations to secure protective measures for all or parts of the Sargasso Sea. Bermuda and the United Kingdom have taken steps to achieve fisheries conservation and management measures in two RFMOs with some competence in the area. The Sargasso Sea Alliance has worked through the Convention on Biological Diversity (CBD) to have the area declared as an ecologically or biologically sensitive area (EBSA). Work is being undertaken to build a case at the IMO around the impact of shipping and pollution on the Sargasso Sea to achieve protective measures.[47] There are eleven signatories to the 2014 Hamilton Declaration, a non-binding instrument that is intended to build a framework for international collaboration and to establish a Commission to support stewardship in the Sargasso Sea.[48] Coastal States could use the Sargasso Sea Alliance as a model for building cooperation in places where the regional infrastructure is not present.

Some of the strengths of the Sargasso Sea model include the ability to raise awareness of the need for protection of an area within competent organizations and promote change in the organizations if necessary.[49] Another strength that has been identified is that there has been a champion State willing to support the initiative, also a feature of the OSPAR MPAs.[50] If the model was adopted for coastal States wishing to protect the extended continental shelf, the coastal State would obviously play this role. However, the process has been time consuming and lengthy so it does not provide a quick solution to a coastal State's needs.

(b) Cooperation and coordination problems

OSPAR has recognized the advantages of cooperation across a range of organizations and gone to lengthy efforts to achieve cooperation to ensure that its measures to protect certain areas are not undermined by activities over which it has no control. OSPAR and the NEAFC are continuing with their efforts by seeking to have

---

of the High Seas Marine Protected Areas' *International Journal of Marine and Coastal Law* 27 (2012): pp. 849–57, p. 854; Elisabeth Druel et al., *Governance of Marine Biodiversity in Areas Beyond National Jurisdiction at the Regional Level: Filling the Gaps and Strengthening the Framework for Action* (Paris: IDDRi and AAMP, 2012): p. 79.

[46] Freestone et al., 'Can Existing Institutions Protect Biodiversity in Areas Beyond National Jurisdiction?', p. 168.

[47] A description of these activities is found in: Freestone et al., 'Can Existing Institutions Protect Biodiversity in Areas Beyond National Jurisdiction?', p. 172; and David Freestone et al., 'Place-based Management of Large-scale Ocean Places: Papahānaumokuākea and the Sargasso Sea' *Stanford Environmental Law Journal* 191 (2013–2014): pp. 191–250.

[48] David Freestone and Kate Killerlain Morrison, 'The Signing of the Hamilton Declaration on Collaboration for the Conservation of the Sargasso Sea: A New Paradigm for High Seas Conservation?' *International Journal of Marine and Coastal Law* 29 (2014): pp. 345–62.

[49] Druel et al., *Governance of Marine Biodiversity in Areas beyond National Jurisdiction at the Regional Level*, p. 84.

[50] Freestone et al., 'Can Existing Institutions Protect Biodiversity in Areas Beyond National Jurisdiction?', p. 168.

the IMO and the ISA join the Collective Arrangement.[51] This has been 'time and labour-intensive', and is an ongoing effort.[52] It has been noted that the cooperative mechanisms are entirely based on non-binding instruments and their success depends on the goodwill of the State parties to those organizations.[53] However, without cooperation and coordination, any MPA risks failing to protect the environment from sectoral activities such as fisheries, shipping and mining.

Even where regional organizations exist in other parts of the world, it may be difficult to persuade a diverse range of States of the need for the measures sought by the coastal State. The Commission for the Conservation of Antarctic Marine Living Resources (CCAMLR Commission) has been attempting over a number of years to get agreement on MPAs in the Southern Ocean. One proposal has been to establish an MPA in the Ross Sea.[54] Since 2012, New Zealand and the United States have been seeking consensus from the State parties to the Convention for the Conservation of Antarctic Marine Living Resources (CCAMLR) to obtain agreement on a fairly substantial MPA.[55] A few States, including China, Russia and Ukraine, have resisted agreeing to such a measure despite ongoing pressure and redrawing of the proposed boundaries of the MPA.[56] Among the objections to the proposals were criticism of the scientific information behind the proposal, a question about the mandate under CCAMLR to create MPAs, and legal ambiguities.[57] Concerns about setting precedents and impacts on fisheries interests are also likely to prompt objections.[58] For the CCAMLR Commission, the key objection appears to be the impact of an MPA on fishing in the area. Although there is no (official) coastal State involved in those negotiations, the fact that one or two States have managed to thwart that process illustrates the practical obstacles that can exist, especially when the proposal will result in the need for the member States to limit the economic activities of their fishing vessels.

International organizations such as the IMO may have competence to implement measures but the difficulty of convincing the membership of the value of those measures could be difficult.[59] There will be even less of an overlap of interests than in regional organizations.

---

[51] Ásmundsson and Corcoran, 'Information Paper', p. 14.
[52] Freestone et al., 'Can Existing Institutions Protect Biodiversity in Areas Beyond National Jurisdiction?', p. 172.
[53] Druel et al., *Governance of Marine Biodiversity in Areas beyond National Jurisdiction at the Regional Level*, p. 27.
[54] Karen N Scott, 'Marine Protected Areas in the Southern Ocean' in Erik J Molenaar, Alex G Oude Elferink and Donald R Rothwell (eds), *The Law of the Sea and the Polar Regions* (Leiden: Martinus Nijhoff, 2013): pp. 113–37.
[55] The initial joint proposal would have covered 2.27 million km² of the Southern Ocean. Karen Scott, 'Protecting the Last Ocean: The Proposed Ross Sea MPA. Prospects and Progress' in Gemma Andreone (ed), *Jurisdiction and Control at Sea: Some Environmental and Security Issues* (Naples: Giannini Editore, 2014): pp. 79–90, p. 82.
[56] The coverage of the proposed MPA was reduced by almost 40 per cent to 1.34 million square kilometres in 2013. Scott, 'Protecting the Last Ocean', p. 82.
[57] Ibid., pp. 82–4.   [58] Scott, 'Marine Protected Areas in the Southern Ocean', p. 133.
[59] Druel et al., *Governance of Marine Biodiversity in Areas beyond National Jurisdiction at the Regional Level*, p. 27.

Although OSPAR and the NEAFC have managed to coordinate some MPAs and closed areas, they do not fully overlap in a geographical sense. This reflects the fact that there are different decision-making processes and interests in each organization. Adding other organizations to the mix introduces a considerable level of complexity and increases the chances that measures sought by the coastal State will be watered down or refused.[60]

(c) Scientific and legal uncertainty

A further obstacle to cooperation is scientific uncertainty and lack of information about the seabed environment. Scientific exploration of the deep-sea environment is expensive and difficult. This means that comprehensive information about biodiversity and the impacts of human activity is slow to obtain and out of reach for many States in the short term.[61] In both OSPAR and the Sargasso Sea, the strong scientific case played a pivotal role in building a persuasive argument for protection.[62] The application of the precautionary approach in the absence of evidence of the need for protection may not be enough to convince States to restrict their activities if asked to do so by the coastal State.

An additional factor identified by the Sargasso Sea Alliance is that scientists from different sectoral bodies can be reluctant to accept the results of scientific research originating outside their organizations.[63] For example, the fact that the CBD approved the Sargasso Sea as an EBSA had little impact on dealings with regional bodies who insisted on using their own scientific processes.

Legal uncertainty over the final limits of the extended continental shelf can also create obstacles. Some of the States in the OSPAR region have made overlapping claims to extended continental shelves and resolution of the matters will take a long time. Some States have been reluctant to agree to marine protection on their continental shelves beyond 200 nm because of this uncertainty, or because national policies have not been developed.[64] Where the proposed sites for protection in the OSPAR area are located on extended shelves claimed by at least one member State, this has slowed down the process of designating MPAs.[65]

Another form of legal uncertainty arises in the case where the regional or international organization does not have clear competence to undertake the measures requested by the coastal State. Both OSPAR and the NEAFC questioned their

---

[60] Ibid.
[61] O'Leary et al., 'The First Network of Marine Protected Areas (MPAs) in the High Seas', p. 601.
[62] Freestone et al., 'Can Existing Institutions Protect Biodiversity in Areas beyond National Jurisdiction?', p. 170.
[63] Ibid., p. 173.
[64] O'Leary et al., 'The First Network of Marine Protected Areas (MPAs) in the High Seas', p. 602.
[65] For example, one proposed protected area in the Charlie-Gibbs area overlapped with Iceland's submission for its extended continental shelf. In the absence of agreement from Iceland for a MPA, the MPA was designated only in respect of the area outside Iceland's continental shelf. See O'Leary et al., 'The First Network of Marine Protected Areas (MPAs) in the High Seas', p. 602; WWF, 'High Seas MPAs: Regional Approaches and Experiences' Global Meeting of the Regional Seas Conventions and Action Plans, 2010 UNEP (DEPI)/RS.12/INF.6.RS, p. 30.

own competence to create high seas MPAs and closed areas. OSPAR concluded that it did have competency to create the MPAs, while the NEAFC States amended the NEAFC treaty to clarify that it could impose area closures.[66] Opponents of the proposed Ross Sea MPA also raised questions about the legal competence of the CCAMLR Commission to establish MPAs on the high seas.[67] This may be a problem for other organizations as well.

(d) Application to third-party States

The problem of the cooperation of third-party States is significant. Although the OSPAR Commission has undergone extensive consultation with other organizations, States that are not party to OSPAR and the NEAFC are not bound by the MPA and fisheries closures. A basic principle of international law reflected in the Vienna Convention on the Law of Treaties is that a treaty does not create obligations or rights for a third-party State without its consent.[68] Because other States can exercise freedoms of navigation and fishing under the LOSC, it is arguable that they will not be bound by the management restrictions imposed by regional bodies.[69]

The main exception is found in the 1995 United Nations Fish Stocks Convention.[70] Article 8(4) provides that: 'Only those States which are members of [a RFMO] or participants in such an arrangement, or which agree to apply the conservation and management measures established by such organization or arrangement, shall have access to the fishery resources to which those measures apply'. If a State is not party to the NEAFC but is a party to the United Nations Fish Stocks Agreement, it will be under an obligation to comply with the closed areas if it wishes to access the fisheries resources in the North-east Atlantic managed by the NEAFC. Although the United Nations Fish Stocks Agreement does not apply directly to sedentary species, the fishing that is being limited may well be for straddling or highly migratory species. Therefore, the agreement is applicable in such a situation.

However, even article 8(4) of the United Nations Fish Stocks Agreement will not be sufficient to impose obligations on States that are not party to that agreement.[71] This highlights a significant flaw in relying on a cooperative approach through

---

[66] Freestone et al., 'Can Existing Institutions Protect Biodiversity in Areas beyond National Jurisdiction?', p. 170.
[67] Scott, 'Protecting the Last Ocean'.
[68] Article 34, Vienna Convention on the Law of Treaties (opened for signature 23 May 1969, entered into force 27 January 1980) 1155 UNTS 331.
[69] Yoshifumi Tanaka, 'Reflections on High Seas Marine Protected Areas: A Comparative Analysis of the Mediterranean and the North-east Atlantic Models' *Nordic Journal of International Law* 8 (2012): pp. 295–326, p. 316.
[70] Agreement for the Implementation of the Provisions of the United Nations Convention on the Law of the Sea of 10 December 1982 relating to the Conservation and Management of Straddling Fish Stocks and Highly Migratory Fish Stocks (opened for signature 4 December 1995, entered into force 11 December 2001) 2167 UNTS 88 (United Nations Fish Stocks Agreement).
[71] At the time of writing, eighty-three States were party to the United Nations Fish Stocks Convention.

regional organizations to achieve protection for the continental shelf beyond 200 nm. However, if a coastal State has managed to secure agreement on protective measures, this would provide a strong basis to make representations to the third-party State and may support any subsequent decision to rely on unilateral expression of its sovereign rights, as discussed in earlier chapters.

(e) Conclusion

Portugal has made significant and noteworthy efforts to pursue cooperative mechanisms through regional organizations to advance its interests in establishing MPAs on its continental shelf beyond 200 nm. Factors influencing this success include a commonality of membership within OSPAR and the NEAFC with shared interests and, in the case of European member States, with the European Union providing additional incentives to move towards protection of marine habitats. Portugal has taken a leadership role in recognizing the importance of establishing MPAs on its extended continental shelf and making the case to the regional bodies for protection. Portugal was fortunate that OSPAR member States were willing to accept it taking such action where it had not yet completed its submissions to the CLCS. Although this book has argued that sovereign rights over the continental shelf beyond 200 nm do not depend on the CLCS process, opposition could have made it very difficult for Portugal politically. It is encouraging that the United Kingdom is following suit with the protection of areas of its extended continental shelf through the NEAFC. The development of a body of practice in the North-east Atlantic can provide guidance for coastal States in other areas.

The advantage of adopting this approach is that measures that enhance protection of the extended continental shelf are negotiated politically in an international forum, allowing the coastal State to persuade others of their necessity. The legal character of the decisions offer reinforcement and, potentially, avenues for ensuring compliance with the measures. Decisions of regional or international organizations will be difficult for third-party States to ignore. This is because it may demonstrate that any unilateral steps taken by the coastal State against the third-party State are justifiable under article 78(2) of the LOSC.

B. Example 2: Bilateral Cooperation Between Mauritius and the Seychelles

Coastal States may choose to enter into cooperative arrangements with just one or two other States where those States would be specially affected by restrictions on uses of the high seas above the continental shelf.[72] One option is to move towards

---

[72] For example, the agreement between the United States and the USSR on how king crab would be fished on the United States continental shelf. Agreement between the Government of the United States of America and the Government of the Union of Soviet Socialist Republics Relating to Fishing for King Crab (5 February 1965) 4 ILM 359.

the joint exercise of sovereign rights. In Chapter 5, it was noted that joint development arrangements can assist in managing rights in the context of a dispute over boundaries. The second example in this chapter is a bilateral arrangement between two States for the joint development of a shared area of continental shelf beyond 200 nm.

Mauritius and the Seychelles are located in the South-west Indian Ocean. In 2008, the two States submitted a joint application to the CLCS in respect of the continental shelf in the Mascarene Shelf Plateau. The CLCS issued recommendations confirming the outer limits of this joint area in 2011. The Mascarene Plateau is a large shallow area of continental shelf situated to the north-east of Mauritius and south-west of the Seychelles. The recommendations of the CLCS resulted in a shared extended continental shelf of 396,000 square kilometres.[73]

The joint submission was an attempt to avoid delay in receiving the recommendations of the CLCS, which will not consider a submission where there is an objection from other States.[74] The States decided that in addition to making a joint submission, that they would go one step further and proceed with co-management of the joint zone. Two treaties were signed between them.

The first was the Treaty Concerning the Joint Exercise of Sovereign Rights over the Continental Shelf in the Mascarene Plateau Region.[75] This treaty confirmed the intention to exercise sovereign rights jointly in the joint extended continental shelf area known as the joint management area (JMA). The second was the Treaty Concerning the Joint Management of the Continental Shelf in the Mascarene Plateau Region (Joint Management Treaty).[76] This treaty established the principles and mechanisms for cooperation between them. It reflects a balanced approach between facilitating exploration and exploitation of the joint shelf area and a desire to protect the marine environment.[77]

Key to the Joint Management Treaty is the establishment of an institutional structure to facilitate joint decision-making. This involves a Ministerial Council, a Joint Commission and a Designated Authority. The Ministerial Council consists of political representatives who consider matters at a higher policy level and matters that have been referred to it by the Commission. The Commission is made up of appointees from Mauritius and the Seychelles to 'establish policies and regulations relating to petroleum and other natural resource activities in the JMA' and oversee the work of the Authority.[78] Both the Council and the Commission will make decisions by consensus. The Designated Authority has legal personality and will carry on the regulation and management of natural resource activities in the JMA.

---

[73] See http://www.un.org/depts/los/clcs_new/submissions_files/submission_musc.htm.
[74] Second Reading, The Maritime Zones Act (Amendment) Bill (No. V of 2012), Speech by the Prime Minister.
[75] (13 March 2012) *Law of the Sea Bulletin* 79 (2012): p. 26.
[76] Ibid., p. 41 (Joint Management Treaty).
[77] Ibid. The preamble refers to the desire for sustainable development as well as the need to manage the resources of the area in a manner that is sustainable and consistent with the precautionary principle, the protection of the marine environment and the biological diversity of the continental shelf.
[78] Ibid., article 4(c)(i).

The Joint Management Treaty provides that the parties will share revenue equally.[79] Provision is made for taxation and natural resources codes to be developed and it sets out expectations in relation to pipelines, unitization of transboundary resources and the right to conduct services. The Treaty covers other matters, including employment, health and safety, customs, safety, surveillance and security. A dispute settlement clause provides for submission of disputes at the request of either party to an arbitral tribunal.

The treaty contains reasonably extensive provisions relating to the protection of the environment, biodiversity and bioprospecting.[80] The emphasis is on cooperation to protect natural resources, seabed habitats and associated ecological communities from risks arising from natural resource activities in the JMA. The Authority can issue regulations to protect the living natural resources and seabed environment and develop contingency plans for combating pollution from natural resource activities. Each party has the right to conduct biological surveys and bioprospecting in the JMA.

One interesting provision relates to activities in the high seas above the extended continental shelf. The Treaty states that parties must apply the precautionary principle. This requirement is accompanied by the following statement:[81]

This shall include measures concerning fishing activity in the waters superjacent to the seabed in the JMA where such activity is having a direct impact upon, or poses a significant risk to, the natural resources of the seabed and subsoil in the JMA.

No mention is made as to the target of these measures. On one reading it can be expected that the parties would be assuming responsibility to take action in relation to problems caused by their own fishing vessels or nationals. However, the text does not rule out the parties taking unilateral or cooperative action in respect of fishing by other States above the JMA.

The two nations have been extremely productive in the years following the conclusion of the treaties. Many meetings of the Ministerial Council and the Joint Commission have been held as they work towards establishing a framework that will allow for prospecting of the resources in the JMA. Among other things, the work programme has resulted in:

- a Joint Ministerial Declaration Concerning the Sustainable Development of the JMA
- an environmental code of practice
- harmonization of legislation
- a model petroleum agreement
- an offshore safety code
- a fiscal and taxation code and
- a strategic plan. Strategic objectives include the development of an institutional framework, technical capacity, promotion of the JMA to oil companies, data acquisition and multi-sector, multi-use planning.[82]

---

[79] Ibid., article 5.   [80] Ibid., article 12.   [81] Ibid., article 12(b).
[82] Draft Project Document: Joint Management Support to Mascarene Plateau Region.

Although it is early days, the Mauritius–Seychelles approach can provide valuable lessons for other States who may wish to take a bilateral approach to managing an area of extended continental shelf. By putting aside potential disputes over the maritime boundary, Mauritius and the Seychelles have secured timely recommendations from the CLCS. The Joint Management Treaty balances economic and environmental issues extremely well and is an excellent basis for future management of activities in the JMA.

This initiative is innovative and, if implemented well, could become a model for the management of other joint areas of continental shelf beyond 200 nm. Naturally, the parties will face the usual challenges around lack of information about the seabed environment and the distance from the coastline. However, Mauritius and the Seychelles have received support from the Commonwealth Secretariat in developing the details of the framework. There is also a proposal for funding through the United Nations Development Programme and the Global Environment Facility for financial and technical support. In the future it could be expected that the Joint Commission will look to achieving cooperation with regional organizations to achieve its objectives.

C. Cooperation with International and Regional Organizations

(1) Regional fisheries management organizations

Assuming a coastal State wishes to create restrictions on fishing that could detrimentally affect sedentary species, it is highly advisable for it to cooperate with the relevant RFMO, if one exists, to establish conservation measures in respect of the shelf.[83] Many RFMOs have taken steps in the last decade to protect vulnerable marine ecosystems from destructive fishing practices as a result of United Nations General Assembly Resolutions and work in the Food and Agriculture Organization (FAO). UN General Assembly Resolution 61/105 called on States and RFMOs to take immediate action to protect vulnerable marine ecosystems, including at seamounts, hydrothermal vents and cold-water reefs, from destructive fishing practices.[84] RFMOs were called upon to identify vulnerable marine ecosystems, to not allow fishing to proceed in areas of vulnerable marine ecosystems in the absence of conservation and management measures, and to establish rules requiring fishing vessels to move on when vulnerable marine ecosystems were encountered.[85]

The United Nations General Assembly resolutions have expressly preserved the position of coastal States and their continental shelves under article 77.[86]

---

[83] Daniel Owen, 'Interactions Between Management of a Water Column Marine Protected Area in the High Seas of the OSPAR Maritime Area and the Exercise of Sovereign Rights Regarding Subjacent Outer Continental Shelf', Report for WWF Germany (2010): p. 46.
[84] UN General Assembly Resolution 61/105 (2006), para. 80.
[85] UN General Assembly Resolution 61/105 (2006), para. 83. See also UN General Assembly Resolution 64/72 (2009), para. 77; UN General Assembly Resolution 66/68 (2011), paras. 121–38.
[86] As an example, see UN General Assembly Resolution 64/72 (2009), para. 115; UN General Assembly Resolution 70/75 (2015), para. 164.

This is an acknowledgement that responsibility for benthic marine ecosystems will fall under coastal State jurisdiction where the ecosystem involves sedentary species. However, this does not mean that the Resolutions are irrelevant for coastal States. Instead, it is likely that RFMOs with competence for areas of the high seas above an extended continental shelf are likely to have begun to implement processes for the protection of vulnerable marine ecosystems on the sea floor. Coastal States may be able to use these RFMO processes to obtain measures that will allow them to protect important ecosystems on their continental shelves beyond 200 nm.

The FAO agreed on International Guidelines for the Management of Deep-Sea Fisheries in the High Seas in 2008.[87] These voluntary Guidelines were endorsed by the UN General Assembly, which called on States and RFMOs to implement them.[88] The Guidelines provide guidance on identifying vulnerable marine ecosystems and assessing whether fishing is having significant adverse impacts on these ecosystems. Suggestions are made for how RFMOs can respond to the need to strengthen regulation of bottom fishing. The FAO has launched a database of vulnerable marine ecosystems in collaboration with RFMOs.[89]

Implementation of the Guidelines has been patchy across RFMOs. Some RFMOs have taken steps to close areas to fishing on the sea floor. These include the NEAFC, North Atlantic Fisheries Organization (NAFO), South-east Atlantic Fisheries Organization (SEAFO), South Pacific Regional Fisheries Management Organization (SPRFMO) and the CCAMLR.[90] However, other RFMOs have been criticized for being slow to act, ignoring scientific advice, and not applying the precautionary principle.[91]

Some RFMOs have limited fishing effort to the existing 'footprint' of fishing activities. This means that vessels will have to fish only in waters that have been subject to fishing effort in the past.[92] This is done in SPRFMO, NAFO, CCAMLR and others. The effectiveness of these measures depends on the extent and intensity of previous effort as some unfished areas may still be found within areas that fall within the 'footprint'.[93]

---

[87] www.fao.org/docrep/011/i0816t/i0816t00.HTM.

[88] UN General Assembly Resolution 64/72 (2009), para. 117.

[89] Report of the Secretary General of the United Nations on Oceans and the Law of the Sea, A/70/74/Add.1, 1 September 2015, para. 80.

[90] Glen Wright et al., 'Advancing Marine Biodiversity Protection Through Regional Fisheries Management: A Review of Bottom Fisheries Closures in Areas Beyond National Jurisdiction' *Marine Policy* 61 (2015): pp. 134–48.

[91] Ibid, p. 146.

[92] J Anthony Koslow et al., 'Chapter 51: Biological Communities on Seamounts and Other Submarine Features Potentially Threatened by Disturbance' in *First Global Integrated Marine Assessment* (New York: United Nations, 2015): p. 10 www.un.org/depts/los/global_reporting/WOA_RegProcess.htm.

[93] Wright et al., 'Advancing Marine Biodiversity Protection Through Regional Fisheries Management', p. 146.

'Move-on' rules require vessels that encounter a certain volume of sensitive benthic species such as coral to immediately cease fishing and move a specified distance before resuming fishing. These rules vary considerably between RFMOs.[94] Fishing vessels may be required to move on between one and five nm from the encounter and the area may or may not be closed to other vessels. Criticisms of move-on rules include a lack of globally available data about vulnerable species, a failure to identify with precision the species that trigger the move-on rule, setting too high a volume of by-catch that will trigger the rule and failure to distinguish between different types of fishing gear.[95]

RFMOs in the process of being formed and not yet in effect may also have processes that can be put in place to protect seabed habitats. For example, parties to the SPRFMO Convention agreed on interim measures to be applied before the Convention came into force. These included limiting fishing to the existing footprint, closing vulnerable marine ecosystems to bottom trawling, and a move-on rule.[96] Parties to the North Pacific Ocean Convention have agreed to limit the catch and the footprint of the fishing effort as interim measures.[97]

Finally, where there is no RFMO, individual flag States may voluntarily restrict their fishing effort. The European Union passed a regulation in 2008 requiring States to identify vulnerable marine ecosystems where their vessels operate and restrict the operation of those vessels.[98] To comply with this regulation, Spain closed nine areas to bottom fishing in 2009.[99]

One problem is that RFMOs may not consider themselves competent to implement conservation and management measures for the benefit of a State's continental shelf. As mentioned, it is common for RFMO conventions to state that nothing in the convention shall prejudice the rights of States under the LOSC. Some conventions go further: the NAFO and SEAFO Conventions specify that they do not apply to sedentary species.[100] However, these types of provisions should not pose a

---

[94] S Hansen, P Ward and A Penney, 'Identification of Vulnerable Benthic Taxa in the Western SPRFMO Convention Area and Review of Move-On Rules for Different Gear Types', SPRFMO Document SC-01-09 (2013).
[95] Peter Auster et al., 'Definition and Detection of Vulnerable Marine Ecosystems on the High Seas: Problems with the "Move-on" Rule' *ICES Journal of Marine Science* (2010): doi:10.1093/icesjms/fsq074.
[96] 'Interim Measures Adopted by Participants in Negotiations to Establish South Pacific Regional Fisheries Management Organization' *International Legal Materials* 46 (2007): pp. 894–6; Howard S Schiffman, 'The South Pacific Regional Fisheries Management Organization (SPRFMO): an Improved Model of Decision-making for Fisheries Conservation?' *Journal of Environmental Studies and Science* 3 (2013): pp. 209–16.
[97] North Pacific Fisheries Commission http://nwpbfo.nomaki.jp/Interim-measures.html.
[98] Council Regulation (EC) No. 734/2008, Protection of Vulnerable Marine Ecosystems in the High Seas from the Impact of Bottom Fishing Gears (2008).
[99] Wright et al., 'Advancing Marine Biodiversity Protection Through Regional Fisheries Management', p. 140.
[100] Convention on Future Multilateral Cooperation in the North-west Atlantic Fisheries (opened for signature 24 October 1978, entered into force 1 January 1979) 1135 UNTS 369, Article 10(4); Convention on the Conservation and Management of Fishery Resources in the South East Atlantic Ocean (opened for signature 20 April 2001, entered into force 13 April 2003) 2221 UNTS 189, article 1(l).

legal impediment to RFMOs exercising jurisdiction over fishing activities aimed at high seas species in proximity to the continental shelf. The provisions are designed to prevent interference with coastal States' rights, not to prevent coastal States working with the RFMO to secure protection of their rights.

Another problem is the necessity of convincing fishing States to limit their effort above an area of the extended continental shelf. It can often be difficult to obtain agreement on closing areas, especially where there is already commercial effort taking place. For example, States creating the South Orkney Islands MPA under CCAMLR required the boundaries to be redrawn to avoid a commercial fishery in the area.[101] Similar issues are arising in relation to the designation of MPAs in the Ross Sea and elsewhere.[102]

In situations where the target species of the RFMO is a straddling or highly migratory stock, article 7(1) and (2) of the United Nations Fish Stocks Agreement may provide support for a coastal State seeking to limit fishing on the continental shelf. These articles require coastal States and other States to cooperate to agree on conservation measures and to seek to achieve compatibility between measures for areas within national jurisdiction and those on the high seas. In a situation where a coastal State wishes to prevent bottom trawling on parts of its continental shelf beyond 200 nm, an argument could be made that the RFMO should close an area that the coastal State has closed to its own fishing vessels for the protection of a vulnerable marine ecosystem. The argument is strained because the closure to fishing is not for the protection of the straddling stock, which is within the competence of the RFMO, but for the sake of sedentary species belonging to the coastal State. In addition, the United Nations Fish Stocks Agreement does not provide a solution where the coastal State and other States disagree about the appropriate conservation and management measures.[103] It is possible for the dispute settlement procedures to apply when agreement cannot be reached, but this not been invoked to date.

Despite the potential obstacles, it is clearly advisable for a coastal State to work with an RFMO where possible to seek to agree on restricting fishing activities if that is a goal, rather than relying on unilateral measures. If an RFMO implements a conservation and management measure restricting fishing above an area of the

---

[101] Karen N Scott, 'Conservation on the High Seas: Developing the Concept of the High Seas Marine Protected Areas' *International Journal of Marine and Coastal Law* 27 (2012): pp. 849–57, p. 855.

[102] Laurence Cordonnery, Alan D Hemmings and Lorne Kriwoken, 'Nexus and Imbroglio: CCAMLR, the Madrid Protocol and Designating Marine Protected Areas in the Southern Ocean' *International Journal of Marine and Coastal Law* 30 (2015): pp. 727–64; CM Brooks, 'Competing Values on the Antarctic High Seas: CCAMLR and the Challenge of Marine Protected Areas' *The Polar Journal* 3 (2013): pp. 277–300.

[103] See, e.g. Olav Schram Stokke, 'Managing Straddling Stocks: The Interplay of Global and Regional Regimes' *Ocean and Coastal Management* 43 (2000): pp. 205–34; Lawrence Juda, 'The 1995 United Nations Agreement on Straddling Fish Stocks and Highly Migratory Fish Stocks: A Critique' *Ocean Development and International Law* 28 (1997): pp. 147–66, at pp. 154–5; and Peter Örebech, Ketill Sigurjonsson and Ted L McDorman, 'The 1995 United Nations Straddling and Highly Migratory Fish Stocks Agreement: Management, Enforcement and Dispute Settlement' *International Journal of Marine and Coastal Law* 13 (1998): pp. 119–41, pp. 127–9.

extended continental shelf, then all States party to the Fish Stocks Agreement will be required to observe the measure or refrain from fishing that stock.[104] Another key advantage for a coastal State in working with an RFMO to achieve an area closure or similar conservation and management measure would be that the measure would be enforceable through compliance and enforcement measures authorized by the RFMO and not reliant on unilateral action by the coastal State.[105]

(2) International Maritime Organization

The IMO has competence to establish rules to mitigate the environmental consequences of shipping, including vessel-sourced pollution. Many coastal States already work with the IMO to implement rules in respect of sensitive areas in their territorial sea and EEZ. These processes could also apply to issues in respect of the continental shelf. It will be necessary to build a clear case that demonstrates that shipping is having a detrimental impact on a particular area, which may be difficult in some cases. However, if this is possible, there are two potential options for States.

The first option is to establish a special area under the MARPOL Convention.[106] Special areas are sea areas that have special oceanological and ecological conditions deserving of protection. The designation of special areas allows the imposition of restrictions on discharges from vessels, including oily waste, noxious liquid substances, sewage and garbage.[107] Existing special areas include the Southern Ocean, the Baltic Sea, the Black Sea, the Mediterranean, the North Sea and the wider Caribbean region.[108] However, a special area must be adopted by an amendment to the MARPOL annexes, which can be a significant obstacle.

A second option would be to establish a particularly sensitive sea area (PSSA). According to the IMO, a PSSA is 'an area that needs special protection through action by IMO because of its significance for recognised ecological, socio-economic, or scientific attributes where such attributes may be vulnerable to damage by international shipping activities'.[109] States must provide a detailed justification for the designation of a PSSA based on ecological, social and economic, and scientific and educational criteria. The attributes of the area should be at risk from international shipping activities, taking into account vessel traffic and natural factors such as hydrographic and oceanographic conditions. As part of the application for a PSSA, States must identify associate protective measures that can be implemented by the IMO. These include designating the area as a special area under MARPOL, or

---

[104] Article 8(4), UN Fish Stocks Agreement.
[105] Owen, 'Interactions', p. 49. See ch 8 for a discussion of the coastal State's unilateral right of enforcement.
[106] International Convention for the Prevention of Pollution from Ships, as Modified by the Protocol of 1978 Relating Thereto (MARPOL 73/78) (concluded 2 November 1973, entered into force 2 October 1983) 1340 UNTS 62.
[107] See Annexes I, II, IV and V to MARPOL.
[108] Coordinates of these areas are set out in Annexes I, II, IV and V to MARPOL.
[109] Revised Guidelines for the Identification and Designation of Particularly Sensitive Sea Areas, IMO Resolution A.24/Res. 982, 6 February 2006, para. 1.2.

adoption of ship routeing and reporting systems near or in the area. Inclusion of the area on United Nations Educational, Scientific and Cultural Organization's (UNESCO's) list of World Heritage Sites or another international mechanism for recognition is encouraged. Applications are considered by the Marine Environment Protection Committee, which may refer the application to other parts of the IMO if their approval is needed to implement the PSSA. The PSSA designation is reviewed by the Assembly of the IMO.

Existing PSSAs primarily focus on areas close to the coastlines of States. Examples include the Great Barrier Reef in Australia, the Wadden Sea, the Canary Islands, the Galapagos archipelago and the Baltic Sea area.[110] In the Papahānaumokuākea Marine National Monument PSSA,[111] the IMO has established six areas to be avoided (ATBA) to protect coral reef areas and created a ship reporting area around the PSSA boundaries.[112] One of the largest PSSAs is the Western European Waters PSSA, which extends northward from the northern coast of Scotland, into the Atlantic and south to Spain.[113] A mandatory ship reporting system is in place in this PSSA. PSSAs could be designated in areas beyond national jurisdiction, which would include the high seas above the extended continental shelf.[114]

There are limitations on the ability of PSSAs to achieve significant additional environmental protection. It has been pointed out that PSSAs are not legally binding.[115] Depending on the measures established under the PSSA, they may not add significantly to existing protective measures. To date, no PSSA has been established on the high seas, which will be a challenge for coastal States seeking to protect an area of extended continental shelf. Political will to limit navigation on the high seas may be hard to establish.[116] However, as the PSSA must be agreed to by IMO members, a successful PSSA will receive a high degree of support and the PSSA is marked on charts used by seafarers.

(3) International Seabed Authority

Although the ISA has responsibility for the management of mining on the seabed beyond national jurisdiction ('the Area') it is still advisable for the coastal State to cooperate with the ISA. It is possible that mining activity, either by the coastal State or authorized by the ISA, could take place near areas under the jurisdiction of the other party.

---

[110] hwww.imo.org/en/OurWork/Environment/PSSAs.
[111] The Papahānaumokuākea Marine National Monument is located around the South-eastern Hawaiian Islands in the United States. See www.papahanaumokuakea.gov/.
[112] Freestone et al., 'Place-based Dynamic Management of Large-scale Ocean Places', p. 227.
[113] http://pssa.imo.org/w-europe/w-europe.htm. [114] Owen, '*Interactions*' (2010): p. 7.
[115] Hélène Lefebvre-Chalain, 'Fifteen Years of Particularly Sensitive Sea Areas: A Concept in Development' *Ocean and Coastal Law Journal* 13 (2007): pp. 47–69, p. 60.
[116] Julian Roberts, Aldo Chircop and Siân Prior, 'Area-based Management on the High Seas: Possible Application of the IMO's Particularly Sensitive Sea Area Concept' *International Journal of Marine and Coastal Law* 25 (2010): pp. 483–522, p. 521.

In a case where resource deposits lie across limits of national jurisdiction, article 142 requires activities in the Area to be conducted with due regard to the rights and interests of the relevant coastal States. Consultation must take place and, where activities in the Area are likely to result in exploitation of the resources within national jurisdiction, the prior consent of the coastal State is required. The latter situation is unlikely to occur when the exploitation is focused on polymetallic nodules that sit on the seabed. Hydrocarbon exploitation in the Area is the sort of activity most likely to require coastal State consent under article 142(2) and there is no indication that this is likely to occur in the short to medium term.

The other interest that a coastal State has is in the protection of the marine environment. A more likely situation is where a contract for exploitation in the Area is awarded close to an area under coastal State jurisdiction and the coastal State has concern about the environmental impact that the activities could pose to the EEZ or extended continental shelf. As an example, in 2015, the ISA signed a contract with Ocean Mineral Singapore for a fifteen-year exploration right in an eastern part of the Clarion-Clipperton Fracture Zone in the Pacific Ocean. This area is very close to the outer limit of Mexico's EEZ—approximately twenty kilometres away. Because the mineral resource subject to the authorization is not going to overlap with resources under Mexico's jurisdiction, the need to obtain prior consent from Mexico in accordance with article 142(2) does not arise. However, the proximity of the activity to Mexico's EEZ does raise particular issues for the contractor and the ISA to consider the impact of the activities on the marine environment. This is an obligation under customary international law[117] and article 142 also indicates the need to consider the interests of the coastal State.

Finally, coastal States that are exploiting the non-living resources of the continental shelf beyond 200 nm will need to cooperate with the ISA in order to carry out their obligations under article 82.

(4) Other relevant international organizations

A coastal State may be able to use environmental regimes to build the case for protection of vulnerable marine ecosystems. The best example is the CBD, which has established a process for identifying EBSAs.[118] In 2008, the meeting of the Conference of the Parties (COP) to the CBD approved scientific criteria for identifying EBSAs in need of protection in open ocean and deep-sea habitats.[119] The work was intended to support the identification of areas beyond national jurisdiction that need protection, but it can also be used to identify areas within national jurisdiction.[120] In

---

[117] See ch 4.
[118] Daniel C Dunn et al., 'The Convention on Biological Diversity's Ecologically or Biologically Significant Areas: Origins, Development, and Current Status' *Marine Policy* 49 (2014): pp. 137–45.
[119] Decision IX/20 on Marine and Coastal Biodiversity, COP 9 (2008), para. 14. See www.cbd.int/ebsa/.
[120] Decision X/29 on Marine and Coastal Biodiversity, COP 10 (2010) para. 21. Dunn et al., 'The Convention on Biological Diversity's Ecologically or Biologically Significant Areas', p. 142.

2010, the COP reinforced that EBSA identification is a matter for States and competent international organizations based on the best available scientific and technical information. The Subsidiary Body on Scientific, Technical and Technological Advice prepares reports setting out details of areas that meet the scientific criteria, for consideration by the COP. Details of the EBSAs that have been approved by the COP or that are waiting for approval can be found on the CBD website.

Designation of a site as an EBSA does not carry legal consequences or require particular management outcomes such as the creation of a MPA.[121] However, EBSA designation can be used to promote protective mechanisms in international organizations or domestic legal systems. Although the criteria for identifying an EBSA will not be identical to mechanisms in other international or regional bodies, there should be similarities between the information required in each case.[122] Despite this fact, it will still be a challenge to convert the EBSA into a protective measure in another forum such as an RFMO or the IMO.[123] It has been suggested that the designation of an area as an EBSA has not been sufficient to persuade organizations to adopt protective measures in the past. Freestone et al., refer to the fact that the NEAFC required areas classified as EBSAs under the CBD to be peer-reviewed by ICES. Not all of the EBSAs were accepted by the NEAFC and OSPAR and some are under ongoing review.[124]

Another option is to seek trade restrictions under the Convention on the International Trade in Endangered Species of Wild Flora and Fauna (CITES). If a coastal State is concerned about inappropriate exploitation of a living resource, it could seek listing on an appendix to CITES. However, this will be of use only if there is a specific species that the coastal State wishes to protect and international trade is undermining the conservation of the species. One difficulty is that there has been some resistance to the listing of marine species, especially where the species is a commercial fish stock.[125]

The World Heritage Convention[126] (WHC) allows States to identify sites of cultural or natural heritage that require protection under domestic laws. Parties are obliged to do all they can to identify, protect and conserve cultural and natural heritage.[127] Other States should assist in the identification, protection and conservation of the sites and must not take deliberate measures that might damage directly or indirectly cultural and natural heritage sites located on the territory of other States.[128] Funding through the UNESCO's World Heritage Fund may be available to States to help them maintain or restore their World Heritage Sites.

---

[121] Ibid., p. 143.
[122] The CBD and FAO have concluded that similar data can be used to establish EBSAs and vulnerable marine ecosystems for the purposes of the International Guidelines for the Management of Deep-sea Fisheries in the High Seas. Jeff A Ardron et al., 'A Systematic Approach Towards the Identification and Protection of Vulnerable Marine Ecosystems' *Marine Policy* 49 (2014): pp. 146–54, p. 147.
[123] Jeff A Ardron et al., 'The Sustainable Use and Conservation of Biodiversity in ABNJ: What Can be Achieved Using Existing International Agreements?' *Marine Policy* 49 (2014): pp. 98–108, p. 102.
[124] Freestone et al., 'Can Existing Institutions Protect Biodiversity in Areas Beyond National Jurisdiction?', p. 173.
[125] Ardron et al., 'The Sustainable Use and Conservation of Biodiversity in ABNJ', p. 102.
[126] Convention for the Protection of the World Cultural and Natural Heritage (opened for signature 16 November 1972, entered into force 17 December 1975) 1037 UNTS 151 (WHC).
[127] Ibid., article 4.   [128] Ibid., article 6.

Natural heritage can include natural features that are of outstanding universal value from an aesthetic or scientific point of view, geological and physiographical formations and areas that constitute the habitat of threatened species of animals and plants of outstanding universal value, and natural sites or areas of outstanding universal value from the point of view of science, conservation or natural beauty.[129] Marine areas can be covered by the Convention as demonstrated by the fact that forty-six natural World Heritage Sites have included aspects of marine values in their listings.[130] For example, the Papahānaumokuākea site in Hawaii includes islands and the surrounding ocean. The listing cites the multitude of habitats—including deep-sea habitats—in the area, the degree of endemism in the marine species, and the importance of the area as habitat to endangered or vulnerable species.[131] The Phoenix Islands Protected Area is an area of over 400,000 square kilometres in the waters of Kiribati that extends to the outer limit of the Kiribati EEZ. The protected area includes an oceanic coral archipelago system, fourteen seamounts and other deep-sea habitats.[132]

One question for a coastal State would be whether the fact that the extended continental shelf is located beneath the high seas would prevent it seeking a listing for World Heritage status. The LOSC is clear that the coastal State exercises sovereign rights over the resources of the shelf, and the WHC has listed marine sites in the past. Theoretically, the fact that the site lies underneath an area beyond national jurisdiction should not prevent the State from seeking recognition for it under the WHC.

A project has been underway through the UNESCO to identify potential marine World Heritage Sites in the Western Indian Ocean.[133] The Mascarene Plateau, part of which is under the joint management of Mauritius and the Seychelles, has been identified as a possible contender for an application. The project has suggested that it could be a transboundary single site nominated by both States.

Nothing in the WHC currently provides a mechanism for World Heritage areas to be created in areas beyond national jurisdiction.[134] Some authors have pointed out that the definitions of cultural and natural heritage do not exclude areas beyond national jurisdiction.[135] Instead, they argue that the silence in the Convention does not prevent its application to areas beyond national jurisdiction. There is a growing recognition that this is a direction that the parties to the WHC may need to consider in the future.[136]

---

[129] Ibid., article 2.
[130] Armeer Awad Abdulla et al., *Marine Natural Heritage and the World Heritage List* (Gland, Switzerland: IUCN, 2013): p. 4.
[131] See http://whc.unesco.org/en/list/1326.
[132] See http://whc.unesco.org/en/list/1325.
[133] www.vliz.be/projects/marineworldheritage/.
[134] Abdulla et al., *Marine Natural Heritage and the World Heritage List*, p. 47.
[135] Freestone et al., 'Place-based Dynamic Management of Large-scale Ocean Places', p. 237.
[136] Abdulla et al., *Marine Natural Heritage and the World Heritage List*, p. 47; Freestone et al., 'Place-based Dynamic Management of Large-scale Ocean Places', p. 238; Abdulla et al., 'Marine World Heritage: Creating a Globally More Balanced and Representative List' *Aquatic Conservation: Marine and Freshwater Ecosystems* 24 (Suppl 2) (2014): pp. 59–74, p. 72.

D. Conclusion

Despite the obvious obstacles for a coastal State taking a cooperative approach along the lines of the OSPAR/NEAFC model, there are considerable advantages. Although it may take time, securing measures that protect State interests in the extended continental shelf through regional and international organizations will result in several important advantages.

First, the process will build an understanding of the reasons for the measures that are sought by the coastal State. In the case of OSPAR, Portugal was able to build on earlier work done by NGOs and other States to convince the organization of the need to protect the Rainbow hydrothermal vent field and certain seamounts. Building awareness of the coastal State's concerns among the international community makes it more likely that the State will obtain agreement within the relevant organizations. If the requested measure is not adopted for some reason, this awareness-building is still useful for future steps the coastal State may take within the organization or in other organizations.

Secondly, the coastal State gains considerable legal certainty around the rules relating to activities in the high seas above the continental shelf. If there is a question whether action taken by the coastal State is an unjustifiable interference with high seas freedoms (for example, by closing part of the sea floor to bottom trawling), the adoption of similar measures by an international organization will support the argument that it is not. This could be important in a situation when a third State, not party to the organization, attempts to conduct activities in the high seas above the extended continental shelf.

Thirdly, depending on the organization, the measures will be binding on a large portion of the international community. An example of this is the coverage that the IMO has in respect of the world's shipping. An area to be avoided under MARPOL will obligate most flag States to comply. Measures taken by a RFMO may be applicable to States outside that RFMO if they are party to the United Nations Fish Stocks Agreement.

Fourth, any compliance and enforcement mechanisms available to the organization can be used to promote observance of the organization's rules. One example would be a case where a RFMO closed a part of the continental shelf to bottom trawling in response to a request from a coastal State. The coastal State could potentially use a RFMO high seas boarding regime to board and inspect vessels flagged to other State parties.[137]

There are limitations on the ability of a cooperative approach to completely meet the objectives of a coastal State. Some of these include: an international or regional organization with competence to regulate the high seas activities may not exist; it

---

[137] See article 21, United Nations Fish Stocks Convention.

will take time and considerable effort to persuade States to adopt the requested measures; a paucity of scientific information about the area will make it difficult to convince other States of the need for the measures; economic reliance on the high seas activities may make States reluctant to agree to restrictions on the activities; and the measures will not bind States that are not party to the organization.

Another option for a coastal State is to undertake bilateral negotiations with States that operate in the region. The joint Commission established by Mauritius and the Seychelles is a case in point. Another example is where, under the Continental Shelf Convention, the United States entered into treaties with Japan and the USSR to establish agreed measures for the exploitation of king crabs on the United States continental shelf.[138]

These bilateral arrangements minimize the potential for conflict to arise between two or more States whose nationals are conducting activities on, or in the vicinity of, the extended continental shelf. The coastal State and the flag State will be able to address the concerns that the other has and reach a compromise if necessary. An advantage is that the agreement can be concluded quickly if the States are able to reach consensus on the appropriate issues. It is most useful in cases where there are a low number of States with vessels operating above the extended continental shelf. Of course, if there are vessels from many States then the time needed to negotiate with each individually would be extensive. In this case it may be more efficient to seek cooperation through a regional or international organization to minimize transaction costs.

---

[138] Exchange of Notes Constituting an Agreement between the United States of America and Japan Relating to the King Crab Fishery in the Eastern Bering Sea (25 November 1964) 1965 UNTS 32; Agreement between the Government of the United States of America and the Government of the Union of Soviet Socialist Republics Relating to Fishing for King Crab (5 February 1965) 4 ILM 359. See ch 3.

10
Looking Ahead

Although the extended continental shelf is an area of the seabed that is far from shore, it would be a mistake to think that it is not a current issue for coastal States. On extended shelves less than 1500 metres below the surface, fishing activities may already be taking place, which will have an impact on sedentary species, including corals and sponges. Exploration for hydrocarbons beyond 200 nm is currently occurring in some parts of the world, and is being actively pursued by coastal States in other places. Scientists are exploring the unique and fascinating ecosystems that can be found on the sea floor. In the future these activities are likely to intensify and expand as technology overcomes the challenges of depth and distance. Inevitably, there will be conflicts between the interests of users of the high seas and the coastal State's rights to the resources of the shelf.

As States work through the process of producing their submissions to the Commission on the Limits of the Continental Shelf (CLCS), they will need to turn their attention to the activities occurring on their extended continental shelves. There are a number of different ways that States can choose to exercise their sovereign rights and this book does not presuppose any particular outcome. However, at a minimum, coastal States that are party to the Convention on Biological Diversity (CBD) will need to turn their minds to the issue of the biodiversity on the continental shelf beyond 200 nm. Other environmental obligations also exist in the Law of the Sea Convention (LOSC) and customary international law, including the obligation to conduct environmental impact assessments (EIAs) for activities that could pose a risk to areas under another State's jurisdiction or areas beyond national jurisdiction. These environmental considerations will need to be integrated into any decision-making about exploiting the resources of the extended continental shelf.

Although many of the legal considerations in relation to the extended continental shelf are not significantly different than within 200 nm—for example, the rights in relation to cables and pipelines—there are some important respects in which the legal regime applying to the extended continental shelf needs separate consideration. The primary examples are the obligations in article 82 in relation to payments for exploitation of the extended continental shelf, and article 246(6) which limits the situations in which coastal States can refuse consent for marine scientific research projects outside 200 nm. The fact that the coastal State has no sovereign rights in respect of the high seas above the continental shelf requires coastal States to be conservative in their actions to avoid interfering with high seas freedoms. In particular, the lack of control over fishing in the high seas means that it may be more difficult to protect vulnerable seabed resources. However, this book has made some suggestions as to how coastal States can respond to this risk.

Coastal States do not need to wait until the outer limits can be set in accordance with the recommendations of the CLCS. In fact, if the progress of the CLCS proceeds at its current pace, it will be several decades before it will finish its work.[1] However, so long as an extended shelf exists in fact, then so too do rights under article 77 of the LOSC. Of course, coastal States will need to act with restraint in the exercise of those rights where there is a possibility that there may be overlapping interests in the high seas, with neighbouring States and in relation to the Area.

A. Disputed Areas

One of the goals of the LOSC was to create certainty about the extent of various maritime zones and the rights and obligations of States in those zones. However, the extension of maritime jurisdiction has resulted in an expansion of the potential delimitation disputes between States. Less than half of the possible global maritime boundaries have been delimited.[2] One estimate was that, of the 30 million km^2 of potential extended continental shelf areas subject to submissions to the CLCS, more than 2.7 million km^2 are subject to overlapping claims.[3] The CLCS has stated that it is not prepared to consider submissions to the extended shelf when there is a dispute as to land or maritime boundaries, unless all parties agree.[4] Therefore, uncertainty as to the precise boundaries of a coastal State's continental shelf beyond 200 nm with its neighbours and the Area is likely to continue for many years.

Article 83 of the LOSC establishes that, in the absence of agreement on maritime boundaries, States should make every effort to enter into provisional arrangements of a practical nature and not to hamper the reaching of the final agreement. There is already considerable practice on joint development agreements for areas within the exclusive economic zone (EEZ) and this practice can be applied to the shelf beyond 200 nm as well.[5] Mauritius and the Seychelles have applied the concept of joint development to the area of their joint continental shelf.[6] The Joint Management Area covers an area that was subject to a joint submission to the CLCS and which

---

[1] Clive Schofield, 'Securing the Resources of the Deep: Dividing and Governing the Extended Continental Shelf' *Berkeley Journal of International Law* 33 (2015): pp. 274–93, p. 286.
[2] Clive Schofield, 'Parting the Waves: Claims to Maritime Jurisdiction and the Division of Ocean Space' *Penn State Journal of Law and International Affairs* 1 (2012): pp. 40–58, p. 48.
[3] Robert Van de Poll and Clive Schofield, 'A Seabed Scramble: A Global Overview of Extended Continental Shelf Submissions', Proceedings of the Advisory Board on the Law of the Sea (ABLOS) Conference on 'Contentious Issues in UNCLOS: Surely Not?' (2010): pp. 1–11, pp. 3–4.
[4] Øystein Jensen, *Commission on the Limits of the Continental Shelf: Law and Legitimacy* (Leiden: Martinus Nijhoff Publishers, 2014): pp. 65–6.
[5] See, e.g. Vasco Becker-Weinberg, *Joint Development of Hydrocarbon Deposits in the Law of the Sea* (Heidelberg: Springer, 2014); and R Beckman et al., (eds), *Beyond Territorial Disputes in the South China Sea: Legal Frameworks for the Joint Development of Hydrocarbon Resources* (Cheltenham: Edward Elgar Publishers, 2013). See also ch 5.
[6] See ch 9.

has been established pursuant to recommendations from the CLCS. Rather than engaging in a potentially divisive delimitation process between the two States, they have developed an innovative solution that involves managing the extended shelf jointly and sharing the profits of any exploitation equally. Although this model will not suit every situation where there is a non-defined boundary, it emphasizes the utility of creative responses under the framework of article 83. It can be much easier to reach agreement on boundaries when little or no exploration or exploitation has occurred because there will be fewer vested interests on either side. This will be the case for most extended shelves at present. Therefore, there is a window of opportunity for coastal States to pursue cooperative approaches towards resolving boundary disputes.

B. Developing States and the Information Gap

Although the provisions in the LOSC regarding the continental shelf do not explicitly refer to developing States, these States face greater challenges than those faced by developed countries, primarily owing to the practicalities of exercising jurisdiction in remote and deep parts of the ocean. A problem for all coastal States is the lack of information about the extended continental shelf, and developing States with limited resources are particularly challenged in this respect. Information is central to the exercise of coastal State rights on the continental shelf beyond 200 nm. If States do not know about the presence of resources on the shelf, then it is very difficult to begin to make management decisions in relation to them. If a coastal State needs to build consensus for protective measures through an international organization, it will need to have a well-developed, scientifically-based case to present. Many coastal States have gathered substantial amounts of data to build a case for the establishment of a right to an extended shelf. This data could be used to draw initial conclusions about the sorts of resources that could be present, although it is likely that much more research will be required for each State to be even modestly informed about the extent and nature of those resources.

Obtaining information about the extended continental shelf and its resources is expensive and time-consuming. In terms of hydrocarbons, it is likely that commercial incentives mean that private actors will be interested in conducting research into the possible presence of non-living resources. States can require companies to provide environmental information as part of the decision-making process, which may add to the coastal State's knowledge base.

Fishing activities are often less strictly regulated than the hydrocarbon industry in terms of their environmental consequences but they have an impact over a much larger area. Therefore, it will be important for the international community to support initiatives to expand knowledge about the areas of continental shelf that extend beyond 200 nm in partnership with those States that wish to learn more. This might be done by funding projects aimed at the identification of biodiversity, that would be conducted by experts and the results shared with the coastal State, for example under the CBD. Another option is for the coastal State to facilitate marine scientific research by foreign researchers on its extended continental shelf that also builds local research capacity.

C. The Intersection Between the Extended Continental Shelf and an International Agreement for Biodiversity Beyond National Jurisdiction

The international community is considering whether to proceed with negotiations for an international agreement (IA) to the LOSC that addresses the conservation and sustainable use of biological diversity in areas beyond national jurisdiction—referred to colloquially as the 'BBNJ' process. The IA will not directly address the issue of the biological resources of the continental shelf on the basis that this is an area within national jurisdiction. However, any IA would cover activities in the high seas above the continental shelf and, because such activities may have an impact on the continental shelf, there will be some important overlaps.

In June 2015, the General Assembly agreed to develop a legally binding instrument under the LOSC on the conservation and sustainable use of marine biological diversity in areas beyond national jurisdiction.[7] This decision was the result of more than a decade of informal discussions at the United Nations about the issues that underpin the need for an agreement.[8] The topics that may be included in the IA comprise:[9]

> [t]he conservation and sustainable use of marine biological diversity of areas beyond national jurisdiction, in particular, together and as a whole, marine genetic resources, including questions on the sharing of benefits, measures such as area-based management tools, environmental impact assessments and capacity-building and the transfer of marine technology.

The General Assembly has instructed that the IA should not undermine existing legal instruments and relevant global, regional and sectoral bodies. Therefore, the treaty is likely to be negotiated as an implementing agreement to the LOSC in a similar manner to the 1995 United Nations Fish Stocks Agreement.[10]

An obvious issue that coastal States will need to consider during the IA negotiation process is whether the IA can address the way in which conflicting high seas and continental shelf activities can be managed cooperatively. A question

---

[7] General Assembly Resolution 69/292.
[8] Information about the informal discussions can be found on the website of the Division for Ocean Affairs and the Law of the Sea www.un.org/Depts/los/biodiversityworkinggroup/biodiversityworkinggroup.htm.
[9] GA Resolution 69/292, para. 2.
[10] Agreement for the Implementation of the Provisions of the United Nations Convention on the Law of the Sea of 10 December 1982 relating to the Conservation and Management of Straddling Fish Stocks and Highly Migratory Fish Stocks (opened for signature 4 December 1994, entered into force 11 December 2001) 2167 UNTS 88.

is whether the IA will contain any special mention of the role of the coastal State when high seas activities potentially impact on the continental shelf beyond 200 nm.

Of course, any discussion about activities in areas beyond national jurisdiction will need to consider the coordination between flag and relevant coastal States where there are issues that affect them horizontally, for example where high seas activities impact on an EEZ. However, with the continental shelf beyond 200 nm there is a direct vertical interaction between the water column and the resources of the continental shelf, which will require special attention. Coastal States will not want the IA to infringe upon or limit their choices in relation to the continental shelf, but establishing processes by which the interactions can be considered and addressed would be useful.

One way to do this is to consider the process for EIAs for activities conducted in the high seas. Warner has suggested that one stage of this process might include public notification and consultation.[11] Questions might arise including whether the coastal State should be informed if the proposed activity will occur close to its continental shelf. What ability does a coastal State have to be included in the consideration of the EIA conducted by another State? In relation to the scope of the assessment, should the EIA assess the potential impact on a continental shelf and take into account the interests of the coastal State? Existing customary international law would suggest that States have obligations to conduct an EIA and notify and consult with affected coastal States if their activities will have a significant impact on the extended continental shelf. These considerations should be incorporated into any EIA process in the IA.

Any consideration of marine spatial planning or the establishment of marine protected areas (MPAs) in the high seas will also need to consider the rights of coastal States when the MPA lies above an extended continental shelf. The MPA should not contain conditions that infringe the coastal State's rights in relation to the living or non-living resources of the continental shelf. An example of this might be a prohibition on loud marine noise above a shelf that would technically prevent seismic surveys. Another example might be a prohibition on fishing that interferes with the harvest of a sedentary species. However, the ability to create MPAs and other spatial measures may be a benefit for coastal States that want to protect vulnerable marine ecosystems from the activities of other States. In this case, the process for establishing MPAs should acknowledge the special interest that coastal States have in relation to activities above the extended continental shelf.

Another important aspect of the IA will be the discussion about the use of genetic resources. One of the contentious issues that arose during the informal discussions in the BBNJ process was the status of genetic resources on the deep

---

[11] Robin Warner, 'Environmental Assessment in Marine Areas beyond National Jurisdiction' in Rosemary Rayfuse (ed), *Research Handbook on International Marine Environmental Law* (Cheltenham: Edward Elgar, 2015): pp. 291–312, p. 310.

seabed and what legal regime applies to their exploitation. A growing interest in biotechnology has raised the prospect that living organisms found in areas beyond national jurisdiction could be a source of revenue. Marine genetic resources are not referred to explicitly in Part XI of the LOSC and a range of views has been expressed about what legal regime, if any, applies to them. One group of States has taken the position that Part XI of the LOSC, which applies the common heritage of mankind to mineral resources on the seabed beyond national jurisdiction, does not cover living resources. Therefore, the appropriate governing principle in relation to genetic resources would be the freedom of the high seas. In that case, any private actor who retrieved a sample of an organism found on the deep seabed and derived a commercial product from it would be free to keep the entire profit for itself.[12]

The opposing view was that the common heritage of mankind principle originated in General Assembly Resolution 25/2749, which declared that the resources of the seabed beyond national jurisdiction are the common heritage of mankind. Although the LOSC defines 'resources' as minerals,[13] no distinction between living and non-living resources appears in Resolution 25/2749.[14] Therefore, it was argued, the benefits of developing these resources should be governed by the principle of the common heritage of mankind. In that case, benefits from exploitation of the resources would be returned to the international community to recognize that the source of the product came from the Area. However, the LOSC makes no provision for such a process—there are significant differences between the issues facing exploitation of minerals and the development of biotechnology from living organisms. Although some have suggested that the International Seabed Authority (ISA) could play a role in a new regime, this will not be possible under the current legal framework.

Over time it has become clear that some middle ground will have to be found between these positions. It is likely that innovative new approaches may need to be used to achieve a balanced system for access to deep seabed resources and benefit sharing.[15] This solution is likely to draw on the experiences under the CBD in relation to access and benefit sharing.[16]

---

[12] Dire Tladi, 'Conservation and Sustainable Use of Marine Biodiversity in Areas beyond National Jurisdiction: Towards an Implementing Agreement' in Rosemary Rayfuse (ed), *Research Handbook on International Marine Environmental Law* (Cheltenham: Edward Elgar, 2015): pp. 259–71, p. 261.

[13] Article 133, LOSC.

[14] Tladi, 'Conservation and Sustainable Use of Marine Biodiversity in Areas beyond National Jurisdiction', p. 261; de La Fayette, 'A New Regime for the Conservation and Sustainable Use of Marine Biodiversity and Genetic Resources', p. 267; Tullio Scovazzi, 'Mining, Protection of the Environment, Scientific Research and Bioprospecting: Some Considerations on the Role of the International Sea-bed Authority' *International Journal of Marine and Coastal Law* 19 (2004): pp. 383–409, p. 386.

[15] Morten Walløe Tvedt and Ane E Jørem, 'Bioprospecting in the High Seas: Regulatory Options for Benefit Sharing' *Journal of World Intellectual Property* 16 (2013): pp. 150–67; Petra Drankier et al., 'Marine Genetic Resources in Areas beyond National Jurisdiction: Access and Benefit Sharing' *International Journal of Marine and Coastal Law* 27 (2012): pp. 375–433.

[16] Arianna Broggiato, 'Marine Genetic Resources Beyond National Jurisdiction: Coordination and Harmonization of Governance Regimes' *Environmental Policy and Law* 41 (2011): pp. 35–41.

One aspect of the negotiations will be particularly relevant for the extended continental shelf. Few people have seriously challenged that living resources in the water column of the high seas will remain subject to the freedom of the high seas. Therefore, two different legal regimes will apply to the resources of the water column and the resources of the seabed, and this will need to be reconciled by the IA.

One possible approach is that the IA could use the 'sedentary species' definition as a starting point for the distinction because it bears similarities to the high seas/Area situation, where different regimes apply to the seabed and the superjacent water column.[17] This would be an attractive analogy because it might seem straightforward. In a situation where the 'sedentary species' definition is used successfully in the IA, it is possible that the balancing of rights between the genetic resources of the seabed and the water column may assist coastal States to manage the intersection between the extended continental shelf genetic resources and the water above through application by analogy. It may provide some legal certainty about the extent of the rights to sedentary species and how they can be exercised in regard to activities in the water column.

However, as discussed in Chapter 4, the 'sedentary species' definition becomes very difficult to apply in relation to ecosystems that may have mixed sedentary and non-sedentary species. Any stark division between seabed and water resources is arbitrary and does not reflect the sorts of ecosystems found on or near the seabed.[18] In addition, there are real questions about the 'harvestable stage' of an organism that is targeted for its genetic properties, rather than for consumption. The negotiations may result in a different approach being taken to distinguish between the high seas and seabed resources—if, indeed, any distinction is necessary. For example, the negotiations may conclude that the seabed regime applies to any genetic resources found within a certain distance from the sea floor. Or the legal regime could focus on types of ecosystems.[19] In this case the IA will have less relevance for coastal States and their management of sedentary species.

Coastal States have an opportunity to consider whether the relationship between the extended continental shelf and activities in the high seas should be addressed in the new IA. It may be that coastal States would be concerned that their sovereign rights over the extended continental shelf will be diminished if the IA refers in any way to the shelf. However, as demonstrated in relation to a possible process for EIA, there may be a case for protecting their interests through inclusion of procedural requirements to consider coastal States' interests.

---

[17] Frida M Armas-Pfirter, 'How Can Life in the Deep Sea Be Protected?' *International Journal of Marine and Coastal Law* 24 (2009): pp. 281–307, p. 303.
[18] De La Fayette, p. 258; David Leary, *International Law and the Genetic Resources of the Deep Sea* (Leiden: Martinus Nijhoff, 2007): p. 55.
[19] As proposed by Leary, who argues that it would be appropriate to create legal regimes to apply to hydrothermal vents, for example, rather than following a sedentary species approach. Leary, *International Law and the Genetic Resources of the Deep Sea*, p. 94.

D. Conclusion

As discussed in Chapter 2, there are a range of possible human activities that could occur in relation to the exploration, exploitation and conservation of the resources of the continental shelf beyond 200 nm. Developments in technology are increasing the depths at which resources can be exploited. Increasingly, attention is turning to offshore areas for the supply of minerals and hydrocarbons, as demand continues to rise and known deposits are exhausted.[20] The willingness of private actors to invest in distant offshore exploitation facilities will in part depend on the price of the resource, the availability of onshore reserves, the cost of extraction, the state of technology and legal certainty in relation to the rights to exploit the resource. Schofield is correct to caution that 'a sense of perspective is advisable'.[21] It is true that there will not be an economic bonanza for coastal States in the near future from exploiting resources on the extended shelf. However, eventually, exploitation of non-living resources on the extended continental shelf will take place.

The issue of living resources may need to be addressed by coastal States earlier than non-living resources in light of international obligations in relation to the preservation of marine biodiversity and interest in bioprospecting. While non-living resources are unlikely to be impacted on by activities of other States, living resources are more vulnerable. Bottom trawling for high seas species is already occurring on some extended continental shelves, with significant impacts on the benthic environment. Of course, living resources can be vulnerable to natural events as well as human perturbation and so establishing a representative network of MPAs where possible will meet a coastal State's obligations under international law, as well as protecting its interests in the genetic material found there.

As it is early days, State practice in relation to coastal State rights and obligations on the continental shelf beyond 200 nm is extremely limited. The extent to which the international community can reach a common viewpoint on the appropriate intersection between coastal and flag State rights will become clearer over time. However, disputes that have arisen in relation to the intersection of rights in other maritime zones, such as the EEZ, indicate that tension between positions is likely to occur. This book has endeavoured to flesh out some of the legal issues that may arise and provide a framework for evaluating competing interests.

When considering the intersection between the continental shelf beyond 200 nm and the high seas, State practice under the Continental Shelf Convention provides some guidance as to how articles 77 and 78 of the LOSC should be dealt with. That Convention created a situation where the coastal State had rights over the resources of the seabed but not the superjacent high seas. Today, coastal States are facing a similar position with the part of their continental shelf that extends beyond 200 nm.

---

[20] Schofield, 'Securing the Resources of the Deep', p. 287.   [21] Ibid., p. 293.

Of course, Part VI of the LOSC is different in many respects from the Continental Shelf Convention. It has introduced new limitations on coastal State rights regarding the shelf. These provisions must be seen in light of the 'package' that was negotiated. Broad margin States were allowed to claim rights over the extended continental shelf but, in return, had to accept limitations on those rights. Article 82 in particular was central to the agreement that was reached. The obligation to make payments in relation to exploitation on the continental shelf beyond 200 nm reflected the compromise that was made by States wanting that area to belong to the international seabed area and be covered by the common heritage of mankind.

Any interpretation of the rights of coastal States must take into account the character of that compromise. At the present time, more than thirty years after the LOSC was concluded, decision-makers may not remember the policy behind articles 82 and 246(6). However, it is important for current interpretation and practice to reflect the history and development of the law of the sea. This is particularly important in the case of States not party to the LOSC, but who wish to make use of article 76 to justify their exercise of jurisdiction over the entire continental shelf. If such States refuse to implement article 82 this could precipitate a serious dispute.

The intersection of the high seas and the continental shelf regimes is another possible point of friction. Where high seas activities take place close to, or even on, the seabed, the coastal State has an interest in the impact that these activities have on the seabed. The requirement not to infringe or unjustifiably interfere with high seas freedoms in article 78(2) limits coastal States' ability to prevent or regulate high seas activities to situations where the high seas activities have a serious impact on the coastal State's rights over the resources of the seabed and where there is no less restrictive option. This does not mean that the coastal State can never interfere with high seas freedoms, but it must be a last resort and only in the most serious cases.

It is important to acknowledge that any attempts by coastal States to unilaterally prevent or interfere with high seas freedoms will be controversial. Even if the coastal State action can be justified in terms of the protection of sovereign rights, it will not easily be accepted by flag States. If it proposes to undertake unilateral action, the coastal State would be strongly advised to consult States that might be affected with a view to reaching agreement. The history of the law of the sea is filled with examples of how unilateral State action can create disputes between States. Coastal States are always under an obligation to seek to resolve such problems peacefully.

Flag States should also understand that there will be limits on what their vessels can do in the high seas as a result of coastal State rights to the continental shelf beyond 200 nm. High seas freedoms are not absolute under the LOSC. Where their vessels are undertaking operations on or near another State's extended continental shelf, it should be expected that some limitations on those activities may be required in order to protect coastal State rights. On the other hand, they should be able to expect that such interference is only made in exceptional circumstances. The focus of this book has been on the coastal State, but the analysis is relevant to all States involved in high seas activities. It can be expected that there will be disagreements over matters lying at the intersection of coastal State and flag State interests.

A preferable approach would be for coastal States to seek to cooperate in multilateral or bilateral fora to provide the best opportunity to achieve their goals in a non-confrontational manner. The case of the North-east Atlantic is an excellent exemplar. Given that activities on remote parts of continental shelves are developing relatively slowly, obtaining support in international or regional organizations for limitations on high seas activities that might impact on the extended continental shelf may be easier if approached sooner rather than later.

The continental shelf beyond 200 nm feels like the 'final frontier' in the LOSC. In respect of all other zones, State practice and international instruments have fleshed out the areas of agreement and disagreement. However, the extended continental shelf still feels like the wild west. States have, legitimately, sought to maximize their control over an area about which the possible economic returns are uncertain. There is much we do not know about the extended continental shelf and its resources. The process for submissions to the CLCS has taken much longer than expected when the LOSC was originally concluded. The legal situation, where the coastal State has exclusive rights to the seabed and the superjacent water is the high seas, is unique in the LOSC and we have to look back to the situation that existed prior to 1972 to find any parallels.

Philip Allott, in his erudite article, observed that legal relations and delegation of powers are layered in the LOSC so that 'my right may begin where someone else's right ends. The more extensive my power, the more limited someone else's freedoms'.[22] He observed that the rights in the LOSC are shared powers, 'shared between the holder of the power and the community of States, in which regard for the interests of other States and of all States is of the essence'.[23] In the context of the relationships established by the LOSC, it is little wonder that the intersection of power, rights and freedoms on the continental shelf beyond 200 nm is complex. It is not just the coastal State that has an interest in the interplay in relation to the continental shelf beyond 200 nm—so too does the international community. Therefore, the first instances of State practice in this area will be heavily scrutinized and analysed. It is hoped that this book provides some framework for the discussion.

---

[22] Philip Allott, 'Power Sharing in the Law of the Sea' *American Journal of International Law* 77 (1983): pp. 1–30, p. 10.
[23] Ibid., p. 27.

Bibliography

Abecasis, Rita C, et al., 'Marine Conservation in the Azores: Evaluating Marine Protected Area Development in a Remote Island Context' *Frontiers in Marine Science* 2 (2015): Article 104, doi: 10.3389/fmars.2015.00104.

Abdel-Aal, Hussein K and Mohammed A Alsahlawi, *Petroleum Economics and Engineering*, 3rd edn (Boca Raton: CRC Press, 2014).

Adede, AO, *The Systems for Settlement of Disputes Under the United Nations Convention on the Law of the Sea* (Leiden: Martinus Nijhoff, 1987).

Adede, AO, 'The System for Exploitation of the "Common Heritage of Mankind" at the Caracas Conference' *American Journal of International Law* 69 (1975): pp. 31–49.

Affholder, M and F Valiron, *Descriptive Physical Oceanography* (Boca Raton: CRC Press, 2001).

Allen, Craig H, 'Protecting the Oceanic Gardens of Eden: International Law Issues in Deep-sea Vent Resource Conservation and Management' *Georgetown International Environmental Law Review* 13 (2001): pp. 563–660.

Allott, Philip, 'Power Sharing in the Law of the Sea' *American Journal of International Law* 77 (1983): pp. 1–30.

Althaus, F, et al., 'Impacts of Bottom Trawling on Deep-coral Ecosystems of Seamounts are Long-lasting' *Marine Ecology Progress Series* 397 (2009): pp. 279–94.

Amin, Amin, et al., 'Subsea Development from Pore to Process' *Oilfield Review* 17 (1999): pp. 4–17.

Anand, Ram Prakash, *Legal Regime of the Sea-bed and the Developing Countries* (Leiden: Brill, 1976).

Anderson, Owen F and Malcolm R Clark, 'Analysis of Bycatch in the Fishery for Orange Roughy, *Hoplostethus atlanticus*, on the South Tasman Rise' *Marine and Freshwater Research* 54 (2003): pp. 643–52.

Andrassy, J, *International Law and the Resources of the Sea* (Columbia University Press, 1970).

Anton, Donald K, 'Law for the Sea's Biological Diversity' *Columbia Journal of Transnational Law* 36 (1998): pp. 341–71.

Anton, Donald K, 'Case Concerning Pulp Mills on the River Uruguay *(Argentina v Uruguay)* (Judgment)' *Australian International Law Journal* 17 (2010): pp. 213–23.

Ardron, Jeff A, et al., 'A Systematic Approach Towards the Identification and Protection of Vulnerable Marine Ecosystems' *Marine Policy* 49 (2014): pp. 146–54.

Ardron, Jeff A, 'The Challenge of Assessing Whether the OSPAR Network of Marine Protected Areas Is Ecologically Coherent' *Hydrobiologia* 606 (2008): pp. 45–53.

Arico, S and C Salpin, *Bioprospecting of Genetic Resources in the Deep Seabed: Scientific, Legal and Policy Aspects* (UNU-IAS 2005) 34 http://moderncms.ecosystemmarketplace.com/repository/moderncms_documents/DeepSeabed.pdf.

Armas-Pfirter, Frida M, 'Submissions on the Outer Limit of the Continental Shelf: Practice to Date and Some Issues of Debate' in Davor Vidas (ed), *Law, Technology and Science for Oceans in Globalisation* (Leiden: Brill, 2010): pp. 477–98.

Arnaud-Haond, Sophie, Jesús M Arrieta and Carlos M Duarte, 'Marine Biodiversity and Gene Patents' *Science* 331 (2011): pp. 1521–2.

Arrieta, Jesús M, Sophie Arnaud-Haond and Carlos M Duarte, 'What Lies Underneath: Conserving the Ocean's Genetic Resources' *Proceedings of the National Academy of Sciences* 107 (2010): pp. 18318–24.

Asebey, Edgar J and Jill D Kempenaar, 'Biodiversity Prospecting: Fulfilling the Mandate of the Biodiversity Convention' *Vanderbilt Journal of Transnational Law* 28 (1995): pp. 703–54.

Ásmundsson, Stefán and Emily Corcoran, 'Information Paper on the Process of Forming a Cooperative Mechanism between NEAFC and OSPAR: From the First Contact to a Formal Collective Arrangement' (2016) www.neafc.org.

Ásmundsson, Stefán, *The Development of NEAFC's Protection of Vulnerable Marine Ecosystems* (2015) www.neafc.org.

Asmus, David and Jacqueline Lang Weaver, 'Unitizing Oil and Gas Fields Around the World: A Comparative Analysis of National Laws and Private Contracts' *Houston Journal of International Law* 28(3) (2006): pp. 1–197.

Attard, David, *The Exclusive Economic Zone in International Law* (Oxford: Clarendon Press, 1987).

Auster, Peter, et al., 'Definition and Detection of Vulnerable Marine Ecosystems on the High Seas: Problems with the "Move-on" Rule' *ICES Journal of Marine Science* (2010): doi:10.1093/icesjms/fsq 074.

Azzam, Issam, 'The Dispute Between France and Brazil Over Lobster Fishing in the Atlantic' *International and Comparative Law Quarterly* 13 (1964): pp. 1453–9.

Bailey, DM, et al., 'Long-term Changes in Deep-water Fish Populations in the Northeast Atlantic: a Deeper Reaching Effect of Fisheries?' *Proceedings of the Royal Society of Britain* 276 (2009): doi:10.1098/rspb.2009.0098.

Bailey, Kenneth, 'Australia and the Law of the Sea' *Adelaide Law Review* 1 (1960–1962): pp. 1–22.

Baird, SJ, BA Wood and NW Bagley, *Nature and Extent of Fishing Effort on or Near the Seafloor within the New Zealand 200 Nautical Mile Exclusive Economic Zone 1989–90 to 2004–05*, NZ Aquatic Environment and Biodiversity Report No. 73 (Wellington: Ministry of Fisheries, 2011).

Baker, Betsy B, 'Law, Science, and the Continental Shelf: The Russian Federation and the Promise of Arctic Cooperation' *American University International Law Review* 25 (2010): pp. 251–81.

Barry, James P, et al., 'Effects of Direct Ocean CO_2 Injection on Deep-sea Meiofauna' *Journal of Oceanography* 60 (2004): pp. 759–66.

Baslar, Kemal, *The Concept of the Common Heritage of Mankind in International Law* (The Hague: Martinus Nijhoff Publishers, 1998).

Bastida, Ana E, et al., 'Cross-border Unitization and Joint Development Agreements: An International Law Perspective' *Houston Journal of International Law* 29 (2006–2007): pp. 355–422.

Becker-Weinberg, Vasco, 'Portugal's Legal Regime on Marine Spatial Planning and Management of the National Maritime Space' *Marine Policy* 61 (2015): pp. 46–53.

Becker-Weinberg, Vasco, *Joint Development of Hydrocarbon Deposits in the Law of the Sea* (Heidelberg: Springer, 2014).

Beckman, R., CH Schofield, I Townsend-Gault, T Davenport and L Bernard (eds), *Beyond Territorial Disputes in the South China Sea: Legal Frameworks for the Joint Development of Hydrocarbon Resources*, (Cheltenham: Edward Elgar Publishers, 2013).

Bernal, Patricio and Alan Simcock, 'Chapter 30: Marine Scientific Research' in *First Global Integrated Marine Assessment* (New York: United Nations, 2015) http://www.un.org/Depts/los/global_reporting/WOA_RegProcess.htm.

Bernal, Patricio, et al., 'Chapter 55: Overall Value of the Oceans to Humans' in *First Global Integrated Marine Assessment* (New York: United Nations, 2015) http://www.un.org/Depts/los/global_reporting/WOA_RegProcess.htm.

Bjørlykke, Knut, 'Introduction to Petroleum Geology' in Knut Bjørlykke (ed), *Petroleum Geoscience: From Sedimentary Environments to Rock Physics* (Berlin: Springer-Verlag, 2010): pp. 1–26.

Bjørn, Kunoy, 'The Terms of Reference of the Commission on the Limits of the Continental Shelf: A Creeping Legal Mandate' *Leiden Journal of International Law* 25 (2012): pp. 109–30.

Boczek, Boleslaw Adam, 'Peacetime Military Activities in the Exclusive Economic Zone of Third Countries' *Ocean Development and International Law* 19 (1988): pp. 445–68.

Bomkamp, RE, HM Page and JE Dugan, 'Role of Food Subsidies and Habitat Structure in Influencing Benthic Communities of Shell Mounds at Sites of Existing and Former Offshore Oil Platforms' *Marine Biology* 146 (2004): pp. 201–11.

Bors, EK, et al., 'Patterns of Deep-sea Genetic Connectivity in the New Zealand Region: Implications for Management of Benthic Ecosystems' *PLOS One* 7(11) (2012): e49474. doi:10.1371/journal.pone.0049474.

Boschen, RE, AA Rowden, MR Clark and JPA Gardner, 'Mining of Deep-sea Seafloor Massive Sulfides: A Review of the Deposits, Their Benthic Communities, Impacts from Mining, Regulatory Frameworks and Management Strategies' *Ocean and Coastal Management* 84 (2013): pp. 54–67.

Bowden, David A, et al., 'Cold Seep Epifaunal Communities on the Hikurangi Margin, New Zealand: Composition, Succession, and Vulnerability to Human Activities' *PLOS One* 8 (2013): e76869.

Boyle, Alan, 'Further Development of the Law of the Sea Convention: Mechanisms for Change' *International and Comparative Law Quarterly* 54 (2005): pp. 563–84.

Boyle, Alan, 'Dispute Settlement and the Law of the Sea Convention: Problems of Fragmentation and Jurisdiction' *International and Comparative Law Quarterly* 46 (1997): pp. 37–54.

Brekke, Harald and Philip Symonds, 'Submarine Ridges and Elevations of Article 76 in Light of Published Summaries of Recommendations of the Commission on the Limits of the Continental Shelf' *Ocean Development and International Law* 42 (2011): pp. 289–306.

Brooks, CM, 'Competing Values on the Antarctic High Seas: CCAMLR and the Challenge of Marine Protected Areas' *The Polar Journal* 3 (2013): pp. 277–300.

Brown, ED, *The Legal Regime of Hydrospace* (London: Stevens & Sons, 1971).

Burke, William T, ' "Customary Law as Reflected in the LOS Convention": A Slippery Formula' in John P Craven, Jan Schneider and Carol Stimson (eds), *The International Implications of Extended Maritime Jurisdiction in the Pacific* (Honolulu: Law of the Sea Institute, 1987): pp. 402–9.

Caddell, Richard, 'Platforms, Protesters and Provisional Measures: The Arctic Sunrise Dispute and Environmental Activism at Sea' in M Ambrus and RA Wessel (eds), *Netherlands Yearbook of International Law 2014*, (TMC Asser, 2015): pp. 359–84.

Calado, H, et al., 'Introducing a Legal Management Instrument for Offshore Marine Protected Areas in the Azores—The Azores Marine Park' *Environmental Science and Policy* 14 (2011): pp. 1175–87.

Caminos, Hugo, and Michael R Molitor, 'Progressive Development of International Law' *American Journal of International Law* 79 (1985): pp. 871–90.

Cater, Carl and Erlet Cater, *Marine Ecotourism: Between the Devil and the Deep Blue Sea* (Cambridge, MA: CABI, 2007).

Charney, Jonathan I, 'The United States and the Law of the Sea after UNCLOS III—The Impact of General International Law' *Law and Contemporary Problems* 46 (1983): pp. 37–54.

Chircop, Aldo, 'Managing Adjacency: Some Legal Aspects of the Relationship Between the Extended Continental Shelf and the International Seabed Area' *Ocean Development and International Law* 42 (2011): pp. 307–16.

Chircop, Aldo and Bruce A Marchand, 'International Royalty and Continental Shelf Limits: Emerging Issues for the Canadian Offshore', *Dalhousie Law Journal* 26 (2003): pp. 273–302.

Churchill, RR and AV Lowe, *The Law of the Sea*, 3rd edn (Manchester: Manchester University Press, 1999).

Clark, Malcolm R, et al., 'The Impacts of Deep-sea Fisheries on Benthic Communities: a Review' *ICES Journal of Marine Science* 73 (Supp. 1) (2016): doi: 10.1093/icesjms/fsv 123.

Clark, Malcolm and Samantha Smith, 'Environmental Management Considerations' in Elaine Baker and Yannick Beaudoin (eds), *Cobalt-rich Ferromanganese Crusts: A Physical, Biological, Environmental, and Technical Review* (Arendal: SPC-Grid Arendal, 2013): pp. 23–40.

Clark, Malcolm and Samantha Smith, 'Environmental Management Considerations' in Elaine Baker and Yannick Beaudoin (eds) *Sea-Floor Massive Sulphides: A Physical, Biological, Environmental and Technical Review* (Arendal: SPC-Grid Arendal, 2013): pp. 27–42.

Clark, Malcolm, et al., 'The Ecology of Seamounts: Structure, Function and Human Impacts' *Annual Review of Marine Science* 2 (2010): pp. 253–78.

Clark, Malcolm, 'Deep-sea Seamount Fisheries: A Review of Global Status and Future Prospects' *Latin American Journal of Aquatic Research* 37(3) (2009): pp. 501–12.

Clark, Malcolm R, et al., 'Large-scale Distant-water Trawl Fisheries on Seamounts' in Tony J Pitcher et al., (eds), *Seamounts: Ecology, Fisheries and Conservation* (Oxford: Blackwell Publishing, 2007): pp. 361–99.

Clark, Malcolm R and J Anthony Koslow, 'Impacts of Fisheries on Seamounts' in Tony J Pitcher et al., (eds), *Seamounts: Ecology, Fisheries and Conservation* (Oxford: Blackwell Publishing, 2007), pp. 413–41.

Cohen, Harlan, 'Some Reflections on Bioprospecting in the Polar Regions' in Davor Vidas (ed), *Law, Technology and Science for Oceans in Globalisation: IUU Fishing, Oil Pollution, Bioprospecting, Outer Continental Shelf* (Leiden: Martinus Nijhoff, 2010): pp. 339–52.

Collins, PC, et al., 'A Primer for the Environmental Impact Assessment of Mining at Seafloor Massive Sulphides' *Marine Policy* 42 (2013): pp. 198–209.

Collins, PC, R Kennedy, CL Van Dover, 'A Biological Survey Method Applied to Seafloor Massive Sulphides (SMS) with Contagiously Distributed Hydrothermal-vent Fauna' *Marine Ecology Progress Series* 452 (2012): pp. 89–107. doi: 10.3354/meps 09646.

Cook, PJ and CM Carleton (eds), *Continental Shelf Limits: The Scientific and Legal Interface* (Oxford: Oxford University Press, 2000).

Cordonnery, Laurence, Alan D Hemmings and Lorne Kriwoken, 'Nexus and Imbroglio: CCAMLR, the Madrid Protocol and Designating Marine Protected Areas in the Southern Ocean' *International Journal of Marine and Coastal Law* 30 (2015): pp. 727–64.

Costello, Mark J and Bill Ballantine, 'Biodiversity Conservation Should Focus on No-take Marine Reserves: 94% of Marine Protected Areas Allow Fishing' *Trends in Ecology and Evolution* 30 (2015): pp. 507–9.

Cruickshank, Michael J, 'Mineral Resources Potential of Continental Margins' in Creighton A Burk and Charles L Drake (eds), *The Geology of Continental Margins* (New York: Springer-Verlag, 1974): pp. 965–1000.

Currie, DR and Leanne R Isaacs, 'Impact of Exploratory Offshore Drilling on Benthic Communities in the Minerva Gas Field, Port Campbell, Australia' *Marine Environmental Research* 59(3) (2005): pp. 217–33.

Curtis, Charles E, 'Alaska's Regulation of King Crab on the Outer Continental Shelf' *UCLA Alaska Law Review* 6 (1976–1977): pp. 375–410.

da Silva, Alexandre Pereira, 'Dealing with Articles 76 and 82 of the United Nations Convention on the Law of the Sea: Legal and Political Challenges for Brazil' *Ocean Yearbook* 28 (2014): pp. 145–72.

Danovaro, Roberto, 'Exponential Decline of Deep-sea Ecosystem Functioning Linked to Benthic Biodiversity Loss' *Current Biology* 18(1) (2008): pp. 1–8.

Davenport, Tara, 'The Exploration and Exploitation of Hydrocarbon Resources in Areas of Overlapping Claims' in R Beckman, CH Schofield, I Townsend-Gault, T Davenport and L Bernard (eds), *Beyond Territorial Disputes in the South China Sea: Legal Frameworks for the Joint Development of Hydrocarbon Resources* (Cheltenham: Edward Elgar Publishers, 2013): pp. 93–113.

Davies, Andrew J, J Murray Roberts and Jason Hall-Spencer, 'Preserving Deep-sea Natural Heritage: Emerging Issues in Offshore Conservation and Management' *Biological Conservation* 138 (2007): pp. 299–312.

de Fontaubert, A Charlotte, David R Downes and Rundi S Agardy, 'Biodiversity in the Seas: Implementing the Convention on Biological Diversity in Marine and Coastal Habitats' *Georgetown International Environmental Law Review* 10 (1997–1998): pp. 753–854.

de Groot, SJ, 'The Impact of Laying and Maintenance of Offshore Pipelines on the Marine Environmental and the North Sea Fisheries' *Ocean Management* 8 (1982): pp. 1–27.

de La Fayette, LA, 'A New Regime for the Conservation and Sustainable Use of Marine Biodiversity and Genetic Resources Beyond the Limits of National Jurisdiction' *International Journal of Marine and Coastal Law* 24 (2009): pp. 221–80.

De Leo, Danielle M, et al., 'Response of Deep-water Corals to Oil and Chemical Dispersant Exposure' *Deep Sea Research Part II: Topical Studies in Oceanography* 129 (2016): pp. 137–47.

Derraik, Jose GB, 'The Pollution of the Marine Environment by Plastic Debris: A Review' *Marine Pollution Bulletin* 44 (2002): pp. 842–52.

de Ronde, Cornel EJ, et al., 'Evolution of a Submarine Magmatic-hydrothermal System: Brothers Volcano, Southern Kermadec Arc, New Zealand' *Economic Geology* 100 (2005): pp. 1097–133.

de Ronde, Cornel EJ, et al., 'Intra-oceanic Subduction-related Hydrothermal Venting, Kermadec Volcanic Arc, New Zealand' *Earth and Planetary Science Letters* 193 (2001): pp. 359–69.

Donaldson, John W, 'Oil and Water: Assessing the Link Between Maritime Boundary Delimitation and Hydrocarbon Resources' in Clive Schofield, Seokwoo Lee and Moon-Dang Kwon (eds), *The Limits of Maritime Jurisdiction* (Leiden: Martinus Nijhoff, 2014): pp. 127–43.

Downey, Morgan, *Oil 101* (Los Angeles: Wooden Table Press, 2009).

Druel, Elisabeth, et al., *Governance of Marine Biodiversity in Areas Beyond National Jurisdiction at the Regional Level: Filling the Gaps and Strengthening the Framework for Action* (Paris: IDDRi and AAMP, 2012).

Dunn, Daniel C, et al., 'The Convention on Biological Diversity's Ecologically or Biologically Significant Areas: Origins, Development, and Current Status' *Marine Policy* 49 (2014): pp. 137–45.

Fedder, Bevis, *Marine Genetic Resources, Access and Benefit Sharing: Legal and Biological Perspectives* (London: Routledge, 2013): p. 46.

Fidell, ER, 'Ten Years Under the Bartlett!Act: A Status Report on the Prohibition on Foreign Fishing' *Boston University Law Review* 54 (1974): pp. 703–56.

Fleischer, Carl August, 'The Continental Shelf Beyond 200 Nautical Miles—a Crucial Element in the 'Package Deal': Historic Background and Implications for Today' in Davor Vidas (ed), *Law, Technology and Science for Oceans in Globalisation* (Leiden: Brill, 2010): pp. 429–48.

Franckx, Erik, 'The International Seabed Authority and the Common Heritage of Mankind: The Need for States to Establish the Outer Limits of Their Continental Shelf' *International Journal of Marine and Coastal Law* 25 (2010): pp. 543–67.

Franssen, Herman T, 'Developing Country Views of Marine Science and Law' in Warren S Wooster (ed), *Freedom of Oceanic Research* (Russak, New York: Crane, 1973): pp. 137–78.

Franssen, Maureen N, 'Oceanic Research and the Developing Nation Perspective' in Warren S Wooster (ed), *Freedom of Oceanic Research* (Russak, New York: Crane, 1973): pp. 179–200.

Freestone, David, et al., 'Can Existing Institutions Protect Biodiversity in Areas Beyond National Jurisdiction? Experiences from Two Ongoing Processes' *Marine Policy* 49 (2014): pp. 167–275.

Freestone, David and Kate Killerlain Morrison, 'The Signing of the Hamilton Declaration on Collaboration for the Conservation of the Sargasso Sea: A New Paradigm for High Seas Conservation?' *International Journal of Marine and Coastal Law* 29 (2014): pp. 345–62.

Freestone, David et al., 'Place-based Management of Large-scale Ocean Places: Papahānaumokuākea and the Sargasso Sea' *Stanford Environmental Law Journal* 191 (2013–2014): pp. 191–250.

Freestone, David, 'Responsibilities and Obligations of States Sponsoring Persons and Entities with Respect to Activities in the Area' *American Journal of International Law* 105 (2011): pp. 755–60.

French, Duncan, 'From the Depths: Rich Pickings of Principles of Sustainable Development and General International Law on the Ocean Floor—the Seabed Disputes Chamber's 2011 Advisory Opinion' *International Journal of Marine and Coastal Law* 26 (2011): pp. 525–68.

Froescheis, Oliver, et al., 'The Deep-sea as a Final Global Sink of Semivolatile Persistent Organic Pollutants? Part I: PCBs in Surface and Deep-sea Dwelling Fish of the North and South Atlantic and the Monterey Bay Canyon (California)' *Chemosphere* 40 (2000): pp. 651–60.

Gardiner, Richard, 'The Vienna Convention Rules on Treaty Interpretation' in Duncan B Hollis (ed), *The Oxford Guide to Treaties* (Oxford: Oxford University Press, 2014): pp. 475–506.

Gena, Kaul, 'Deep Sea Mining of Submarine Hydrothermal Deposits and Its Possible Environmental Impact in Manus Basin, Papua New Guinea', *Procedia Earth and Planetary Science* 6 (2013): pp. 226–33.

Genin, Amatzia and John F Dower, 'Seamount Plankton Dynamics' in Tony J Pitcher et al., (eds), *Seamounts: Ecology, Fisheries and Conservation* (Oxford: Blackwell Publishing, 2007): pp. 85–100.

Gerbruck, Andrey, 'Locomotory Organs in the Elasipodid Holothurians: Functional-morphological and Evolutionary Approaches' in Roland Emerson, Andrew Smith and Andrew Campbell (eds), *Echinoderm Research 1995* (Rotterdam: AA Balkema, 1995): pp. 95–101.

Gislason, Sigurdur R and Eric H Oelkers, 'Carbon Storage in Basalt' *Science* 344 (2014): pp. 373–4.

Glavovic, Bruce C, 'Disasters and the Continental Shelf: Exploring New Frontiers of Risk' in Myron H Nordquist et al., (eds), *The Regulation of Continental Shelf Development: Rethinking International Standards* (Leiden: Martinus Nijhoff, 2013): pp. 225–58.

Glover, Adrian and Craig R Smith, 'The Deep-sea Floor Ecosystem: Current Status and Prospects of Anthropogenic Change by the Year 2025' *Environmental Conservation* 30(3) (2003): pp. 219–41.

Glowka, Lyle, 'Putting Marine Scientific Research on a Sustainable Footing at Hydrothermal Vents' *Marine Policy* 27 (2003): pp. 303–12.

Glowka, Lyle, 'The Deepest of Ironies: Genetic Resources, Marine Scientific Research, and the Area' *Ocean Yearbook* 12 (1996): pp. 154–78.

Goldberg, David S, Dennis C Kent and Paul E Olsen, 'Potential Onshore and Offshore Reservoirs for CO_2 Sequestration in Central Atlantic Magmatic Province Basalts' *PNAS* 107 (2010): pp. 1327–32.

Goldie, LFE, 'The Ocean's Resources and International Law: Possible Developments in Regional Fisheries Management' *Columbia Journal of Transnational Law* 8 (1969): pp. 1–53.

Goldie, LFE, 'Australia's Continental Shelf: Legislation and Proclamations' *International and Comparative Law Quarterly* 3 (1954): pp. 535–75.

Gorina-Ysern, Monserrat, *An International Regime for Marine Scientific Research* (New York: Transnational Publishers, 2003).

Paul Gragl, 'Marine Scientific Research' in Malgosia Fitzmaurice and Norman A Martínez Gutiérrez (eds), *The IMLI Manual on International Maritime Law: Volume I Law of the Sea* (Oxford: Oxford University Press, 2014): pp. 396–429.

Grant, Alastair and Andrew D Briggs, 'Toxicity of sediments from around a North Sea oil platform: are metals or hydrocarbons responsible for ecological impacts?' *Marine Environmental Research* 53(1) (2002): pp. 95–116.

Grunstein, Miriam, 'Unitized we Stand, Divided we Fall: A Mexican Response to Karla Urdaneta's Analysis of Transboundary Petroleum Reservoirs in the Deep Waters of the Gulf of Mexico' *Houston Journal of International Law* 33 (2011): pp. 345–67.

Gutteridge, JAC, 'The 1958 Geneva Convention on the Continental Shelf' *British Yearbook of International Law* 35 (1959): pp. 102–23.

Halfar, Jochen and Rodney M Fujita, 'Precautionary Management of Deep-sea Mining' *Marine Policy* 26 (2002): pp. 103–6.

Handl, Günther, 'Transboundary Impacts' in Daniel Bodansky, Jutta Brunnée and Ellen Hey (eds), *The Oxford Handbook of International Environmental Law* (Oxford: Oxford University Press, 2008): pp. 531–49.

Hannington, M, et al., 'The Abundance of Seafloor Massive Sulfide Deposits' *Geology* 39 (2011): pp. 1155–8.

Hansen, S, P Ward and A Penney, '*Identification of Vulnerable Benthic Taxa in the Western SPRFMO Convention Area and Review of Move-on Rules for Different Gear Types*', SPRFMO Document SC-01-09 (2013).

Harel, Assaf, 'Preventing Terrorist Attacks on Offshore Platforms: Do States Have Sufficient Legal Tools?' *Harvard National Security Journal* 4 (2012): pp. 131–84.

Harris, Peter, et al., 'Chapter 21: Offshore Hydrocarbon Industries' in *First Global Integrated Marine Assessment* (New York: United Nations, 2015) http://www.un.org/Depts/los/global_reporting/WOA_RegProcess.htm.

Harrison, James, 'The *Arctic Sunrise* Arbitration (Netherlands v Russia)' *International Journal of Marine and Coastal Law* 31 (2016): pp. 145–59.

Hayashi, Moritaka, 'Military and Intelligence Gathering Activities in the EEZ: Definition of Key Terms' *Marine Policy* 29 (2005): pp. 123–37.

Heezen, Bruce C, 'Atlantic-type Continental Margins' in Creighton A Burk and Charles L Drake (eds), *The Geology of Continental Margins* (New York: Springer-Verlag, 1974): pp. 13–24.

Hein, James R, et al., 'Deep-ocean Mineral Deposits as a Source of Critical metals for High- and Green-technology Applications: Comparison with Land-based Resources' *Ore Geology Reviews* 51 (2013): pp. 1–14.

Hoagland, Porter, et al., 'Deep-sea Mining of Seafloor Massive Sulfides' *Marine Policy* 34 (2010): pp. 728–32.

Hubert, Anna-Maria, 'The New Paradox in Marine Scientific Research: Regulating the Potential Environmental Impacts of Conducting Ocean Science' *Ocean Development and International Law* 42 (2011): pp. 329–55.

Hui Zhang, 'The Sponsoring State's "Obligation to Ensure" in the Development of the International Seabed Area' *International Journal of Marine and Coastal Law* 28 (2013): pp. 681–99.

Hyne, Norman J, *Nontechnical Guide to Petroleum Geology, Exploration, Drilling and Production*, 3rd edn (Tulsa: Penn Well, 2012).

Imhoff, JF, A Labes and J Wiese, 'Bio-mining the Microbial Treasures of the Ocean: New Natural Products' *Biotechnology Advances* 29 (2011): pp. 468–82.

International Law Association, *Report on Article 82 of the 1982 UN Convention on the Law of the Sea* (2008).

International Law Association, *Legal Issues of the Outer Continental Shelf*, Second Report, Toronto Conference (2006).

International Seabed Authority, *Environmental Management of Deep-sea Chemosynthetic Ecosystems: Justification of and Considerations for a Spatially-based Approach*, Technical Study No. 9 (2011).

International Seabed Authority, *Non-living Resources of the Continental Shelf Beyond 200 Nautical Miles: Speculations on the Implementation of Article 82 of the United Nations Convention on the Law of the Sea*, Technical Study No. 5 (2010).

International Seabed Authority, *Issues Associated with the Implementation of Article 82 of the United Nations Convention on the Law of the Sea*, Technical Study No. 4 (2009).

International Seabed Authority, *Global Non-living Resources on the Extended Continental Shelf: Prospects at the Year 2000*, Technical Study No. 1 (2001).

IPCC, *Special Report on Carbon Dioxide Capture and Storage* (Cambridge: Cambridge University Press, 2005).

Jabour-Green, Julia and Dianne Nichol, 'Bioprospecting in Areas Outside National Jurisdiction: Antarctica and the Southern Ocean' *Melbourne Journal of International Law* 4 (2003): pp. 76–111.

Jacobson, Jon L, 'Marine Scientific Research Under Emerging Ocean Law' *Ocean Development and International Law* 9 (1981): pp. 187–218.

Jares, Vladimir, 'Continental Shelf Beyond 200 Nautical Miles: The Work of the Commission on the Limits of the Continental Shelf and the Arctic' *Vanderbilt Journal of Transnational Law* 42 (2009): pp. 1265–305.

Jennings, RY, 'The Limits of Continental Shelf Jurisdiction: Some Possible Implications of the North Sea Case Judgment' *International and Comparative Law Quarterly* 18 (1969): pp. 819–32.

Jensen, Øystein, *Commission on the Limits of the Continental Shelf: Law and Legitimacy* (Leiden: Martinus Nijhoff Publishers, 2014).

Jensen, Øystein, 'The Commission on the Limits of the Continental Shelf: An Administrative, Scientific, or Judicial Institution?' *Ocean Development and International Law* 45 (2014): pp. 171–85.

Jørem, Ane and Morten Walløe Tvedlt, 'Bioprospecting in the High Seas: Existing Rights and Obligations in View of a New Legal Regime for Marine Areas Beyond National Jurisdiction' *International Journal of Marine and Coastal Law* 29 (2014): pp. 321–43.

Joyner, Christopher C, 'Legal Implications of the Concept of the Common Heritage of Mankind' *The International and Comparative Law Quarterly* 35 (1986): pp. 190–9.

Juda, Lawrence, 'The 1995 United Nations Agreement on Straddling Fish Stocks and Highly Migratory Fish Stocks: A Critique' *Ocean Development and International Law* 28 (1997): pp. 147–66.

Kachel, Markus J, *Particularly Sensitive Sea Areas: The IMO's Role in Protecting Vulnerable Marine Areas* (Berlin: Springer, 2008).

Kamau, EC, B Fedder and G Winter, 'The Nagoya Protocol on Access to Genetic Resources and Benefit Sharing: What is New and What are the Implications for Provider and User Countries and the Scientific Community?' *Law, Environment and Development Journal* 6 (2010): pp. 246–62.

Kanehara, Atsuko, 'The Revenue Sharing Scheme with Respect to the Exploitation of the Outer Continental Shelf Under Article 82 of the United Nations Convention on the Law of the Sea—A Plethora of Entangling Issues', Seminar on the Establishment of the Outer Limits of the Continental Shelf Beyond 200 Nautical Miles Under UNCLOS, Ocean Policy Research Foundation, 2008 https://www.sof.or.jp/en/topics/pdf/aca.pdf.

Katin, Ernest, *The Legal Status of the Continental Shelf as Determined by the Conventions Adopted at the 1958 United Nations Conference on the Law of the Sea: An Analytical Study of an Instance of International Law Making* (Minneapolis: University of Minnesota, 1962).

Kato, Yasuhiro, 'Deep-sea mud in the Pacific Ocean as a Potential Source for Rare-earth Elements' *Nature Geoscience* 3 July 2011: doi: 10.1038/NGEO 1185.

Kaye, Stuart, 'Joint Development in the Timor Sea' in R Beckman, CH Schofield, I Townsend-Gault, T Davenport and L Bernard (eds), *Beyond Territorial Disputes in the South China Sea: Legal Frameworks for the Joint Development of Hydrocarbon Resources* (Cheltenham: Edward Elgar Publishers, 2013): pp. 249–67.

Kaye, Stuart, 'International Measures to Protect Oil Platforms, Pipelines, and Submarine Cables from Attack' *Tulane Maritime Law Journal* 31 (2006–2007): pp. 377–424.

Kingston, Paul F, 'Long-term Environmental Impact of Oil Spills' in *Spill Science and Technology Bulletin* 7 (2002): pp. 53–61.

Kingston, PF, 'Impact of Offshore Oil Production Installations on the Benthos of the North Sea' *ICES Journal of Marine Science* 49 (1992): pp. 45–53.

Kiss, Alexandre, 'The Common Heritage of Mankind: Utopia or Reality?' *International Journal* 40 (1985): pp. 423–41.

Klein, Natalie, *Dispute Settlement in the UN Convention on the Law of the Sea* (Cambridge: Cambridge University Press, 2005).

Kong, Lingjie, 'Environmental Impact Assessment Under the United Nations Convention on the Law of the Sea' *Chinese Journal of International Law* 10 (2011): pp. 651–69.

Koslow, J Anthony, et al., 'Chapter 51: Biological Communities on Seamounts and Other Submarine Features Potentially Threatened by Disturbance' in *First Global Integrated Marine Assessment* (New York: United Nations, 2015) www.un.org/depts/los/global_reporting/WOA_RegProcess.htm.

Koslow, Tony, *The Silent Deep: The Discovery, Ecology and Conservation of the Deep Sea* (Sydney: UNSW Press, 2007).

Koslow, Tony, et al., 'Seamount Benthic Microfauna off Southern Tasmania: Community Structure and Impacts of Trawling' *Marine Ecology Progress Series* 213 (2001): pp. 111–25.

JA Koslow et al., 'Continental Slope and Deep-sea Fisheries: Implications for a Fragile Ecosystem' *ICES Journal of Marine Science* 57 (2000): pp. 548–57.

Kovalev, AA, *Contemporary Issues of the Law of the Sea: Modern Russian Approaches*, translated by WE Butler (Utrecht: Eleven International Publishing, 2004).

Kunz, Josef L, 'Continental Shelf and International Law: Confusion and Abuse' *American Journal of International Law* 50 (1956): pp. 828–53.

Kwiatkowska, Barbara, 'Creeping Jurisdiction Beyond 200 Miles in the Light of the 1982 Law of the Sea Convention and State Practice' *Ocean Development and International Law* 22 (1991): pp. 153–87.

Kwiatkowska, Barbara, *The 200 Mile Exclusive Economic Zone in the New Law of the Sea* (Dordrecht: Martinus Nijhoff, 1989).

Kvalvik, Ingrid, 'Managing Institutional Overlap in the Protection of Marine Ecosystems on the High Seas: The Case of the North East Atlantic' *Ocean and Coastal Management* 56 (2012): pp. 35–43.

Lagoni, Rainer, 'Interim Measures Pending Maritime Delimitation Agreements' *American Journal of International Law* 78 (1984): pp. 345–68.

Lagoni, Rainer, 'Oil and Gas Deposits Across National Frontiers', *American Journal of International Law* 73 (1979): pp. 215–43.

Lauterpacht, H, 'Sovereignty Over Submarine Areas' *British Yearbook of International Law* 27 (1950): pp. 376–433.

Leary, David and S Kim Juniper, 'Addressing the Marine Genetic Resources Issue: Is the Debate Headed in the Wrong Direction?' in Clive Schofield, Seokwoo Lee and Moon-Sang Kwon (eds), *The Limits of Maritime Jurisdiction* (Leiden: Martinus Nijhoff, 2014): pp. 769–85.

Leary, David, 'International Law and the Genetic Resources of the Deep Sea' in Davor Vidas (ed), *Law, Technology and Science for Oceans in Globalisation: IUU Fishing, Oil Pollution, Bioprospecting, Outer Continental Shelf* (Leiden: Martinus Nijhoff, 2010): pp. 353–69.

Leary, David, et al., 'Marine Genetic Resources: A Review of Scientific and Commercial Interest' *Marine Policy* 33 (2009): pp. 183–94.

Leary, David, *International Law and the Genetic Resources of the Deep Sea* (Leiden: Martinus Nijhoff, 2007).

Le Bris, Nadine, et al., 'Chapter 45: Hydrothermal Vents and Cold Seeps' in *First Integrated Global Marine Assessment* (New York: United Nations, 2015) http://www.un.org/depts/los/global_reporting/WOA_RegProcess.htm.

Lefebvre-Chalain, Hélène, 'Fifteen Years of Particularly Sensitive Sea Areas: A Concept in Development' *Ocean and Coastal Law Journal* 13 (2007): pp. 47–69.

Lee, Luke T, 'The Law of the Sea Convention and Third States' *American Journal of International Law* 77 (1983): pp. 541–68.

Leung, Dennis YC, Giorgio Caramanna and M Mercedes Maroto-Valer, 'An Overview of Current Status of Carbon Dioxide Capture and Storage' *Renewable and Sustainable Energy Reviews* 39 (2014): pp. 426–43.

Levin, Lisa A and Paul K Dayton, 'Ecological Theory and Continental Margins: Where Shallow Meets Deep' *Trends in Ecology and Conservation* 24 (2009): pp. 606–17.

Liquete, Camino, 'Current Status and Future Prospects for the Assessment of Marine and Coastal Ecosystem Services: A Systematic Review' *PLOS One* 8(7) (3 July 2013): doi: 10.1371/journal.pone.0067737.

Lodge, Michael, 'The Deep Seabed' in Donald R Rothwell et al., (eds), *Oxford Handbook of the Law of the Sea* (Oxford: Oxford University Press, 2015): pp. 226–53.

Lodge, Michael W, 'The Common Heritage of Mankind' *International Journal of Marine and Coastal Law* 27 (2012): pp. 733–42.

Lodge, Michael W, 'The International Seabed Authority and Article 82 of the UN Convention on the Law of the Sea' *International Journal of Marine and Coastal Law* 21 (2006): pp. 323–33.

Lowe, Vaughan, 'Jurisdiction' in Malcolm Evans (ed), *International Law* (Oxford: Oxford University Press, 2003): pp. 329–55.

Lösekann, TA, et al., 'Endosymbioses Between Bacteria and Deep-sea Siboglinid Tubeworms from an Arctic Cold Seep (Haakon Mosby Mud Volcano, Barents Sea)' *Environmental Microbiology* 10(12) (2008): pp. 3237–54.

Lumb, RD 'Australian Legislation on Sedentary Resources of the Continental Shelf' *University of Queensland Law Journal* 7 (1970–1971): pp. 111–4.

Madin, Laurence P, et al., 'The Unknown Ocean' in LK Glover and SA Earle (eds), *Defying Ocean's End: An Agenda for Action* (Washington, DC: Island Press, 2004): pp. 213–36.

Magnússon, Bjarni Már, *The Continental Shelf Beyond 200 Nautical Miles* (Leiden: Brill Nijhoff, 2015).

McClain, Craig R and Sarah Mincks Hardy, 'The Dynamics of Biogeographic Ranges in the Deep Sea' *Proceedings of the Royal Society*, 277 (2010): pp. 3533–46.

McClain, Craig R, 'Seamounts: Identity Crisis or Split Personality?' *Journal of Biogeography* 34 (2007): pp. 2001–8.

McDorman, Ted L, 'The Continental Shelf Beyond 200 Nautical Miles: A First Look at the Bay of Bengal (*Bangladesh v Myanmar*) Case' in Myron H Nordquist, John Norton Moore, Aldo Chircop and Ronán Long (eds), *The Regulation of Continental Shelf Development: Rethinking International Standards* (Leiden: Martinus Nijhoff, 2013): pp. 89–103.

McDorman, Ted L, 'The Continental Shelf Beyond 200 NM: Law and Politics in the Arctic Ocean' *Journal of Transnational Law and Policy* 89 (2008): pp. 155–93.

McDorman, Ted L, 'The Role of the Commission on the Limits of the Continental Shelf: A Technical Body in a Political World' *International Journal of Marine and Coastal Law* 17 (2002): pp. 301–24.

McDorman, Ted L, 'The Entry into Force of the 1982 LOS Convention and the Article 76 Outer Continental Shelf Regime' *International Journal of Marine and Coastal Law* 10 (1995): pp. 165–87.

McDorman, Ted L, 'The New Definition of "Canada Lands" and the Determination of the Outer Limit of the Continental Shelf' *Journal of Maritime Law and Commerce* 14 (1983): pp. 195–223.

McDougal, MS and WT Burke, *The Public Order of the Oceans: A Contemporary International Law of the Sea* (New Haven: New Haven Press, 1987).

Mendelssohn, Irving A, et al., 'Oil Impacts on Coastal Wetlands: Implications for the Mississippi River Delta Ecosystem After the Deepwater Horizon Oil Spill' *Bioscience* 62(6) (2012): pp. 562–74.

Menot, Lenaick, et al., 'New Perceptions of Continental Margin Biodiversity' in Alasdair D McIntyre (ed), *Life in the World's Oceans: Diversity, Distribution, and Abundance* (Chichester: Wiley-Blackwell, 2010): pp. 79–102.

Miyoshi, Masahiro, 'The Basic Concept of Joint Development of Hydrocarbon Resources on the Continental Shelf' *International Journal of Estuarine and Coastal Law* 3 (1988): pp. 1–18.

Molenaar, EJ and AG Oude Elferink, 'Marine Protected Areas in Areas Beyond National Jurisdiction: The Pioneering Efforts Under the OSPAR Convention' *Utrecht Law Review* 5 (2009): pp. 5–20.

Moore, JD and ER Trueman, 'Swimming of the Scallop, Chlamys Operculis' *Journal of Exp Mar Biol Ecol* 6 (1971): pp. 179–85.

Morin, Michael D, 'Jurisdiction Beyond 200 Miles: A Persistent Problem' *California Western International Law Journal* 10 (1980): pp. 514–35.

Mossop, Joanna, 'Protests against Oil Exploration at Sea: Lessons from the *Arctic Sunrise* Arbitration' *International Journal of Marine and Coastal Law* 31 (2016): pp. 60–87.

Mossop, Joanna, 'Marine Bioprospecting' in Donald Rothwell et al., (eds), *The Oxford Handbook on the Law of the Sea* (Oxford: Oxford University Press, 2015): pp. 825–42.

Mossop, Joanna, 'Regulating Uses of Marine Biodiversity on the Outer Continental Shelf' in Davor Vidas (ed), *Law, Technology and Science for Oceans in Globalisation* (Leiden: Brill, 2010): pp. 319–37.

Mossop, Joanna, 'Protecting Marine Biodiversity on the Continental Shelf Beyond 200 Nautical Miles' *Ocean Development and International Law* 38 (2007): pp. 283–304.

Mukherjee, PK, 'The Consent Regime of Oceanic Research in the New Law of the Sea' *Marine Policy* 5 (1981): pp. 98–113.

National Research Council, *Oil in the Sea III: Inputs, Fates and Effects* (Washington, DC: National Academies Press, 2003).

Nordquist, Myron H, (ed), *United Nations Convention on the Law of the Sea 1982: A Commentary, Volume II* (The Hague: Martinus Nijhoff, 1993).

Nordquist, Myron H, (ed), *United Nations Convention on the Law of the Sea 1982: A Commentary, Volume IV* (The Hague: Martinus Nijhoff, 1991).

Norse, Elliott A, et al., 'Sustainability of Deep-sea Fisheries' *Marine Policy* 36 (2012): pp. 307–20.

Noyes, John E, 'The Common Heritage of Mankind: Past, Present and Future' *Denver Journal of International Law and Policy* 40 (2011–2012): pp. 447–71.

O'Connell, DP, *The International Law of the Sea: Volume II*, edited by Ivan Shearer (Oxford: Clarendon Press, 1984).

O'Connell, DP, *The International Law of the Sea: Volume I*, edited by Ivan Shearer (Oxford: Clarendon Press, 1982).

O'Connell, DP, 'Sedentary Fisheries and the Australian Continental Shelf' *American Journal of International Law* 49 (1955): pp. 185–209.

Oda, Shigeru, *Fifty Years of the Law of the Sea, With a Special Section on the International Court of Justice* (Dordrecht: Martinus Nijhoff, 2003).

Oda, Shigeru, *International Control of Sea Resources*, reprint with new Introduction (Dordrecht: Martinus Nijhoff, 1989).

Oda, Shigeru, 'The Ocean: Law and Politics' *Netherlands International Law Review* 25 (1978): pp. 149–58.

Oda, Shigeru, 'Proposals for Revising the Convention on the Continental Shelf' *Columbia Journal of Transnational Law* 7 (1968): pp. 1–31.

Oda, Shigeru, 'The Reconsideration of the Continental Shelf Doctrine' *Tulane Law Review* 32 (1957–1958): pp. 21–36.
Odek, James O, 'Bio-piracy: Creating Proprietary Rights in Plant Genetic Resources' *Journal of Intellectual Property Law* 2 (1) (1994–1995): pp. 141–81.
O'Leary, BC, et al., 'The First Network of Marine Protected Areas (MPAs) in the High Seas: The Process, the Challenges and Where Next' *Marine Policy* 36 (2012): pp. 598–605.
Olsgard, Frode and John S Gray, 'A Comprehensive Analysis of the Effects of Offshore Oil and Gas Exploration and Production on the Benthic Communities of the Norwegian Continental Shelf' *Marine Ecology Progress Series* 122 (1995): pp. 277–306.
Ong, David M, 'A Legal Regime for the Outer Continental Shelf? An Inquiry as to the Rights and Duties of Coastal States within the Outer Continental Shelf', Paper presented to the Third ABLOS Conference, 2003, http://www.iho.int/mtg_docs/com_wg/ABLOS/ABLOS_Conf3/PAPER7-4.PDF.
Ong, David, 'The New Timor Sea Arrangement 2001: Is Joint Development of Common Offshore Oil and Gas Deposits Mandated Under International Law?' *International Journal of Marine and Coastal Law* 17 (2002): pp. 79–122.
Ong, David M, 'Joint Development of Common Offshore Oil and Gas Deposits: 'Mere' State Practice or Customary International Law?' *American Journal of International Law*, 93 (1999): pp. 771–804.
Örebech, Peter, Ketill Sigurjonsson and Ted L McDorman, 'The 1995 United Nations Straddling and Highly Migratory Fish Stocks Agreement: Management, Enforcement and Dispute Settlement' *International Journal of Marine and Coastal Law* 13 (1998): pp. 119–41.
Oude Elferink, Alex G, 'The *Arctic Sunrise* Incident: A Multi-faceted Law of the Sea Case with a Human Rights Dimension' *International Journal of Marine and Coastal Law* 29 (2014): pp. 244–89.
Oude Elferink, Alex G, 'Environmental Impact Assessment in Areas Beyond National Jurisdiction' *International Journal of Marine and Coastal Law* 27 (2012): pp. 449–80.
Oude Elferink, Alex G, 'Causes, Consequences and Solutions Relating to the Absence of Final and Binding Outer Limits of the Continental Shelf' in Clive R Symmons (ed), *Selected Contemporary Issues in the Law of the Sea* (Leiden: Martinus Nijhoff, 2011): pp. 253–72.
Oude Elferink, Alex, 'The Establishment of Outer Limits of the Continental Shelf Beyond 200 Nautical Miles by the Coastal State: The Possibilities of Other States to Have an Impact on the Process' *International Journal of Marine and Coastal Law* 24 (2009): pp. 535–56.
Oude Elferink, Alex, 'Article 76 of the Law of the Sea Convention on the Definition of the Continental Shelf: Questions Concerning its Interpretation from a Legal Perspective' *International Journal of Marine and Coastal Law* 21 (2006): pp. 269–85.
Owen, Daniel, *Interactions Between Management of a Water Column Marine Protected Area in the High Seas of the OSPAR Maritime Area and the Exercise of Sovereign Rights regarding Subjacent Outer Continental Shelf*, Report for WWF Germany (2010).
Owen, Daniel, 'The Powers of the OSPAR Commission and Coastal State Parties to the OSPAR Convention to Manage Marine Protected Areas on the Seabed Beyond 200 nm from the Baseline', Report for WWF Germany (2006).
Oxman, Bernard, 'The Territorial Temptation: A Siren Song at Sea' *American Journal of International Law* 100 (2006): pp. 830–51.
Oxman, Bernard, 'The Preparation of Article 1 of the Convention on the Continental Shelf' *Journal of Maritime Law and Commerce* 3 (1971–1972): pp. 245–305.

Pardo, A and CQ Christol, 'The Common Interest: Tension Between the Whole of the Parts' in R St J Macdonald and D M Johnston (eds), *The Structure and Process of International Law: Essays in Legal Philosophy Doctrine and Theory* (The Hague: Martinus Nijhoff, 1983): p. 654.

Paskal, Cleo and Michael Lodge, 'A Fair Deal on Seabed Wealth: The Promise and Pitfalls of Article 82 on the Outer Continental Shelf', Chatham House Briefing Paper, February 2009.

Penick, FVW, 'The Legal Character of the Right to Explore and Exploit the Natural Resources of the Continental Shelf' *San Diego Law Review* 22 (1985).

Pen-Yuan Hsing et al., 'Evidence of Lasting Impact of the Deepwater Horizon Oil Spill on a deep Gulf of Mexico Coral Community' *Elementa: Science of the Anthropocene* (2013): doi: 10.12952/journal.elementa.000012.

Peterson, Charles H, et al., 'Ecological Consequences of Environmental Perturbations Associated with Offshore Hydrocarbon Production: A Perspective on Long Term Exposures in the Gulf of Mexico' *Canadian Journal of Fisheries and Aquatic Sciences* 53 (1996): pp. 2637–54.

Peterson, Charles H, et al., 'Long-term Ecosystem Response to the Exxon Valdez Oil Spill' *Science* 302 (5653) (19 December 2003): pp. 2082–6.

Pham, Christopher K, et al., 'Deep-water longline fishing has reduced impact on Vulnerable Marine Ecosystems' *Scientific Reports* 4 (2014): doi: 10.1038/srep 04837.

Phillips, JC, 'The Exclusive Economic Zone as a Concept in International Law' *International and Comparative Law Quarterly* 26 (1977): pp. 585–618.

Pitcher, Tony J, et al., *Seamounts: Ecology, Fisheries and Conservation* (Chichester: John Wiley & Sons, 2008).

Poisel, Tim, 'Implications of Seabed Disputes Chamber's Advisory Opinion' *Australian International Law Journal* 19 (2012): pp. 213–33.

Porteiro, Filipe M and Tracey Sutton, 'Midwater Fish Assemblages and Seamounts' in Tony J Pitcher et al., (eds), *Seamounts: Ecology, Fisheries and Conservation* (Oxford: Blackwell Publishing, 2007): pp. 101–16.

Poulantzas, NM, *The Right of Hot Pursuit in International Law* (Leiden: AW Sijthoff, 1969).

Prip, Christian, et al., *The Australian ABS Framework: A Model Case for Bioprospecting?* (Oslo: Fridtjof Nansens Institute, 2014).

Pulvenis, Jean-François, 'The Continental Shelf Definition and Rules Applicable to Resources' in René-Jean Dupuy and Daniel Vignes (eds), *A Handbook on the New Law of the Sea, Volume 1* (Dordrecht: Martinus Nijhoff, 1991): pp. 315–81.

Ramirez-Llodra, E, et al., 'Man and the Last Great Wilderness: Human Impact on the Deep Sea' *PLOS One* 6(7) (2011): e22588. doi:10.1371/journal.pone.0022588.

Ramirez-Llodra, E, et al., 'Deep, Diverse and Definitely Different: Unique Attributes of the World's Largest Ecosystem' *Biogeosciences* 7 (2010): pp. 2851–99.

Reagan, Matthew T and George J Moridis, 'Modeling of Oceanic Gas Hydrate Instability and Methane Release in Response to Climate Change' (Berkeley, CA: Lawrence Berkeley National Laboratory, 2008), http://escholarship.org/uc/item/1b11w8ns#.

Redford, Michael, 'Legal Framework for Oceanic Research' in Warren S Wooster (ed), *Freedom of Oceanic Research* (Russak, New York: Crane, 1973): pp. 41–96.

Ribeiro, Marta Chantal, 'Marine Protected Areas: The Case of the Extended Continental Shelf' in Marta Chantal Ribeiro (ed), *30 Years after the Signature of the United Nations Convention on the Law of the Sea: the Protection of the Environment and the Future of the Law of the Sea* (Coimbra: Coimbra Editora, 2014): pp. 179–208.

Ribeiro, Marta Chantal, 'The "Rainbow": The First National Marine Protected Area Proposed Under the High Seas' *International Journal of Marine and Coastal Law* 25 (2010): pp. 183–207.

Rife, Alexis N, et al., 'When Good Intentions are not Enough: Insights on Networks of "Paper Park" Marine Protected Areas' *Conservation Letters* 6 (2013): pp. 200–12.

Roach, J Ashley and Robert W Smith, *Excessive Maritime Claims*, 3rd edn (Leiden: Martinus Nijhoff, 2012).

Roach, J Ashley, 'Dispute Settlement in Specific Situations' *Georgetown International Law Review* 7 (1994–1995): pp. 775–89.

Roberts, Julian, Aldo Chircop and Siân Prior, 'Area-based Mangement on the High Seas: Possible Application of the IMO's Particularly Sensitive Sea Area Concept' *International Journal of Marine and Coastal Law* 25 (2010): pp. 483–522.

Robertson Jr, Horace B, 'Navigation in the Exclusive Economic Zone' *Virginia Journal of International Law* 24 (1983–1984): pp. 865–915.

Rochette, Julien, et al., 'The Regional Approach to the Conservation and Sustainable Use of Marine Biodiversity in Areas Beyond National Jurisdiction' *Marine Policy* 49 (2014): pp. 109–17.

Rogers, Alex D, et al., 'Corals on Seamounts' in Tony J Pitcher et al., (eds), *Seamounts: Ecology, Fisheries and Conservation* (Oxford: Blackwell Publishing, 2007): pp. 141–69.

Rona, Peter A, 'Resources of the Seafloor' *Science* 299 (2003): pp. 673–4.

Rooker, Jay R, 'Spatial, Temporal and Habitat-Related Variation in Abundance of Pelagic Fishes in the Gulf of Mexico: Potential Implications of the Deepwater Horizon Oil Spill' *PLOS One* (10 October 2013): doi: 10.1371/journal.pone.0076080.

Rothwell, Donald R and Tim Stephens, *The International Law of the Sea* (Oxford: Hart, 2010).

Roughton, Dominic and Colin Trehearne, 'The Continental Shelf' in DJ Attard, M Fitzmaruice and NA Martinez Gutierrez (eds), *The IMLI Manual on International Maritime Law: Volume I: The Law of the Sea* (Oxford: Oxford University Press, 2014): pp. 137–76.

Rovine, AW, 'Contemporary Practice of the United States Relating to International Law' *American Journal of International Law* 69 (1975): pp. 141–53.

Rowden, Ashley A, et al., 'A Test of the Seamount Oasis Hypothesis: Seamounts Support Higher Epibenthic Megafaunal Biomass Than Adjacent Slopes' *Marine Ecology* 31 (Supp. 1) (2010): pp. 95–106.

Rozakis, CL, 'Compromises of States Interests and Their Repercussions upon the Rules on the Delimitation of the Continental Shelf: From the Truman Proclamation to the 1982 Convention on the Law of the Sea' in CL Rozadkis and CA Stephanou (eds), *The New Law of the Sea* (Amsterdam: Elsevier Science Publishers, 1983): pp. 155–84.

Ryder, AA and SC Rapson, 'Pipeline Technology' in Stefan T Orszulik (ed), *Environmental Technology in the Oil Industry* (Berlin: Springer, 2008): pp. 229–80.

Sanderson, Hans, et al., 'Environmental Hazards of Sea-dumped Chemical Weapons' *Environmental Science and Technology* 44 (2010): pp. 4389–94.

Samadi, Sarah, Thomas Schlacher and Bertrand Richer de Forges, 'Seamount Benthos' in Tony J Pitcher et al. (eds), *Seamounts: Ecology, Fisheries and Conservation* (Oxford: Blackwell Publishing, 2007): pp. 119–40.

Sands, Phillippe and Jacqueline Peel, *Principles of International Environmental Law*, 3rd edn (Cambridge: Cambridge University Press, 2012).

Schiffman, Howard S, 'The South Pacific Regional Fisheries Management Organization (SPRFMO): an Improved Model of Decision-making for Fisheries Conservation?' *Journal of Environmental Studies and Science* 3 (2013): pp. 209–16.

Schofield, Clive, 'The El Dorado Effect: Reappraising the "Oil Factor" in Maritime Boundary Disputes' in Clive Schofield, Seokwoo Lee and Moon-Dang Kwon (eds), *The Limits of Maritime Jurisdiction* (Leiden: Martinus Nijhoff, 2014): pp. 111–26.

Schofield, Clive, 'Securing the Resources of the Deep: Dividing and Governing the Extended Continental Shelf' *Berkeley Journal of International Law* 33 (2015): pp. 274–93.

Schofield, Clive, 'New Marine Resource Opportunities, Fresh Challenges' *University of Hawaii Law Review* 35 (2013): pp. 715–33.

Schofield, Clive, 'No Panacea? Challenges in the Application of Provisional Arrangements of a Practical Nature' in Myron H Nordquist and John Norton Moore (eds), *Maritime Border Diplomacy* (Leiden: Martinus Nijhoff, 2012): pp. 151–69.

Schofield, Clive, 'Blurring the Lines? Maritime Joint Development and the Cooperative Management of Ocean Resources', Issues in Legal Scholarship, *Berkeley Electronic Press* 8(1) (2009): (Frontier Issues in Ocean Law: Marine Resources, Maritime Boundaries, and the Law of the Sea), Article 3.

Schofield, Clive, 'Minding the Gap: The Australia-East Timor Treaty on Certain Maritime Arrangements in the Timor Sea' *International Journal of Marine and Coastal Law* 22 (2007): pp. 189–234.

Scholz, Wesley S, 'Oceanic Research: International Law and National Legislation' *Marine Policy* 4 (1980): pp. 91–127.

Schram Stokke, Olav, 'Managing Straddling Stocks: The Interplay of Global and Regional Regimes' *Ocean and Coastal Management* 43 (2000): pp. 205–34.

Schrag, Daniel P, 'Storage of Carbon Dioxide in Offshore Sediments' *Science* 325 (2009) pp. 1658–9.

Schroeder, Donna M and Milton S Love, 'Ecological and Political Issues Surrounding Decommissioning of Offshore Oil Facilities in the South California Bight' *Ocean and Coastal Management* 47 (2000): pp. 21–48.

Scott, Karen N, 'Protecting the Last Ocean: The Proposed Ross Sea MPA. Prospects and Progress' in Gemma Andreone (eds), *Jurisdiction and Control at Sea: Some Environmental and Security Issues* (Naples: Giannini Editore, 2014): pp. 79–90.

Scott, Karen N, 'Marine Protected Areas in the Southern Ocean' in Erik J Molenaar, Alex G Oude Elferink and Donald R Rothwell (eds), *The Law of the Sea and the Polar Regions* (Leiden: Martinus Nijhoff, 2013): pp. 113–37.

Scott, Karen N, 'Conservation on the High Seas: Developing the Concept of the High Seas Marine Protected Areas' *International Journal of Marine and Coastal Law* 27 (2012): pp. 849–57.

Scott, Karen N, 'The Day After Tomorrow: Ocean CO_2 Sequestration and the Future of Climate Change' *Georgetown International Environmental Law Review* 18 (2005–2006): pp. 57–108.

Scott, SV 'The Inclusion of Sedentary Fisheries within the Continental Shelf Doctrine' *International and Comparative Law Quarterly* 41 (1992): pp. 788–807.

Scovazzi, Tullio, 'Marine Protected Areas in Waters Beyond National Jurisdiction' in Marta Chantal Ribeiro (ed), *30 Years after the Signature of the United Nations Convention on the Law of the Sea: the Protection of the Environment and the Future of the Law of the Sea* (Coimbra: Coimbra Editora, 2014): pp. 209–38.

Scovazzi, Tullio, 'Mining, Protection of the Environment, Scientific Research and Bioprospecting: Some Considerations on the Role of the International Sea-bed Authority' *International Journal of Marine and Coastal Law* 19 (2004): pp. 383–409.

Seibel, Brad A and Patrick J Walsh, 'Potential Impacts of CO_2 Injection on Deep-sea Biota' *Science* 294 (2001): pp. 319–20.

Serdy, Andrew, 'The Commission on the Limits of the Continental Shelf and its Disturbing Propensity to Legislate' *International Journal of Marine and Coastal Law* 16 (2011): pp. 355–83.

Shaw, Malcolm M, *International Law*, 6th edn (Cambridge: Cambridge University Press, 2008).

Slouka, Zdenek J, *International Custom and the Continental Shelf* (The Hague: Martinus Nijhoff, 1969).

Smith, Robert W and George Taft, 'Legal Aspects of the Continental Shelf' in Peter J Cook and Chris M Carleton (eds), *Continental Shelf Limits: The Scientific and Legal Interface* (Oxford: Oxford University Press, 2000): pp. 17–24.

Smith, Samantha and Robert Heydon, 'Processes Related to the Technical Development of Marine Mining' in Elaine Baker and Yannick Beaudoin (eds), *Cobalt-rich Ferromanganese Crusts: A Physical, Biological, Environmental, and Technical Review* (Arendal: SPC-Grid Arendal, 2013): pp. 41–6.

Socks, Karen I and Paul JB Hart, 'Biogeography and Biodiversity of Seamounts' in Tony J Pitcher et al., (eds), *Seamounts: Ecology, Fisheries and Conservation* (Oxford: Blackwell Publishing, 2007): pp. 255–81.

Soons, Alfred HA, *Marine Scientific Research and the Law of the Sea* (The Hague: TMC Asser Instituut/Kluwer, 1982).

Spalding, Mark D et al., 'Protecting Marine Spaces: Global Targets and Changing Approaches' *Ocean Yearbook* 27 (2013): pp. 213–48.

Stone, Gregory, et al., 'Seamount Biodiversity, Exploitation and Conservation' in LK Glover and SA Earle (eds), *Defying Ocean's End: An Agenda for Action* (Washington, DC: Island Press, 2004): pp. 43–70.

Suarez, Suzette, 'The Commission on the Limits of the Continental Shelf and its Function to Provide Scientific and Technical Advice' *Chinese Journal of International Law* 12 (2013): pp. 301–24.

Suarez, Suzette V, *The Outer Limits of the Continental Shelf: Legal Aspects of Their Establishment* (Berlin: Springer, 2008).

Subedi, Surya P, 'Problems and Prospects for the Commission on the Limits of the Continental Shelf in Dealing with Submissions by Coastal States in Relation to the Ocean Territory Beyond 200 Nautical Miles' *International Journal of Marine and Coastal Law* 26 (2011): pp. 413–31.

Symonds PA, et al., 'Characteristics of Continental Margins' in Peter J Cook and Chris M Carleton (eds), *Continental Shelf Limits: The Scientific and Legal Interface* (Oxford: Oxford University Press, 2000): pp. 25–63.

Tanaka, Yoshifumi, 'Obligations and Liability of Sponsoring States Concerning Activities in the Area: Reflections on the ITLOS Advisory Opinion of 1 February 2011' *Netherlands International Law Review* 60 (2013): pp. 205–30.

Tanaka, Yoshifumi, 'Reflections on High Seas Marine Protected Areas: A Comparative Analysis of the Mediterranean and the North-east Atlantic Models' *Nordic Journal of International Law* 8 (2012): pp. 295–326.

Tanaka, Yoshifumi, *The International Law of the Sea* (Cambridge: Cambridge University Press, 2012).

Taverne, Bernard, *Petroleum, Industry and Governments: A Study of the Involvement of Industry and Governments in the Production and Use of Petroleum*, 2nd edn (The Hague: Wolters Kluwer, 2008).

Teulings, Jasper, 'Peaceful Protests against Whaling on the High Seas—A Human Rights-based Approach' in Clive R Symmons (ed), *Selected Contemporary Issues in the Law of the Sea* (Leiden: Martinus Nijhoff, 2011): pp. 221–49.

Thiel, Hjalmar, 'Anthropogenic Impacts on the Deep Sea' in PA Tyler (ed) *Ecosystems of the World 28, Ecosystems of the Deep Oceans* (Elsevier, 2003): pp. 427–72.

Tho Pesch, Sebastian, 'Coastal State Jurisdiction around Installations: Safety Zones in the Law of the Sea' *International Journal of Marine and Coastal Law* 30 (2015): pp. 512–32.

Townsend-Gault, Ian, 'Rationales for Zones of Cooperation' in R Beckman, CH Schofield, I Townsend-Gault, T Davenport and L Bernard (eds), *Beyond Territorial Disputes in the South China Sea: Legal Frameworks for the Joint Development of Hydrocarbon Resources* (Cheltenham: Edward Elgar Publishers, 2013): pp. 114–38.

Ulfstein, Geir, 'The Conflict Between Petroleum Production, Navigation and Fisheries in International Law' *Ocean Development and International Law* 19 (1988): pp. 229–62.

Urdaneta, Karla, 'Transboundary Petroleum Reservoirs: A Recommended Approach for the United States and Mexico in the Deepwaters of the Gulf of Mexico' *Houston Journal of International Law* 32 (2010): pp. 333–91.

Valencia, Mark and Kazumine Akimoto, 'Guidelines for Navigation and Overflight in the Exclusive Economic Zone' *Marine Policy* 30 (2006): pp. 704–11.

van de Poll, Robert and Clive Schofield, 'A Seabed Scramble: A Global Overview of Extended Continental Shelf Submissions' (Contentious Issues in UNCLOS—Surely Not?, Advisory Board on the Law of the Sea Conference, Monaco, October 2010).

Van Dover, Cindy Lee, et al., 'Designating Networks of Chemosynthetic Ecosystem Reserves in the Deep Sea' *Marine Policy* 36 (2012): pp. 378–81.

Van Dover, Cindy Lee, 'Tighten Regulations on Deep-sea Mining' *Nature* 470 (3 February 2011): pp. 31–3.

Van Dover, Cindy Lee, 'Mining Seafloor Massive Sulphides and Biodiversity: What is at Risk?' *ICES Journal of Marine Science* (2010): doi: 10.1093/icesjms/fsq 086.

Van Dover, Cindy Lee, *The Ecology of Deep-sea Hydrothermal Vents* (Princeton: Princeton University Press, 2000).

Van Dyke, JM, 'Modifying the 1982 Law of the Sea Convention: New Initiatives on Governance of High Seas Fisheries Resources: the Straddling Stocks Negotiations', *International Journal of Marine & Coastal Law* 10 (1995): pp. 219–27.

Vanreusel, Ann, et al., 'Biodiversity of Cold Seep Ecosystems along the European Margins' *Oceanography* 22(1) (2009): pp. 110–27.

Vargas, Jorge A, 'The 2012 US–Mexico Agreement on Transboundary Hydrocarbon Reservoirs in the Gulf of Mexico: A Blueprint for Progress or a Recipe for Conflict?' *San Diego International Law Journal* 14 (2012): pp. 3–70.

Vasciannie, SC, *Land-locked and Geographically Disadvantaged States in the International Law of the Sea* (Oxford: Clarendon Press, 1990).

Vattel, Emer de and Joseph Chitty (eds), *The Law of Nations*, 6th American edn (Philadelphia: T & JW Johnson, 1844).

Verlaan, Philomène A, 'Experimental Activities that Intentionally Perturb the Marine Environment: Implications for the Marine Environmental Protection and Marine Scientific Research Provisions of the 1982 Convention on the Law of the Sea' *Marine Policy* 31 (2007): pp. 210–16.

von Kries, Caroline and Gerd Winter, 'Harmonizing ABS Conditions for Research and Development Under UNCLOS and CBD/NP' in Evanson Chege Kamau, Gerd Winter and Peter-Tobias Stoll (eds), *Research and Development on Genetic Resources: Public Domain Approaches in Implementing the Nagoya Protocol* (Abingdon: Routledge, 2015): pp. 75–90.

von Zharen, WM, 'An Ecopolicy Perspective for Sustaining Living Marine Species' *Ocean Development & International Law* 30 (1999): pp. 1–42.

Waczewski, James, 'Legal, Political and Scientific Response to Ocean Dumping and Subseabed Disposal of Nuclear Waste' *Journal of Transnational Law and Policy* 7 (1997–1998): pp. 97–118.

Warner, Robin, 'Protecting the Diversity of the Depths: Environmental Regulation of Bioprospecting and Marine Scientific Regulation Beyond National Jurisdiction' *Ocean Yearbook* 22 (2008): pp. 411–43.

Watling, Les and Elliott A Norse, 'Disturbance of the Seabed by Mobile Fishing Gear: A Comparison to Forest Clearcutting' *Conservation Biology* 12 (1998): pp. 1180–97.

Watt, Donald Cameron, 'First Steps in the Enclosure of the Oceans: The Origins of Truman's Proclamation on the Resources of the Continental Shelf, 28 September 1945' *Marine Policy* 3 (1979): pp. 211–24.

Weeks, Lewis G, 'Petroleum Resources Potential of Continental Margins' in Creighton A Burk and Charles L Drake (eds), *The Geology of Continental Margins* (New York: Springer-Verlag, 1974): pp. 953–64.

Wegelein, Florian, *Marine Scientific Research: The Operation and Status of Research Vessels and Other Platforms in International Law* (Leiden: Brill, 2005).

Wessel, Paul, 'Seamount Characteristics' in Tony J Pitcher et al., (eds), *Seamounts: Ecology, Fisheries and Conservation* (Oxford: Blackwell Publishing, 2007): pp. 3–25.

Whiteman, Majorie M, 'Conference on the Law of the Sea: Convention on the Continental Shelf' *American Journal of International Law* 52 (1958): pp. 629–59.

Williams, Alan, et al., 'Seamount Megabenthic Assemblages Fail to Recover from Trawling Impacts' *Marine Ecology* 31 (Suppl 1) (2010): pp. 183–99.

Winner, Russ, 'Science, Sovereignty, and the Third Law of the Sea Conference' *Ocean Development and International Law Journal* 4 (1977): pp. 297–342.

Wolfrum, Rüdiger and Nele Matz, 'The Interplay of the United Nations Convention on the Law of the Sea and the Convention on Biological Diversity' *Max Planck Yearbook of United Nations Law* 4 (2000): pp. 445–80.

Wolfrum, Rüdiger, 'The Principle of the Common Heritage of Mankind' *Zaö RV* 44 (1983): pp. 312–37.

Woodall, Lucy C, et al., 'Deep-sea Litter: a Comparison of Seamounts, Banks and a Ridge in the Atlantic and Indian Ocean Reveals Both Environmental and Anthropogenic Factors Impact Accumulation and Composition' *Frontiers in Marine Science* 2 (2015): pp. 3–10.

Wright, Glen, et al., 'Advancing Marine Biodiversity Protection Through Regional Fisheries Management: A Review of Bottom Fisheries Closures in Areas Beyond National Jurisdiction' *Marine Policy* 61 (2015): pp. 134–48.

Wright, Glen, Julien Rochette and Elisabeth Druel, 'Marine Protected Areas in Areas Beyond National Jurisdiction' in Rosemary Rayfuse (ed), *Research Handbook on International Marine Environmental Law* (Cheltenham: Edward Elgar, 2015): pp. 272–90.

Xiaofeng, Ren and Cheng Xizhong, 'A Chinese Perspective' *Marine Policy* 29 (2005): pp. 139–46.

Xiaopin Zhu and Jinlong He, 'International Court of Justice's Impact on International Environmental Law: Focusing on the Pulp Mills Case' *Yearbook of International Environmental Law* 23 (2012): pp. 106–30.

Yesson, Chris, et al., 'The Global Distribution of Seamounts Based on 30 Arc Seconds Bathymetry Data' *Deep-Sea Research Part I Oceanographic Research Papers* 58(4) (2011): pp. 442–53.

Young, Richard, 'The Geneva Convention on the Continental Shelf: A First Impression' *American Journal of International Law* 52 (1958): pp. 733–8.
Young, Richard, 'Sedentary Fisheries and the Convention on the Continental Shelf' *American Journal of International Law* 63 (1969): pp. 359–73.
Young, R, 'Saudi Arabian Offshore Legislation' *American Journal of International Law* 43 (1949): pp. 530–2.
Young, R, 'Recent Developments with Respect to the Continental Shelf' *American Journal of International Law* 42 (1948): pp. 849–57.

Index

Area, the
  see also International Seabed Authority, Common Heritage of Mankind
  boundary between continental shelf and 72
  and the CLCS 74, 79
  common heritage of mankind 9, 60, 86, 89–90, 125
  influence on Part VI 10, 60–1, 67, 125–6
  resources shared with the continental shelf 139, 146–7
  regulations governing mining 44, 133
Arrest see Enforcement rights
Article 82
  see also International Seabed Authority
  ambiguities 130–6
  background 125–7
  carbon sequestration 150
  common heritage of mankind 86, 89–90
  cross-boundary issues 138–47
  disputes 136–8
  importance 13
  International Seabed Authority 89
  interpretation 127–30
  negotiation of 61, 125–7
  payments or contributions in kind 130–2
  status in customary international law 86–91
Artificial islands, see Installations, structures and artificial islands
Australia
  bioprospecting 112
  enforcement rights 208, 210
  negotiation of LOSC 60
  origins of continental shelf doctrine 55, 60
  sedentary species 62–3, 65

Biodiversity
  see also Convention on Biological Diversity (CBD), Living resources, Marine protected areas
  beyond national jurisdiction 244–7
  protection of 107–8
Bioprospecting
  under CBD 115–17
  defined 31
  impacts of 32–3
  under LOSC 113–15
  marine scientific research distinguished 110–13
  regulation of 110–19

Bottom trawling
  balancing high seas freedoms and coastal State rights 120, 194–5
  danger to subsea installations 148
  impacts of 27–9, 120
Broad margin States 10–11, 60, 87, 120, 123, 126, 128, 168, 249

Cables and pipelines
  'due regard' provisions 182
  Part VI LOSC 12
Canada
  disputes over sedentary species 66, 209–10
  enforcement rights 210
  fishing for sedentary species 94
  hydrocarbon exploration and exploitation 38, 124
  negotiation of LOSC 60
  onshore production of gas hydrates 42
  origins of continental shelf doctrine 54
  petroleum exploration licenses 38, 124
  regulation of oil and gas vessels 195, 200
Carbon mitigation measures
  see also Climate change
  carbon sequestration 47–8, 149–50
  injection of carbon into seabed 47
  release of carbon dioxide into deep ocean 48
Charlie-Gibbs Fracture Zone 221
Climate change
  adverse effect on the sea floor 20, 42, 45
  mitigation measures
    carbon sequestration 47–8
    injection of carbon into seabed 47
    release of CO_2 into deep ocean 48
Coastal States
  see also Article 82, Bioprospecting, Enforcement rights, Environmental protection, Fishing, Living resources, Marine Scientific Research, Mining, Non-living resources, Sedentary species, Unjustifiable interference
  cooperation with international organizations 230–8
  expansion of jurisdiction 8, 57, 59–62, 91–2
  marine scientific research under LOSC 156–8
  rights under Part VI 10–14
Cold seeps 25–8, 32–3

Commission on the Limits of the Continental
  Shelf (CLCS)
    completed submissions 1, 73
    defining the outer limits 71–4
    delay in considering submissions 242
    Part VI LOSC 11
    relationship with existence of coastal State
      rights 74–80
Common heritage of mankind
    Art 82 and 86, 89–90, 128–9
    biodiversity beyond national
      jurisdiction 245–6
    customary international law 89–90
    definition 129
    importance 10
    relationship to Pt VI 10, 13, 60–1
Continental shelf
    see also Delimitation of continental shelf;
      Extended continental shelf; Outer limits
      of continental shelf
    biological resources and seabed
      environment 21–7
    exploitation of living resources 27–33
      bioprospecting 31–3
      fishing 27–31
      impact on benthic ecosystems 27–33
    legal v scientific definition 1–2, 20
    meaning and scope 20–1
    modern development of doctrine 59–62
    non-living resources
      hydrocarbon exploration, exploitation and
        impacts 36–42
      hydrocarbon resources 34
      increasing activity 33
      mineral resources 34–6
      mining and impacts 42–5
    origins of doctrine 9, 53–6
    physical description, 1–2, 20–1
    terminology 6–7
Convention on Biological Diversity (CBD)
    bioprospecting 115–17
    coastal State obligations 99
    ecologically and biologically significant areas
      (EBSAS) 100, 236–7
    LOSC and 117–19
    marine protected areas 99–100, 107–8, 237
    Nagoya Protocol 116–17
Convention on the Continental Shelf
    as customary international law 58–9, 64
    basis for Part VI LOSC 11, 59, 176–7,
      205, 248
    enforcement rights under
      academic views 206–7
      State practice 208–10
    influence on Art 78 LOSC 176–80
    marine scientific research 152, 160–2
    negotiations of 52, 56–8
    outer limits under 56–7, 67–8
    sedentary species 62–4

Cooperation
    international and regional organizations
      concluding remarks 239–40
      European Union 221, 222, 227
      International Maritime Organization
        (IMO) 234–5
      International Seabed Authority
        (ISA) 235–6
      other relevant organizations 236–8
      regional fisheries management
        organizations (RFMOs) 230–4
      regional seas organizations 222
      Sargasso Sea Alliance 222–3, 225
    multilateral regional cooperation in NE
      Atlantic
      Iceland 221
      lessons from the experience 222–7
      Portugal 109–10, 216–21
      United Kingdom 221
    obstacles to success
      absence of competent
        organisations 222–3
      coordination problems 223–5
      scientific and legal
        uncertainty 225–6
      third-party States 226–7
    overview 216–17
    preferable approach 202, 250
    Seychelles and Mauritius
      joint application to the CLCS 228
      lessons for other States 230
      precautionary approach 229
      treaty arrangements 228–9
    transboundary hydrocarbon
      resources 139–44
'Creeping jurisdiction' 8
Cross-boundary issues (Art 82)
    fields located across the 200 nm line on the
      continental shelf 139
    fields lying under the continental shelf and
      the Area 146–7
    fields that cross the maritime boundaries of
      two or more States 139–44
    fields where there are disputed
      claims 144–6
    overview 138–9
Customary international law
    environmental principles
      obligation not to cause
        transboundary harm and duty of
        prevention 101–2
      obligation to exercise due
        diligence 103–5
    marine scientific research 152
    status of LOSC
      Art 76 82–6
      Art 82 86–91
      indivisible regime 88–9
      Pt VI 81–2

status of the Continental Shelf
    Convention 58–9

Deep seabed *see* Area
***Deep-Water Horizon* incident** 37, 40–1
Delimitation of continental shelf
  see also Outer limits of continental Shelf
  cases on 59, 74–80
  cooperation prior to 144–6
  coverage of 5, 13–14
  delimitation agreements 141, 144
  need for restraint in absence of 80
  Part VI LOSC
    matter for agreement 13
    provisional arrangements 14, 144–6
  relationship to the rights of coastal
    States 74–80
Developing States
  exemption from Art 82 127, 133–5, 144
  future challenges 243–4
  marine scientific research 152–3, 161
  negotiation of LOSC 61
Dispute settlement
  see also Cooperation
  interpretation of enforcement rights 211–14
  marine scientific research 157, 168
  options under Art 82 136–8
  seabed disputes chamber 137–8
Drilling operations
  hydrocarbon exploration and
    exploitation 36–42
  Part VI LOSC 13
'Due regard' provisions
  compared to Art 78 184
  confusion with Art 78 184–5
  in the LOSC 176, 182–4

**Ecologically and Biologically Significant Areas
  (EBSAs)** 100, 236–7
Ecotourism 50
Enforcement rights
  arguments against 203–5
  arguments for
    academic views 206–7
    State practice 208–10
    overview 205–6
    UN Convention on the Law of the Sea
      (LOSC) 210–14
Environmental impact assessments 98, 99,
    104, 245
Environmental protection *see also* Convention
    on Biological Diversity, Marine protected
    areas, Ecologically and biologically
    significant areas
  activities affecting the sea floor
    carbon mitigation measures 46–8
    climate change 45
    ecotourism 49
    marine scientific research 49

    waste disposal 45–6
  impact of fishing 27–31, 119–20
  impact of hydrocarbon exploration and
    exploitation
    Deep-Water Horizon incident 40–1
    discharges 39
    oil and gas platforms 40
    pipelines 39–40
  impact of mining operations 42–5
  legal obligations of coastal States
    customary international law 101–5
  LOSC
    Part VI 95–7
    Part XII 97–9
    multilateral treaties 99–101
    non-binding instruments 105–6
    summary of obligations 106–7
    UN General Assembly
      Resolutions 105–6
European Union 221, 222, 227
Exclusion zone *see* Safety zones
Exclusive Economic Zones
  agreement on rights under LOSC 9
  'due regard' provisions 182–4
  enforcement rights 3, 174, 204
  extent of State sovereignty 3, 9
  hydrocarbon exploration and exploitation 36
  negotiation of LOSC 60
  relationship with continental shelf 9, 184–5
Extended continental shelf
  balancing high seas freedoms and
      enforcement rights
    application of the framework 193–201
    under Art 78(2) 185–201
    bottom trawling 193–5
    consultation with affected States 192–3
    exclusion zones around survey
      vessels 195–201
    international or regional institutions 192
    likelihood of interference with continental
      shelf resources and level of
      harm 189–90
    minimal interference 191–2
    overview 187–9
    relative importance of the
      interests 187–8
  current issue for coastal States 17–18, 248
  customary international law 81–90
  enforcement rights
    arguments against 203–5
    arguments for 205–14
  environmental protection
    bioprospecting 110–19
    creation of MPAs 107–10
    customary international law 101–5
    fishing 119–20
    LOSC 95–9
    multilateral treaties 99–101
    non-binding instruments 105–6

Extended continental shelf (*cont.*)
  summary of obligations 106–7
  UN General Assembly
    Resolutions 105–6
examples of ecosystems
  cold seeps 24–7
  hydrothermal vents 24–6
  seamounts 22–4
installations, structures and artificial islands
  Art 60/80 13, 147–9, 175–6, 193
  creation of safety zones 147
  jurisdictional principles 147
  'subsea developments' 148–9
marine scientific research
  Art 246 164–70
  Continental Shelf Convention 152, 160–2
  exploitation distinguished 158–9
  LOSC 153, 156–8, 162–3
  overview 152–4
  relevant activities 154–5
outer limits 56–7, 67–71
relationship between CLCS recommendations and rights 174–80

Fishing
  see also Living resources; Regional fisheries management organizations (RFMOs)
  balancing high seas freedoms and coastal State rights 12, 193–5
  bottom fishing 28–30
  danger to subsea installations 149
  expansion of coastal State jurisdiction 9
  impacts of 27–31, 120
  North East Atlantic Fisheries Commission (NEAFC) 109, 219–21
  regional fisheries management organizations (RFMOs) 100, 230–4
  regulation of 119–20
  techniques 27–8, 120
  voluntary restriction of fishing 232
Flag State
  Environmental obligations 98, 107

Gas *see* Hydrocarbons
Gas hydrates 35, 42
Gulf of Mexico
  Deep-Water Horizon incident 37, 40–1
  hydrocarbon exploration and exploitation 37
  severe weather risks 41

High seas
  activity affecting sedentary species 12, 119–20, 173, 176, 194–5
  balancing high seas and coastal State rights under LOSC
    Art 78 176–9, 184–201
    Art 60/80 174, 175, 193
    comparison of balancing provisions 184–5
  decision making framework 187–93
  'due regard' 182–4
  innocent passage 181
  interpretation of Art 78(2) 185–7
biodiversity beyond national jurisdiction 244–7
initial study by ILC 56
original concept of the freedom of the seas 7
possible point of friction 249
Hot pursuit 205, 207, 213, 215
Hydrocarbons
  See also Joint development, Unitization agreements
  Deep-Water Horizon incident 37, 40–1
  environmental impacts of exploitation
    discharges 39
    natural spillage 41
    oil and gas platforms 40
    pipelines 40
    severe weather risks 41
  exploitation on the extended continental shelf
    ability to exploit greater depths 38
    current practice 38, 124
    stages of exploitation 36–45
  obligations of mutual restraint and cooperation 141–2
  oil and gas reserves 34
  potential in the continental shelf 34
Hydrothermal vents
  described 24–7
  discovery of Rainbow hydrothermal vent field 217
  impact of mining operations 43–5
  location 25
  source of mineral deposits 35

Iceland 221
Infringements *see* Unjustifiable interference
Innocent passage 181
Installations, structures and artificial islands
  Exclusive Economic Zones 9
  extended continental shelf
    Art 60/80 13, 147–9, 175–6, 193
    balancing of high seas freedoms and coastal State rights 175, 193
    creation of safety zones 147
    jurisdictional principles 147
    'subsea developments' 148–9
International Council for the Exploration of the Seas (ICES) 219
International Court of Justice (ICJ)
  environmental principles
    obligation not to cause transboundary harm and duty of prevention 101–2
    obligation to exercise due diligence 103–5

existence of rights in extended continental shelf 75–6, 77–80
marine scientific research 161
status of the LOSC as customary international law 82, 83–6, 88–9
terminology 6
International Law Commission (ILC)
balancing high seas freedoms and coastal State rights 178–9, 190, 193
draft articles on the continental shelf 56–7
enforcement rights 205–6, 211, 213, 215
identification of outer limits of continental shelf 56, 67
sedentary species 63
International Marine Minerals Society 50
International Maritime Organization (IMO)
balancing high seas freedoms and coastal State rights 192, 199–200
cooperation with 234–5
cooperation in NE Atlantic 219
environmental protection 100, 107, 109, 234–5
establishment of particularly sensitive sea areas (PSSA) 234–5
special areas under MARPOL Convention 234
International Oceanographic Commission 49
International Seabed Authority (ISA)
Art 82 89, 125–36
deposit of charts 14
dispute settlement 136–8
need for cooperation 146–7, 235–6
protection of common heritage 10
regulations governing mining 44
International Tribunal for the Law of the Sea (ITLOS)
concept of single continental shelf 2
interpretation of enforcement rights 211–13
relationship between CLCS recommendations and coastal State rights 76–7, 79–80
seabed dispute chamber 137
terminology 6
International Whaling Commission (IWC) 219

Joint development
see also Unitization agreements
extended continental shelf and 142–4
no duty to enter into 142
overview 141

Law of the Sea Convention *see* UN Convention on the Law of the Sea (LOSC)
Living resources
see also Bioprospecting; Fishing; Sedentary species
biological resources and seabed environment 21–6

exploitation and regulation
bioprospecting 31–3, 110–19
fishing 27–31, 119–20, 230–4
marine scientific research 165–70
other activities affecting the sea floor
carbon mitigation measures 46–7
climate change 45
ecotourism 50
marine scientific research 49–50
waste disposal 45–6
range of activities on extended continental shelf 19
Ross sea 224, 226, 233

Marine protected areas
see also Convention on Biological Diversity; Environmental protection
biodiversity beyond national jurisdiction 245
CBD 99–100, 107, 236–7
coordination problems 223–5
creation 107–10, 168–9
EBSAs 100, 223, 225
OSPAR 109–10, 218–19, 221
Portugal
incorporation into domestic law 220
MPAs on its extended continental shelf 107–11, 217–21
Marine scientific research
Art 246(6)
application to living resources 165–70
interpretation 164–5
marine protected areas 168–9
bioprospecting distinguished 110–13
coastal State jurisdiction 9
extended continental shelf 158–63
LOSC and 156–8
conditions of granting consent 170–1
conditions on the conduct of 153
customary international law 152
disciplines of 154–5
environmental impacts of 49
exploitation distinguished 159
history of 154
negotiation of LOSC 152–3
'on' the continental shelf 162
Mauritius *see* Seychelles/Mauritius bilateral arrangements
Mexico
agreement with US 124, 142–4
Deep-Water Horizon incident 40–1
hydrocarbon exploration and exploitation 37
severe weather risks 41
Mineral resources
see also Non-living resources, Mining
coastal State rights under LOSC 12
increasing demand 36
types of deposits 34–5

Mining
see also Mineral resources
Art 82
  ambiguities 130–6
  background 125–7
  cross-boundary issues 138–47
  interpretation 127–30
  options for settling disputes 136–8
  payments and contributions 123–5
exploitation 42–5
Part VI LOSC
  drilling operations 13
  tunnelling under seabed 14
'Move-on' rules 100, 232

Nagoya Protocol 116–19
New Zealand
enforcement rights 210
exclusion zones around survey vessels 195–201
exploration and exploitation of non-living resources 38, 43
negotiation of LOSC 59
petroleum exploration licenses 38, 124
Non-interference zone *see* Safety zones
Non-living resources
see also Article 82; Mineral resources; Mining
coastal State rights 13–14
exploitation
  hydrocarbon exploration and exploitation 36–42
  increasing activity 33
  mineral resources 34–6
  mining 42–5
  oil and gas reserves 34
increasing importance 248
influence on the continental shelf doctrine 54
North Atlantic Marine Mammal Commission (NAMMCO) 219
North Atlantic Salmon Conservation Organization (NASCO) 219
North East Atlantic Fisheries Commission (NEAFC)
cooperation in NE Atlantic 109–10, 217–21
marine protected areas 109–10, 218–21
obstacles to cooperation
  coordination problems 223–5
  scientific and legal uncertainty 225–6
Norway
carbon sequestration 47
delimitation agreement with UK 141
hydrocarbon exploration and exploitation 38, 124

Oil *see* Hydrocarbons
OSPAR
establishment 218
limited jurisdiction 109, 219
nomination of MPAs 109, 218, 221
nomination of Rainbow Field as MPA 218
obstacles to cooperation
  coordination problems 223–5
  scientific and legal uncertainty 225–6
Outer limits of Continental Shelf
see also Delimitation of continental shelves
defined
  Commission on the Limits of the Continental Shelf 71–4
  Continental Shelf Convention 67–8
  LOSC 69–71
deposit of charts 14
establishment 11

Package deal 8, 10, 61, 84, 87, 89, 129, 249
Papua New Guinea *see also* Solwara I. 25, 43
Particularly sensitive sea areas (PSSA) 234
Pipelines *see* Cables and pipelines
Polymetallic sulfides 35, 43
Portugal
concluding remarks 227
cooperation with North East Atlantic Fisheries Commission (NEAFC) 219–20
discovery of Rainbow hydrothermal vent field 217
EMEPC 218, 221
incorporation of MPAs into domestic law 220
nomination of MPAs 109–10, 218
Precautionary approach
balancing of high seas rights and coastal State rights 189, 195
environmental protection 104, 106
Seychelles/Mauritius bilateral arrangements 229
Protests
human rights considerations 198
interruption to surveys 197
recognition of right by IMO 199
restrictions on navigation around survey vessels 195
toleration of some nuisance 201
vessels exercising right of navigation 198

Rare earth elements 36
Regional fisheries management organizations (RFMOs) 230–4
see also Fishing, NEAFC, Sedentary species
balancing high seas freedoms and coastal State rights 192
competence over sedentary species 232–3
cooperation in NE Atlantic 219–20
difficulty to obtain agreement on closing areas 233
gaps in coverage of RFMOs 223

International Guidelines for the Management
  of Deep-Sea Fisheries in the High
  Seas 105, 231
'move-on' rules 100, 232
need for cooperation with 230
protection of vulnerable ecosystems 100,
  102, 105, 106, 109, 230, 231
UN General Assembly
  Resolutions 105, 230
Russia *see also* USSR 165

Safety zones *see also* Installations, structures and
  artificial islands 13, 174, 193
  balancing high seas freedoms and coastal State
    rights 175, 195–201
  enforcement of 203, 205, 207, 209, 211–13
  non-interference zones 196–201
  on the extended continental shelf 147–9
  vulnerability of 201
Santiago Declaration 55
Sargasso Sea Alliance 222–3, 225
Seabed, deep *see* Area, the
Seamounts 22–4, 28–31, 35, 51
Sedentary species 12
  biodiversity beyond national jurisdiction
    and 245, 247
  bioprospecting and 113–15
  criticism of 63–4, 67, 93, 113–15
  defined 64–7, 113–15, 247
  disputes about 65–6, 209–10
  environmental protection of the
    extended continental shelf 93–4,
    106, 119–20
  genetic resources and 62, 113–15, 247
  origins of doctrine 62–4
  relationship to the EEZ regime 95–6
  relationship with non-sedentary species 93
Seychelles/Mauritius bilateral arrangements
  joint application to the CLCS 228
  lessons for other States 230
  environmental protection 229
  treaty arrangements 228–30
Solwara 1, 42, 43
Sovereignty *see* Enforcement rights
Structures *see* Installations, structures and
  artificial islands
Submarine cables and pipelines *see* Cables and
  pipelines
Survey vessels 195–201

Terminology
  bioprospecting 31, 111
  'extended continental shelf' 6, 7
  nautical miles (nm) 7
  sedentary species 64–5
Territorial sea
  agreement on rights under LOSC 8
  expansion of coastal State jurisdiction 8
  innocent passage 181

origins of continental shelf
  doctrine 53–4
Tourism 50
Truman Proclamation 54–5

**UN Convention on the Law of the Sea
  (LOSC)**
  See also Article 82, Commission on the Limits
    of the Continental Shelf, Common
    heritage of mankind, Dispute resolution,
    Fishing, Living resources, Marine
    scientific research, Mining, Non-living
    resources, Sedentary Species
  Art 78
    drafting history 176–81
    interpretation of 185–7
  balancing high seas freedoms and coastal State
    rights 182–4
  Art 78 176–81
  Art 60/80 174, 175, 193
  comparison of balancing provisions 184–5
  concluding remarks 201–2
  'due regard' 182–4
  framework for
    decision-making 187–200
  bioprospecting 113–15
  CBD and 117–19
  cooperation prior to delimitation 144–5
  delimitation agreements 141, 144
  delimitation cases on 59, 75–80
  need for restraint in absence of
    delimitation 80
  Part VI LOSC 10–14, 95–7, 175–6, 204
  disputed areas 144–6, 242–3
  drafting history 59–62, 125–7, 162–3,
    180–1
  enforcement rights 203–5, 210–15
  environmental protection 95–9
  major achievements 8
  marine scientific research 152
    Art 246(6) 164–70
    coastal State jurisdiction 156–63
  package deal 8, 10, 61, 84, 87, 89, 129, 249
  Part VI
    cables and pipelines 12
    delimitation of continental shelf 13–14
    development of 9, 59–62
    environmental protection in 95–7
    establishment of outer limits 10–11,
      69–71
    importance of Art 77 11
    importance of Art 82 13
    influence of the Continental Shelf
      Convention 11, 59, 64, 152, 162,
      176–7, 205, 248
    installations, structures and artificial
      islands 12, 147–49
    overview 10–14
  Part XII 97–9, 107

UN Convention on the Law of the Sea
   (LOSC) (*cont.*)
   status in customary international law
      Art 76 82–6
      Art 82 86–91
      Pt VI 81–2
United Kingdom
   Chagos Marine Protected Area case 183
   Delimitation agreement with Norway 141
   'due regard' provisions 182
   enforcement rights 209, 211
   multilateral regional cooperation in NE
      Atlantic 221
   negotiation of LOSC 59
   origins of continental shelf doctrine 52
United States
   agreements on king crabs 66, 227, 240
   definition of sedentary species in domestic
      law 67
   dispute with Canada over scallops 66,
      209–10
   enforcement rights 208, 209–10
   hydrocarbon exploration and
      exploitation 38, 124
   Mexico and 38, 142–3
   regulation of protests 199
   Truman Proclamation 54–5

views on customary status of Part VI 82–4,
   87–8
Unitization agreements *see also* Joint
   Development 140–1, 143, 147
Unjustifiable interference
   Art 78
      drafting history 176–81
      interpretation of 185–7
      balancing high seas freedoms and coastal State
         rights 182–4
      Art 78 177–81
      Art 60/80 174, 175, 193
      comparison of balancing
         provisions 184–5
      concluding remarks 201–2
      'due regard' 182–4
      framework for decision-making 187–200
USSR
   See also Russia
   agreement with US on king crab 66,
      227, 240
   enforcement 208
   marine scientific research 161, 165
   negotiation of LOSC 180

Waste disposal 45–6
World Heritage Convention (WHC) 237–8